Anas N. Al-Rabadi

Reversible Logic Synthesis

Springer-Verlag Berlin Heidelberg GmbH

Anas N. Al-Rabadi

Reversible Logic Synthesis

Springer-Verlag Berlin Heidelberg GmbH

Anas N. Al-Rabadi

Reversible Logic Synthesis

From Fundamentals to Quantum Computing

With 213 Figures

Springer

Dr. Anas N. Al-Rabadi

Post Office Box 85
97207-0085 Portland, OR
USA

E-mail: alra.badi@ece.pdx.edu

ISBN 978-3-642-62325-7 ISBN 978-3-642-18853-4 (eBook)
DOI 10.1007/978-3-642-18853-4

Cataloging-in-Publication Data applied for
Bibliographic infonnation published by Die Deutsche Bibliothek.
Dic Deutsche Bibliothck lists this publication in the Deutsche Nationalbibliografie;
detailed bibliographic data is available in the Internet at <http://dnb.ddb.de>.

http://www.springer.de

Originally published by Springer-Verlag Berlin Heidelberg New York in 2004
Softcover reprint of the hardcover 1st edition 2004

Typesetting: Camera ready by the author
Cover-design: design & production, Heidelberg
Printed on acid-free paper 62 / 3020 hu - 5 4 3 2 1 0 -

To

The Memory of My Grandmother

Preface

Quantum Computing (QC) is the upcoming revolution in computation. The fact that the tendency of current technologies is towards the nano-scale (i.e., dimensions of a single atom in the order of 10^{-10} m) will have, and is already having, disastrous effects on the signal integrity in classical designs for processing and transmitting information bits. The higher packing of devices on increasingly smaller and smaller areas will have, and is already having, tremendous power consumption effects. Thus, one would ask the question: what is the solution? The answer for both problems is simply: Quantum Circuits.

Since QC must be reversible, Reversible Computing (RC) becomes an inseparable intrinsic ingredient of QC. Consequently, the first step towards the implementation of QC is to find methods for RC. It turns out that various methods for RC produce various amounts of "garbage" outputs that are needed only for the purpose of reversibility not more. Thus, one needs to explore efficient synthesis methods using Reversible Logic (RL) to be used in future Computer Aided Design (CAD) tools for the synthesis of RL circuits in analogy to the current advanced CAD tools for the synthesis of classical irreversible circuits. Just this single purpose of Reversible CAD (RCAD) tools will impose an overall re-evaluation of all existing methods that are traditionally used in classical logic synthesis, including decomposition methods, factorization methods, and minimization methods. New benchmarks have to be created for the comparison of efficiency of various RL synthesis techniques. Thus one will ask the important question: what are the available methods that exist for reversible logic synthesis? The answer is almost none! This exact answer was the whole reason behind the gradual development and growth of this Book over the years. Totally new reversible logic synthesis methodologies had to be created and various evaluations had to be conducted. The objective is obvious: design RL circuits with minimum, if not none, garbage outputs. This optimization constraint is reflected in quantum circuits in the form of obtaining quantum registers of minimal size (i.e., length and width).

The body of this Book contains twelve Chapts. These Chapts. evolve gradually from basics towards the contributions. Chapter 1 provides the overall introduction to the subject of RC and QC. Chapter 2 introduces the basic concepts of two-valued and multiple-valued logic systems. An application of the concepts developed in Chapt. 2 is the Shannon/Davio (S/D) trees introduced in Chapt. 3, which will be used later on in Chapt. 8 for minimizing expressions to be realized in reversible cascades. Chapter 4 introduces the synthesis of logic functions using lattice structures, which resemble another important application of the concepts that are developed in Chapt. 2. Basic background and new results are introduced in Chapt. 5. Chapters 6, 7, and 8 introduce new structures and methods for RL synthesis. An initial evaluation of the RL synthesis methods is presented in Chapt. 9. Chapter 10 implements the quantum logic circuits using the new results from the previous Chapts. Two-valued and multiple-valued quantum computing for the quantum circuits from Chapt. 10 is performed in Chapt. 11. Chapter 12 provides conclusions, highlights of new results that were presented in this Book, and future directions of research. The end matter of this Book contains eleven Appendices and a Bibliography.

Around four years ago, at the start up of my research into reversible and quantum computing, many researchers advised me to rethink again about conducting research in this field, since it is a fairly new research specialty and unlike most other research fields has only minimal literature, which meant that I had to create "from scratch" my own solutions to the continuously emerging problems. Another concern was that part of this work especially in QC is more or less futuristic, which means that there is little current use of the results for industry and consequently limited number of accepted money grants and proposals. Nevertheless, I felt that these discouragements I faced, to proceed in this field, are in fact a challenge and motivation for me to conduct research in this field, and the result is this Book which accumulates some of my work for the past four years. This Book tried to fill as much as possible the large missing gaps in previous literature, especially for the synthesis of reversible circuits and their consequent QC. Consequently, the result of this work was an opening of a window of a new research area, and does not mean by any means the end of the road. In fact this work is only one possible beginning, from which serious researchers can start towards achieving the ultimate ambitious goal of fully reliable super-fast power-free nano computing.

Portland, Spring 2003 Anas N. Al-Rabadi

Abstract

The biggest problems in system design today, and in the future, are the high rate of power consumption, and the emergence of quantum effects for highly dense ICs. The real challenge is to design reliable systems that consume as little power as possible and in which the signals are processed and transmitted at very high speeds with very high signal integrity. Current tools are used to design ICs using only classical design methodologies that apply conventional synthesis constraints such as area, delay, and power.

As it was proven, physical processes have to be logically reversible in order to reduce (theoretically eliminate) power consumption. Logical reversibility requires that one can obtain the vector of inputs from the vector of outputs (i.e., backward process) as well as the vector of outputs from the vector of inputs (i.e., forward process). Since only ad-hoc methods were used previously for the reversible synthesis of logic functions, and since systematic and efficient reversible logic synthesis methodologies were significantly missing from previous literature, this Book provides several original contributions to reversible logic synthesis by providing a set of tools that can be systematically used to synthesize and evaluate logic functions using reversible logic. This includes, among other new results, Reversible Lattice Structures, Reversible Modified Reconstructability Analysis, Reversible Nets, Reversible Decision Diagrams, and Reversible Cascades.

To solve the problem of high signal integrity when processing (computing and transmitting) information using extremely high dense circuits one needs to incorporate the physical quantum mechanical effects that are unavoidable in the nano scales. Since quantum circuits are reversible, and since many of the underlying theorems and formalisms for multiple-valued quantum computing were significantly missing from previous literature, new fundamental foundations for such computations had to be established. These new results include, but not limited to, Quantum Chrestenson Operator, new types of Quantum Decision Trees and Quantum Decision Diagrams as efficient representations for quantum computing, new Composite Basis States, new multiple-valued Einstein-Podolsky-Rosen (EPR) Basis States, and new classes of quantum primitives.

Initial evaluations and conclusions for the comparative advantages and disadvantages of the new reversible and quantum computing methodologies are also provided, and applications to Optical Computing and Quantum Neural Networks are presented.

Acknowledgements

I would like to express my deep gratitude to all of the people who never stopped their support to me during the process of preparing this manuscript. Without their complete support, my journey would have been much tougher. In particular, I would like to express my sincere appreciation to my beloved parents who never stopped supporting my endeavors.

Contents

Appendices

Glossary

AC	Ashenhurst-Curtis
AF	Activation Function
ALU	Arithmetic Logic Unit
ASIC	Application Specific Integrated Circuit
BBM	Billiard Ball Model
BBL	Billiard Ball Logic
BD	Bi-Decomposition
BDD	Binary Decision Diagram
C	Complexity
CA	Cellular Automata
CAD	Computer Aided Design
CCD	Charge Coupling Device
CCNOT	Controlled-Controlled-NOT
CCW	Counter Clock Wise
CMOS	Complementary Metal Oxide Semi-Conductor
CNF (POS)	Conjunctive Normal Form
CNOT	Controlled-NOT
CoQDD	Composite Quantum Decision Diagram
CoQDT	Composite Quantum Decision Tree
CQDD	Computational Quantum Decision Diagram
CQDT	Computational Quantum Decision Tree
CRA	Conventional Reconstructability Analysis
CRT	Chinese Remainder Theorem
D	Davio
DD	Decision Diagram
DFC	Decomposed Function Cardinality
DFT	Discrete Fourier Transform
DM	Data Mining
DNF (SOP)	Disjunctive Normal Form
DT	Decision Tree
EC	Evolutionary Computation
EDA	Electronic Design Automation
EPR	Einstein-Podolsky-Rosen
EPRQDD	Einstein-Podolsky-Rosen Quantum Decision Tree
EPRQDT	Einstein-Podolsky-Rosen Quantum Decision Tree

ESOP	Exclusive Sum Of Products
EXNOR	Exclusive Not OR (XNOR)
EXOR	Exclusive OR (XOR)
F	Functionality
FIR	Finite Impulse Response
FPGA	Field Programmable Gate Arrays
FPRM	Fixed Polarity Reed Muller
GA	Genetic Algorithm
GBFM	Generalized Basis Function Matrix
GF	Galois field
GFSOP	Galois field Sum-Of-Product
GIF	Generalized Inclusive Forms
G(P)L	Generalized (Post) Literal
GP	Genetic Programming
GQD	Generalized Quaternary Davio
GRM	Generalized Polarity Reed Muller
IC	Integrated Circuit
ID	Invariant Davio
ID/IS	Invariant Davio Invariant Shannon
ID/S	Invariant Davio Shannon
IF	Inclusive Forms
IfD	Invariant flipped Davio
IfS	Invariant flipped Shannon
$IF_{n,2}$ Triangles	Inclusive Forms of arbitrary radix and two variables triangles
I/O	Input-Output
IPDE	Invariant Permuted Davio Expansion
IPSE	Invariant Permuted Shannon Expansion
IS	Invariant Shannon
IS/D	Invariant Shannon/Davio
ISID	Iterative Symmetry Indices Decomposition
KD	Knowledge Discovery
KDD	Knowledge Discovery in Database
K-Map	Karnaugh Map
LF	Log-Functionality
LI	Linearly Independent
LIL	Linearly Independent Logic
LL	Log-Linear

LOS	Lattice-Of-Structures
LS	Linearly Separable
LUT	Look-Up Table
MAX	Maximum
MC Gate	Maximum Cofactor Gate
MIMO	Multiple-Input Multiple-Output
MIN	Minimum
MIN/MAX	Minimum/Maximum
MISO	Multiple-Input Single-Output
ML	Machine Learning
MRA	Modified Reconstructability Analysis
MUX	Multiplexer
MV	Multiple-Valued, Many-Valued, Multi-Valued
MVL	MV Logic
MVQC	Multiple-Valued Quantum Computing
MvQDD	Multiple-Valued Quantum Decision Diagram
MvQDT	Multiple-Valued Quantum Decision Tree
NC	Neural Computing
ND	Negative Davio
NDDD	Negative Davio Decision Diagram
NLS	Not-Linearly Separable
nMOS	Negative Metal Oxide Semi-conductor
NN	Neural Network
NPN	Negation of variables, Permutation of variables, and Negation of function
OBDD	Ordered Binary Decision Diagram
PD	Positive Davio
PDDD	Positive Davio Decision Diagram
PDF	Probability Density Function
PL	Post Literal
PLA	Programmable Logic Arrays
PLD	Programmable Logic Devices
POS	Product-Of-Sum
PPRM	Positive Polarity Reed-Muller
PSDD	Pseudo-Symmetric Decision Diagram
PTL	Pass Transistor Logic
PUS	Positive Unate Symmetric
QC	Quantum Computing, Quantum Computation

QCA	Quantum Cellular Automata
QCAD	Quantum Computer Aided Design
QChT	Quantum Chrestenson Transform
QCt	Quantum Circuit
QCT	Quantum Chrestenson Transform
QDD	Quantum Decision Diagram
QDT	Quantum Decision Tree
QFT	Quantum Fourier Transform
QIF	Quaternary Inclusive Forms
QL	Quantum Logic
QMRA	Quantum Modified Reconstructability Analysis
QN	Quantum Neuron
QNN	Quantum Neural Network
QS/DT	Quaternary Shannon/Davio Tree
Qubit	Quantum Bit
QuDT	Quaternary Decision Tree
QWHT	Quantum Walsh-Hadamard Transform
RA	Reconstructability Analysis
RC	Reversible Computing
RCAD	Reversible Computer Aided Design
RCt	Reversible Circuit
RDD	Reversible Decision Diagram
RDT	Reversible Decision Tree
RGBFM	Reversible Generalized Basis Function Matrix
RL	Reversible Logic
RM	Reed-Muller
RMRA	Reversible Modified Reconstructability Analysis
RPGA	Reversible Programmable Gate Array
RPL	Reduced Post Literal
ROBDD	Reduced Ordered Binary Decision Diagram
RQDD	Reduced Quaternary Decision Diagram
S	Shannon
S/D	Shannon/Davio
SDD	Shannon Decision Diagram
SE	Schrodinger Equation
SET	Single Electron Transistor
SG	Stern-Gerlach
SIMO	Single-Input Multiple-Output

SISO	Single-Input Single-Output
SJ	Summing Junction
SOP	Sum-Of-Product
SVD	Singular Value Decomposition
TDD	Ternary Decision Diagram
TDSE	Time Dependent Schrodinger Equation
TDT	Ternary Decision Tree
TFKRODT	Ternary Free Kronecker Decision Tree
TGIF	Ternary Generalized Inclusive Forms
TGIGKDT	Ternary Generalized Inclusive Forms Kronecker Decision Tree
TIF	Ternary Inclusive Forms
TISE	Time Independent Schrodinger Equation
TKRODT	Ternary Kronecker Decision Tree
TPRMDT	Ternary Pseudo-Reed-Muller Decision Tree
TPKRODT	Ternary Pseudo-Kronecker Decision Tree
UL	Universal Literal
ULM	Universal Logic Module
WL	Window Literal
1-D	One-Dimensional
2-D	Two-Dimensional
3-D	Three-Dimensional

1 Introduction

Computing structures have been evolving since early times of mankind. Such structures evolved from simple systems that use simple mechanical elements such as ropes and pulleys, to mechanical systems that use carefully designed elements, to machines that use discrete electronic elements, to nowadays computing machinery that uses highly complex integrated electronic elements [256]. The power of the computing machinery has been growing with the growing of the complexity of such machines for processing, storage, and interfacing capabilities. This is observed in the fact that the pre-electronic computing machines were able only to perform basic arithmetic calculations, while modern computing machines that are made up of highly complex electronic integrated circuits (ICs) are capable of performing many more tasks such as three-dimensional graphics (i.e., 3-D visualizations) and networking. This evolution of computing has been driven by the need to fulfill the increasingly demanding design specifications of more speed, less power consumption, smaller size, better testability, better reliability, and more regularity.

If the trends in computing keep going according to Moore's law, by the year 2020 the basic memory components of a computer will be the size of individual atoms, and the ongoing scale down of technology to produce very high dense ICs will reach its limit, where the scale down of technology will evolve from the micro-scale to the nano-scale. At such scales, the current theory of computation becomes invalid, and a new field, called "quantum computing", starts emerging which is about re-inventing the foundations of computer science and information theory in a way that is consistent with quantum physics - the most accurate model of reality that is currently known [81,208].

Remarkably, this new theory predicts that quantum computers can perform certain tasks breathtakingly faster than classical computers and, better yet, can accomplish mind-boggling feats such as teleporting information, breaking supposedly unbreakable codes, generating true random numbers and communicating with messages that betray the presence of eavesdropping [93,167,253,254]. Indeed, a quantum scheme for sending and receiving ultra-secure messages has already been implemented over a distance of 100 km - far enough to wire the financial district of any major city [149,167,253,254].

At the nano-scale, which is the scale of atomic diameter, a different kind of physics emerges which is governed by quantum mechanics of atoms and particles [81,208]. The physical laws will be a driving force for different directions in computing due to the fact that new phenomena are encountered in the quantum level, which were not previously observed. This includes for instance quantum entanglement, where the physical properties of one particle affect one or more other particles and thus particles are *entangled*, and quantum interference where physical phenomenon is interpreted as several waves *interfering* with each other to produce specific physical patterns [107,115,167]. Utilizing such new physical phenomena that emerge in the nano-scale, quantum computations use basic properties of particles that can be performed such as using the spin of such particles to encode logic values [167].

Several powerful features are harnessed using such type of computation. One powerful feature is super-fast computing [268] that can be achieved in the quantum domain, as compared to fast computational speeds, which are achieved in the conventional domains [57]. The speedups of calculations in quantum computing are due to the new physical phenomenon encountered in the nano-scale which is the quantum entanglement in which parallel computations are performed simultaneousely [107,167]. Such computational power could not be observed previously in the micro-scale or in the macro-scale. This speedup in computations can have many implications on highly important applications that can range from consumer products having faster computers to national security issues like the encryption of highly classified information [162,163,167]. For example, it is known in the complexity theory of algorithms (i.e, computational complexity) that an algorithm is classified as a polynomial-time algorithm if that algorithm is guaranteed to terminate within a number of steps which is a polynomial function of the size of the problem. The new speedup of quantum computations can have the ability to solve problems that were previously thought to be unsolvable in polynomial time such as the factoring problem [226,228], and this will lead to new applications in encrypting and decrypting communicated information which will have a direct impact on how much the electronically transferred messages are secured [162,163].

In addition to the need of quantum computing due to the anticipated failure of Moore's law at the quantum level, which predicts that computing will be ultimately performed in hardware in the size of nano-scale (less than 10^{-10} m = 1 Angestrom), the power needed to switch a single bit in future nano-technologies will be much lower than its counterparts in the bigger scales, and according to the fundamental principles of thermodynamics the limit will be K·T·ln(2) [139,140,141], where K is Boltzmann constant ($\approx 1.380658 \cdot 10^{-23}$ Jouls/Kelvin) and T is the operating temperature (Kelvins).

Figure 1.1 illustrates this ongoing trend in power consumption. One can observe from Fig. 1.1 that the energy needed to switch one bit will be decreasing with the advancements of chip manufacturing with highly increasing integration densities according to Moore's law, and that by the year 2020 the energy consumption for switching a single bit will reach the thermodynamical limit of K·T·ln (2) which is the threshold after which Moore's law will not hold due to the emerging quantum nano-scale effects.

Using basic principles of thermodynamics, the thermodynamical limit of K·T·ln (2) can be derived as follows [95]: for a gas compression experiment, the following Eq. holds: $P \cdot V = R \cdot T$, where P is pressure, V is volume, T is temperature (Kelvin), and R = K·N, where K is Boltzmann constant ($\approx 1.380658 \cdot 10^{-23} \dfrac{Joules}{Kelvin}$), and N is the number of molecules in the gas. Then from the basic laws of thermodynamics:

$$T\Delta S = \Delta Q = \int_{V_o}^{\frac{1}{2}V_o} PdV = \frac{RT}{V} dV \Big|_{V_o}^{\frac{1}{2}V_o} = RT\ln(2).$$

Therefore, for N molecules (~ N bits of information) one has:

$$\Delta S = R\ln(2) = KN\ln(2)\frac{Joules}{Kelvin},$$

and for 1 molecule (~ 1 bit of information) one obtains:

$$\frac{\Delta S}{N} = K\ln(2)\ \frac{Joules}{Kelvin}.$$

For 1 molecule, one obtains the following energy dissipation ΔQ:

$$\Delta Q = \Delta S\ \frac{Joules}{Kelvin}\cdot T\quad Kelvin = KT\ln(2)\ Joules.$$

Besides reaching the quantum barrier, trends in computer hardware are leading toward higher density and lower energy dissipation.

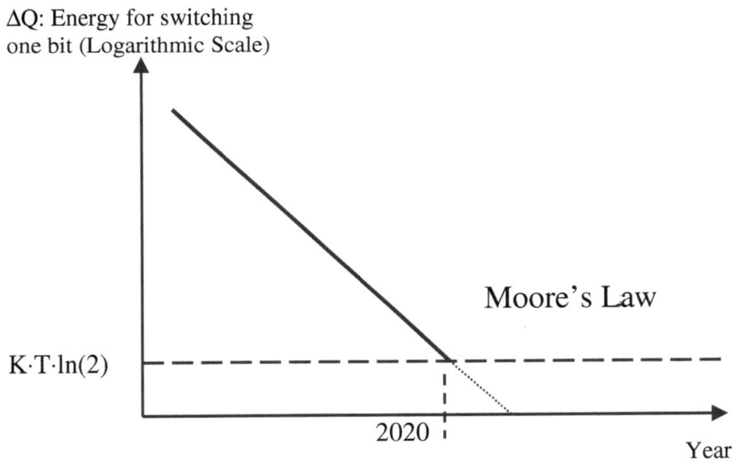

Fig. 1.1. Trend in energy consumption for switching one bit for conventional and quantum computing.

Ultimately, some approaches should result in packing densities in excess of 10^{17} logic devices in a cubiccentimeter [152]. The trend towards higher packing density and higher speeds strongly influence energy dissipation. For example, conventional devices must dissipate more than K·T· ln (2) Joules in switching, so 10^{17}conventional devices operating at room temperature (K·T· ln (2) ~ $3\cdot10^{-21}$ Joules for T = 300 Kelvins) at a frequency of 10 gigaHertz would dissipate more than 3,000,000 Watts; a computer with 1,000 times as many logic elements would still be of reasonable size but would dissipate 3,000,000,000 Watts! [152]. Consequently, a new computational approach has to be invented such as quantum computing.

In quantum computing, low power consumption is needed (theoretically zero) as no power is needed for processing information and the power is only consumed when reading and writing information into and from quantum computing machines [92,93,94,130,139,140]. The minimal power consumption is due to the fact that quantum computations are naturally reversible [37,38,39,40,98,139,140,141], which means that the information entropy must be conserved, since it has been proven using fundamental principles of thermodynamics that reversible computation does not consume power since no logical information is lost [139,140]. Accordingly, the reversibility of computation is a necessary but *not sufficient* step for quantum computing, as more constraints are needed in addition to reversibility, to achieve quantum computations. These additional constraints, that exist in the quantum domain and do not exist in the conventional domain, drive the use of *new* mathematical computational methodologies that map the underlying quantum phenomena such as the quantum entanglement, for which a new type of information representation and the corresponding operations must be used. This includes the use of the information element of quantum bit (qubit), which is a vector of bits, as compared to a (scalar) bit, which exists in the conventional domain [93,95,107,115,167,253,254].

As quantum computation is reversible, reversible computing is an essential ingredient of quantum computing, and reversible logic serves as a mathematical concept to describe the physical reality of quantum logic. Consequently, new methodologies of synthesizing logic functions using reversible structures have to be invented. The reversible computations aspect of quantum computing has its own design constraints (design goals, design objectives) in the form of (1) minimizing the number of garbage outputs that are needed only for the purpose of reversibility (which leads to minimizing area and power), (2) minimizing the number of gates used (which leads to minimizing area and power), and (3) minimizing the delay of signal propagation from inputs to outputs (i.e., more speed). These design constraints, which are encountered in the conventional design of reversible circuits, are reflected as design constraints in the quantum domain in the form of the "size of the quantum register" [93,95,98] (i.e., the width and length of the quantum register), and therefore one wants to design a quantum register of minimal size that would perform the underlying reversible computations. Reversible computing as a mean of low-power computing has been investigated and encouraging results have been reported [35,42,70,71,70,72,131,143,206,262,263]. Figure 1.2 illustrates the set-theoretic inclusion relationships between various computing methodologies, where the shaded areas indicate the types of logic synthesis that are presented in this Book.

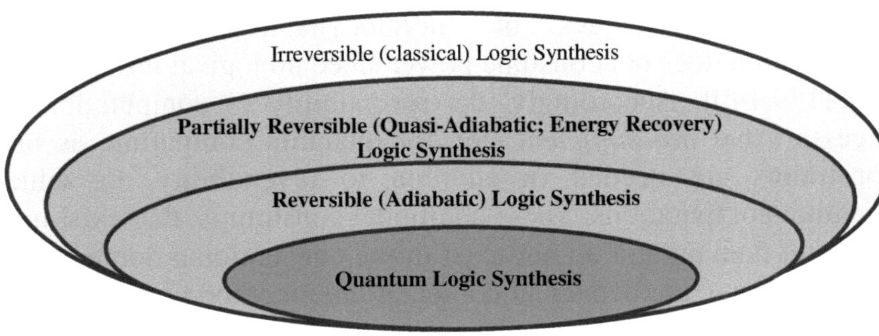

Fig. 1.2. Set-theoretic relationship between various computing methodologies.

The inclusive relationship in Fig. 1.2 reflects the hierarchically increasing design constraints when moving from the domain of classical logic synthesis to the domain of quantum logic synthesis. In Fig. 1.2, the classical (irreversible) logic synthesis includes all methodologies that are developed in the conventional logic synthesis field [136,164,118]. The two lightly shaded areas in Fig. 1.2 are included within the general framework of reversible logic synthesis. The word *adiabatic* comes from a Greek word that describes a process that occurs without any loss or gain of heat (i.e., no heat is injected in or generated out of a system). (Adiabatic system is the opposite of *isothermal* system, where heat is injected in or generated out of a system in order to preserve the constant temperature in that system). In real-life computing, such an ideal process (or hardware) cannot be achieved because of the presence of dissipative elements like resistances in a circuit. However, one can achieve very low energy dissipation by slowing down the speed of operation and only switching the switches (e.g., transistors) under certain conditions. Consequently, the real-life applications of rversible computing are based on quasi-adiabatic (or energy recovery) techniques rather than fully adiabatic techniques [206,262,263].

It has been shown [37,38,139,140,141] that if the physical processes that are associated with computing are nondissipative, the natural laws require that the *physical entropy* must be conserved. Entropy conservation means that the processes must be physically reversible. One of the conclusions from earlier studies is that the abstract logical operations composing the computing tasks must be reversible, that is, the *information entropy* must be conserved in order to be performed by physically nondissipative hardware. Consequently, logical reversibility is the only necessary abstract condition for nondissipative computing. Reversible logic operations can be realized by either reversible or nonreversible hardware. Consequently, the concept of reversibility can be implemented spanning all of the abstraction levels of the conventional logic design.

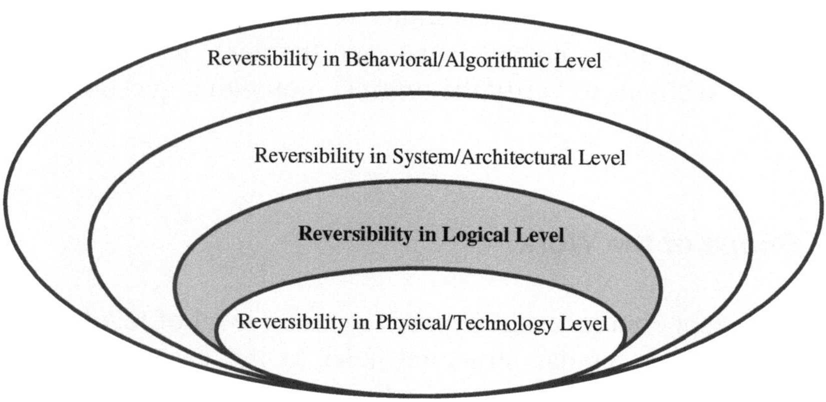

Fig. 1.3. Reversibility in computing system design.

Abstraction levels of reversibility includes the algorithmic level, architectural level, logic level, and the physical level. Figure 1.3 illustrates the implementation of reversibility within all of the design abstraction levels. This Book investigates reversibility in the logic level, which is the shaded area in the abstraction levels in Fig. 1.3, which was largely missing from previous literature.

The original motivating research guideline of this Book was that reversibility and regularity are interrelated and various regularity levels in two-valued and multiple-valued reversible structures will lead to various sizes of two-valued and multiple-valued quantum logic circuits and consequently various complexity levels in two-valued and multiple-valued quantum computing. More specifically, for several classes of small-dimension functions, the more regularities that exist in two-valued and multiple-valued reversible structures will lead to larger two-valued and multiple-valued quantum logic circuits and consequently more operations are needed in two-valued and multiple-valued quantum computing. Consequently, in general, if one relaxes the regularity levels which are imposed as constraints in the synthesis of reversible circuits, one would expect to obtain reversible circuits with smaller size. To prove this point, and since synthesis methodologies for binary and multiple-valued reversible logic synthesis and multiple-valued quantum computing were substantially missing in previous literature, new mathematical formalisms, representations, novel two-valued and multiple-valued reversible logic synthesis methodologies and structures, and new operations for multiple-valued quantum computing had to be invented first, and the next step was to apply these new methods to verify the original motivating questions of the Book.

1.1 Scope of the Work

Since modern circuit design requires a certain level of regularity due to the fact that regular structures lead to the ease of testability [99,124,198,199,204,218], ease of manufacturability, and free-library synthesis, one would like to design reversible structures that are regular, which will produce (1) minimal, (2) universal, (3) regular, and (4) reversible circuits. Minimal means to reduce (or if possible to eliminate) the number of *garbage* outputs that are needed only for the purpose of reversibility, and to reduce the number of gates used. Universal (or complete) means that the structure must be able to realize *all* logic functions for particular radix of logic and particular number of variables. Regular means that the structure must have a fixed number of gate types and interconnect types from which the whole structure is synthesized. Consequently, *full regularity* means that one type of internal nodes and one type of interconnects are used, *semi regularity* means that fixed number of internal node types and fixed number of interconnect types are used, and *non-regularity* means that arbitrary types of internal nodes and arbitrary types of interconnects are used. Synthesis methods to design minimal-size regular reversible circuits that will produce minimal size quantum registers were largely missing from previous research and literature, and this has been the driving force behind the development of reversible and quantum computing methods presented in this Book. To achieve the general goal of reversibility and regularity new reversible logic synthesis methodologies have been developed. Figure 1.4 shows the main ideas (i.e., tree paths) that were the driving force behind the development of this work.

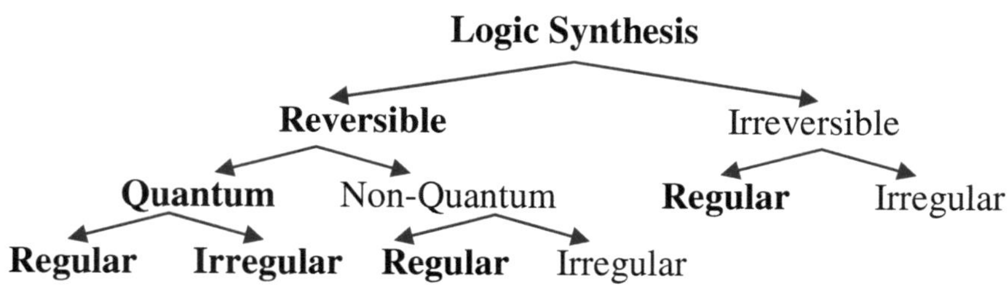

Fig. 1.4. A general characterization of reversibility in logic synthesis.

Since minimal size is one important design specification of reversible and consequently quantum logic structures, functional minimization techniques, which exist in the conventional design tools, can be used to produce minimal size functional expressions, and consequently algorithms can manipulate such expressions to efficiently design reversible and quantum circuits. Conventional ESOP minimizers and other minimization techniques, such as S/D trees, can be used for this purpose [4,9,52,114,157,232,233,235]. Another direction of area minimization of reversible structures is using multiple-valued logic, especially as multiple-valued logic has been efficiently used in conventional hardware for learning [186,187], testing [124], and IC design [86,267]. Similar to the conventional case, using higher radix in multiple-valued logic will minimize the number of wiring used as compared to binary logic to achieve the same functionality of logic structure [86,119,120,155,166,229,267]. Multiple-valued computing becomes important especially as multiple-valued quantum computations are performed on the same atomic structures on which two-valued quantum computations are performed without the need of adding new structural elements as compared to the conventional domain. This is due to the fact that quantum computing is performed using fundamental properties of particles such as spins of electron or polarizations of light [162,163], and these same physical properties are used to perform both two-valued and multiple-valued computations without the need of adding new circuit elements as in the conventional circuit design, especially the fact that multiple-valued quantum devices that perform the corresponding multiple-valued quantum computations have been created using trapped ions [54,165], and tunnel diodes [220]. For example, another way to harness the functional power of performing multiple-valued quantum computations is to perform minimal number of light polarizations to execute the same functionality as compared to using only two-valued quantum computations [163]. (One objective of this Book is to develop a theory for multiple-valued quantum computing that includes the binary case as a special case.) Consequently, the core stream of this Book follows the diagram shown in Fig. 1.5.

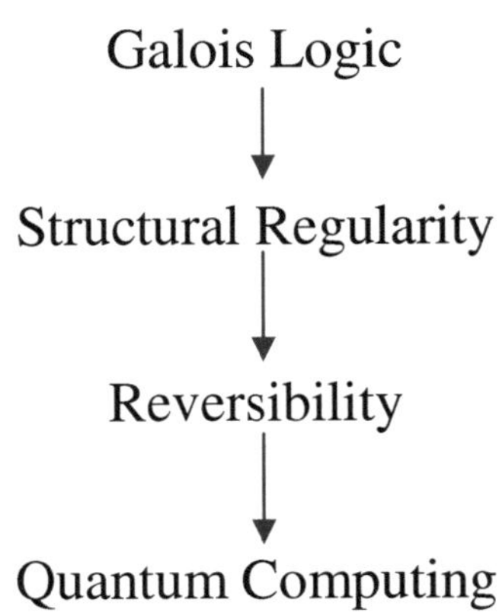

Fig. 1.5. Stream of topic development in this Book.

The flow chart in Fig. 1.5 illustrates the logical build up of most part of this Book. This starts with Galois field as the fundamental algebraic basis from which other components are built upon. Using specific radix of Galois field, logic structures that possess certain amount of regularity are synthesized. Such structures, if not reversible, are generalized to the reversible domain. The applications of two-valued and multiple-valued quantum computations, using the new reversible logic structures, are then performed.

1.2 Organization of the Book

To reach the objective shown in Fig. 1.5, this Book is divided into several intermediate steps that include the general components of: (1) reversibility, (2) multiple-valued logic, (3) minimization, (4) regularity, and (5) quantum computing. These elements of the Book are illustrated using the lattice diagram in Fig. 1.6.

Chapter 2 includes fundamentals and mathematical background that are needed to construct various important reversibility theorems in the next Chapts. This include binary and multiple-valued normal Galois forms, and new types of expansions which constitute a generalization of some basic decompositions that play classically a central role in modern logic synthesis tools.

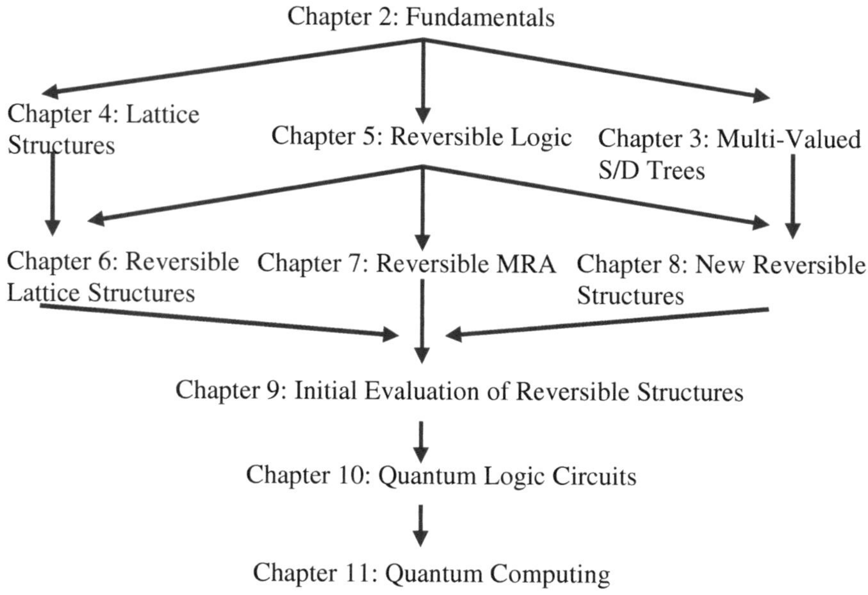

Fig. 1.6. General organization of this Book.

Chapter 3 presents new types of families of multiple-valued trees, their associated properties, and their corresponding canonical forms and hierarchies. These new forms serve as an intermediate step to produce one important minimization methodology of multiple-valued Galois functions that uses the polarity of multiple-valued Inclusive Forms (IFs) which are generated from Shannon/Davio (S/D) trees. The new multiple-valued minimizer will be used for functional minimization in order to realize logic functions in minimal size reversible structures such as reversible Cascades that will be presented in Chapt. 8.

An important class of regular structures that will be used in Chapt. 6 to reversibly realize Boolean and multiple-valued logic functions, which is called lattice structure, is presented in Chapt. 4. New three-dimensional lattices, that are built using the new spectral transforms from Chapt. 2, are introduced. An important methodology that restricts the realization of lattice structures to specific structural boundaries, called Iterative Symmetry Indices Decomposition (ISID), is also introduced.

Chapter 5 introduces the foundations of reversible computing. New reversible logic circuits and the corresponding theorems in reversible logic are introduced. The new theorems in reversible logic produce new reversible primitives from which more complex reversible structures will be synthesized in the following Chapts. The important process of garbage elimination in binary and multiple-valued reversible circuits is also presented.

The first type of the new reversible structures is presented in Chapt. 6. This type of regular reversible circuits is called reversible lattice structure. The binary and multiple-valued reversible lattice structures are used to realize regularly Boolean and multiple-valued logic functions, respectively.

Reversible Modified Reconstructability Analysis (RMRA) is presented in Chapt. 7. This includes the introduction of a novel binary and multiple-valued decomposition called Modified Reconstructability Analysis (MRA), and then the reversible realization of such structure.

New types of binary and multiple-valued reversible structures that present some advantages over previous reversible structures are presented in Chapt. 8. This includes reversible Nets, reversible Cascades, and reversible Decision Diagrams. Certain advantages and disadvantages of such structures are discussed, and examples are provided.

Chapter 9 presents an initial evaluation of the reversible logic synthesis methodologies that have been produced in previous Chapts. This evaluation is conducted on different symmetric and non-symmetric NPN-classified logic functions, and lead to several new insights into the new nature of some of the reversible circuits such as reversible lattice circuits.

The introduction of the physical operational quantum notation and the associated examples of quantum circuits for the previously invented reversible structures is presented in Chapt. 10. Advantages of the usage of such quantum notation and its pragmatic meaning and use are also presented.

Chapter 11 introduces new formalisms, representations, and operations in two-valued and multiple-valued quantum computing that uses theorems of reversible computing and reversible structures from previous Chapts. to compute functionalities using quantum logic. This includes the production of the quantum representation of the previously created reversible primitives in Chapt. 5, and then the performance of quantum computations using such quantum representations. The important concept of multiple-valued quantum entanglement is introduced. Examples for the use of algebraic mathematical decompositions for quantum computing such as the Singular Value Decomposition (SVD), Spectral Theorem, Quantum Fourier Transform (QFT), and Quantum Walsh-Hadamard Transform (QWHT) are presented. The new quantum representations of quantum decision trees and diagrams, as new means of representations for the manipulation of quantum circuits using Computer-Aided Design (CAD) tools, are also introduced.

Chapter 12 presents conclusions of the Book and perspective future work.

This Book is terminated with a series of eleven Appendices that produce new related results, important background, and motivations for many components of the work which were introduced in the previous Chapts. For example, some of these new results include the counts of several theorems that were presented in previous Chapts. Functions of these counts can be incorporated as upper or lower bounds in search heuristics that can be used to search for solutions to solve synthesis problems that have no concrete formal solutions.

Appendix A introduces count results to count the various classes of the new binary and multiple-valued invariant Shannon and Davio expansions from Chapt. 2. Circuits that implement the quaternary Galois field Sum-Of-Products (GFSOP) expressions, which are discussed in Chapt. 2, are introduced in Appendix B. Two novel count results for the count of S/D trees and the corresponding Inclusive Forms, that were the main result of Chapt. 3, are presented in Appendix C. Circuit realizations of multiple-valued S/D trees are introduced in the form of Universal Logic Modules (ULMs) in Appendix D. Background on Evolutionary Computing, which is used in various algorithms in different locations in Chapt. 3, Chapt. 8, and Chapt. 11, is presented in Appendix E. The count of all possible families of binary and multiple-valued reversible Shannon and Davio decompositions that result from Chapt. 5 is introduced in Appendix F. Appendix G presents the NPN classification method of Boolean functions and the complexity measures that are used in Chapt. 7 and Appendix H of this Book. New evaluation results that compare the new Modified Reconstructability Analysis (MRA) structure from Chapt. 7 and Ashenhurst-Curtis and Bi-Decomposition are presented in Appendix H. Appendix I introduces the count for reversible Nets that were introduced in Chapt. 8. Novel optical realizations of two-valued and multiple-valued classical and reversible logics are presented in Appendix J. Appendix K utilizes results in multiple-valued quantum computing from Chapt. 11 to introduce new results in multiple-valued quantum implementation of discrete Artificial Neural Networks.

2 Fundamentals

This Chapt. presents the necessary mathematical background and the fundamental formalisms of the work that will be introduced and further developed in the next Chapts. This includes the main reversible decompositions in Chapt. 5 that will be used to construct reversible primitives, from which reversible structures are built in Chapts. 6, 7 and 8, respectively. Also, the foundations that are introduced in this Chapt. will be used to construct the quantum gates and their associated quantum circuits and computing in Chapt. 11.

Spectral transforms play an important role in synthesis, analysis, testing, classification, formal verification, and simulation of logic circuits. Dyadic families of discrete transforms: Reed-Muller and Green-Sasao hierarchy, Walsh, Arithmetic, Adding, and Haar wavelet transforms and their generalizations to p-adic (multi-valued) transforms, have found a fruitful use in digital system design [120,125]. In this Chapt., we present a specialized framework for the creation, classification, and counts of new non-singular generalized Reed-Muller-like families of expansions for an arbitrary radix of Galois field.

Reed-Muller-like spectral transforms [240] have found a variety of useful applications in minimizing Exclusive Sum-Of-Products (ESOP) and Galois field SOP (GFSOP) expressions [9,76,77,79,80,171,264], creation of new forms [4,78,104,173,265,266], binary decision diagrams [2,45,142], spectral decision diagrams [82,238,239], regular structures [5,7,13,18,50,51,84,177], besides their well-known uses in digital communications [125], digital signal processing [89,257,260], digital image processing [90], and fault detection (testing) [99,124,147,198,199,204,218]. Ternary Reduced Post Galois field Sum-Of-Products (RP-GFSOPs), their generalized Green/Sasao hierarchies, and the extensions of such hierarchies to the case of quaternary Galois field hierarchy were recently developed [4,9].

For higher radices of Galois fields, there exist very large number of nonsingular transforms that still need to be systematically generated and classified. The main contributions in this Chapt. are:

- New methods of Galois Eqs. which are based on multiple-valued Galois logic to produce the corresponding multiple-valued Shannon and Davio expansions.
- A generic methodology of generating new types of multi-valued Shannon and Davio spectral transforms.
- The classification of the new types of multi-valued Shannon and Davio spectral transforms into families. The corresponding counts of the new families are also provided.

The methodology of generating the new families of multi-valued Shannon and Davio spectral transforms is based on the fundamental multiple-valued Shannon expansions (i.e., p-adic) and the fundamental multiple-valued Davio expansions. The new families of multiple-valued decompositions possess many advantages. The first advantage is the comparative ease of generation of the new multiple-valued transforms, since they are very closely related to the fundamental multiple-valued Shannon and Davio transforms. The second advantage is that the new transforms allow for a fast construction of the inverse transform because the basis functions of such transforms are the same basis functions of the fundamental multiple-valued Shannon and Davio transforms just scaled by constants. These constants are the multiplicative inverse of the corresponding constants that scale the rows of the corresponding basic multiple-valued Shannon and Davio transform matrices. This feature is very useful in hardware and software implementation for the fast processing of digital signals. The third advantage is that the flipped Shannon has an important role in the construction of reduced-size lattices [50,51]. Such regular structures have found application in the design for test and design for self-repair of logic circuits [221]. Consequently, the new multiple-valued flipped Shannon can find similar applications in the design for test and self-repair of multiple-valued logic circuits, utilizing more general regular structures such as three-dimensional (3-D) lattices as shown in [5,13,18]. The fourth advantage is that all the new multiple-valued canonical expansions can be used in the implementation of various types of the corresponding lattice structures linking the expansion choice to technology issues, like improvements in area, speed, power, and testability.

The remainder of this Chapt. is organized as follows. The basic definitions of the fundamental binary expansions and their multiple-valued extensions are given in Sect. 2.1. A new methodology for the creation and classification of new Galois-based transforms and examples of such transforms are presented in Sect. 2.2. A Summary of the Chapt. is presented in Sect. 2.3.

2.1 Normal Galois Forms in Logic Synthesis

Normal canonical forms play an important role in the synthesis of logic circuits [113,213,217,219]. This role includes testing, synthesis, and optimization. The main algebraic structure which is used in this work for developing the canonical normal forms is the Galois field (GF) algebraic structure, which is a fundamental algebraic structure in the theory of algebras [56,67,87,146,160,166].

Galois field has proven high efficiency in various applications such as in logic synthesis and computer engineering, communications, information systems and computer science, and mathematics. This includes items like: design for test [124], reversible logic synthesis (cf. Sect. 5.4 in Chapt. 5) [6], error correction codes [48], cryptography, number theory, and proving Fermat's last theorem [251]. The importance of Galois field results from the fact that every finite field is isomorphic to a Galois field [146]. In general, the attractive properties of GF-based circuits, such as the high testability of such circuits, are due to the fact that the GF operators exhibit the Cyclic Group (Latin Square) Property [67]. This property can be explained, for example, using the four-valued (quaternary) GF operators as shown in Figs. 2.1e and 2.1f, respectively. Note that in any row and column of the addition table (Fig. 2.1e), the elements are all different, which is cyclic, and that the elements have a different order in each row and column. Another cyclic group can be observed in the multiplication table; if the zero elements are removed from the multiplication table (Fig. 2.1f), then the remaining elements form a cyclic group. In binary, for example, the GF(2) addition operator, EXOR, has the cyclic group property.

+	0	1
0	0	1
1	1	0

a

*	0	1
0	0	0
1	0	1

b

+	0	1	2
0	0	1	2
1	1	2	0
2	2	0	1

c

*	0	1	2
0	0	0	0
1	0	1	2
2	0	2	1

d

+	0	1	2	3
0	0	1	2	3
1	1	0	3	2
2	2	3	0	1
3	3	2	1	0

e

*	0	1	2	3
0	0	0	0	0
1	0	1	2	3
2	0	2	3	1
3	0	3	1	2

f

Fig. 2.1. Galois field addition and multiplication tables: **a** $GF_2(+)$, **b** $GF_2(*)$, **c** $GF_3(+)$, **d** $GF_3(*)$, **e** $GF_4(+)$, and **f** $GF_4(*)$.

Reed-Muller based normal forms have been classified using the Green/Sasao hierarchy [4,52]. The Green/Sasao hierarchy of families of canonical forms and corresponding decision diagrams is based on three generic expansions: (1) Shannon [43,217], (2) positive Davio [265,266,217], and (3) negative Davio [265,266,217] expansions. Since Shannon and Davio expansions play an essential role in logic synthesis, the corresponding generalized Green/Sasao hierarchy of families of canonical forms and corresponding decision diagrams have been developed [4]. Shannon, positive Davio, and negative Davio decompositions are given below:

$$f(x_1,x_2,\ldots,x_n) = x_1' \cdot f_0(x_1,x_2,\ldots,x_n) \oplus x_1 \cdot f_1(x_1,x_2,\ldots,x_n),$$

$$= \begin{bmatrix} \bar{x}_1 & x_1 \end{bmatrix} \begin{bmatrix} 1 & 0 \\ 0 & 1 \end{bmatrix} \begin{bmatrix} f_0 \\ f_1 \end{bmatrix}, \tag{2.1}$$

$$f(x_1,x_2,\ldots,x_n) = 1 \cdot f_0(x_1,x_2,\ldots,x_n) \oplus x_1 \cdot f_2(x_1,x_2,\ldots,x_n),$$

$$= \begin{bmatrix} 1 & x_1 \end{bmatrix} \begin{bmatrix} 1 & 0 \\ 1 & 1 \end{bmatrix} \begin{bmatrix} f_0 \\ f_1 \end{bmatrix}, \tag{2.2}$$

$$f(x_1,x_2,\ldots,x_n) = 1 \cdot f_1(x_1,x_2,\ldots,x_n) \oplus x_1' \cdot f_2(x_1,x_2,\ldots,x_n),$$

$$= [1 \quad \bar{x}_1] \begin{bmatrix} 0 & 1 \\ 1 & 1 \end{bmatrix} \begin{bmatrix} f_0 \\ f_1 \end{bmatrix}, \tag{2.3}$$

where $f_0(x_1,x_2,\ldots,x_n) = f(0,x_2,\ldots,x_n) = f_0$ is the negative cofactor of variable x_1, $f_1(x_1,x_2,\ldots,x_n) = f(1,x_2,\ldots,x_n) = f_1$ is the positive cofactor of variable x_1, and $f_2(x_1,x_2,\ldots,x_n) = f(0,x_2,\ldots,x_n) \oplus f(1,x_2,\ldots,x_n) = f_0 \oplus f_1$.

An arbitrary n-variable function $f(x_1, x_2, \ldots, x_n)$ can be represented using the Positive Polarity Reed-Muller (PPRM) expansion as follows [217,239]:

$$f(x_1,x_2, \ldots,x_n) = a_0 \oplus a_1x_1 \oplus a_2x_2 \oplus \ldots \oplus a_nx_n \oplus a_{12}x_1x_2 \oplus a_{13}x_1x_3 \oplus a_{n-1,n}x_{n-1}x_n \oplus \ldots \oplus a_{12\ldots n}x_1x_2\ldots x_n. \tag{2.4}$$

For each function f, the coefficients a_i in Eq. (2.4) are determined uniquely, so PPRM is a canonical form. If we use either only the positive literal or only the negative literal for each variable in Eq. (2.4) we obtain the Fixed Polarity Reed-Muller (FPRM) form. There are 2^n possible combinations of polarities and as many FPRMs for any given logic function. If we freely choose the polarity of each literal in Eq. (2.4), we obtain Generalized Reed-Muller (GRM) form. In GRMs, contrary to FPRMs, the same variable can appear in both positive and negative polarities. There are $n.2^{(n-1)}$ literals in Eq. (2.4), so there are $2^{n.2^{(n-1)}}$ polarities for an n-variable function and as many GRMs [217]. Each of the polarities determines a unique set of coefficients, and thus each GRM is a canonical representation of a function. Two other types of expansions result from the flattening of certain binary trees that will produce Kronecker (KRO) forms and Pseudo Kronecker (PKRO) forms for Shannon, positive Davio, and negative Davio expansions [217]. There are 3^n and at most $3^{2^{(n)}-1}$ different KROs and PKROs, respectively.

The good selection of the various permutations in using the Shannon and Davio expansions (in addition to other expansions like Walsh, Arithmetic, etc) as internal nodes in decision trees (DTs) and diagrams (DDs) will result in DTs and DDs, that represent the corresponding logic functions, with smaller sizes in terms of the total number of hierarchical levels used, and the total number of internal nodes needed [217]. The minimization of the size of DD, to represent a logic function, will result in speeding up the manipulations of logic functions using DD as data structure, and the minimization of the use of memory space during the execution of such manipulations.

We call the Shannon and Davio expansions presented in this Sect. the "Fundamental" spectral transforms in order to distinguish them from the more generalized case of the "Invariant" Shannon and Davio expansions that will be presented in Sect. 2.2. One can observe that by going from PPRM to GRM forms, less restrictions (constraints) are imposed on the canonical forms due to the enlarged set of polarities that one can choose from. The gain of more freedom (less constraints) on the polarity of the canonical expansions will provide an advantage of obtaining Exclusive-Sum-Of-Product (ESOP) expressions with less number of terms and literals, and consequently expressing Boolean functions using ESOP forms will produce on average expressions with less size as if compared to Sum-Of-Product (SOP) expressions for example. Table 2.1 illustrates these observations [217].

In general, a literal can be defined as any function of a single variable. Basis functions in the general case of multiple-valued expansions are constructed using literals. Galois field Sum-Of-Products expansions can be performed on variety of literals.

Table 2.1. The number of product terms required to realize some arithmetic functions using different Reed-Muller forms.

Function	PPRM	FPRM	GRM	ESOP	SOP
adr4	34	34	34	31	75
log8	253	193	105	96	123
nrm4	216	185	96	69	120
rdm8	56	56	31	31	76
rot8	225	118	51	35	57
sym9	210	173	126	51	84
wgt8	107	107	107	58	255

For example, one can use, among others: K-Reduced Post literal (K-RPL) to produce K-RPL-GFSOP [4,9], Post literal (PL) to produce PL-GFSOP, Window literal (WL) to produce WL-GFSOP, Generalized (Post) literal (GL) to produce GL-GFSOP, or Universal literal (UL) to produce UL-GFSOP. Figure 2.2 demonstrates set-theoretic relationships beween the various literals, where the shaded Reduced Post literal is the type of literal that will be used through this Book. (Note that RPL is analogous to the delta function in the continuous domain.)

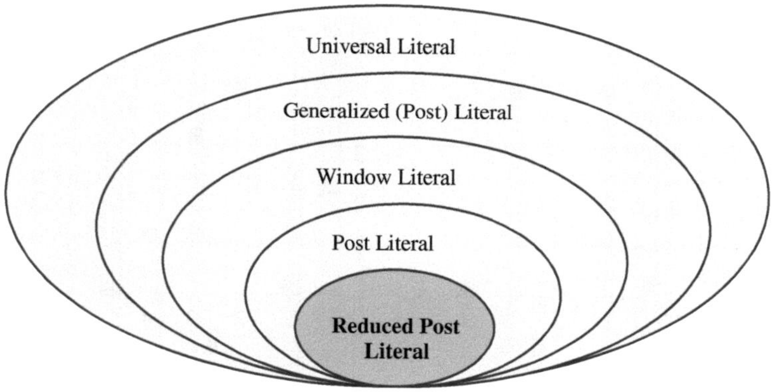

Fig. 2.2. Inclusion relationship of various types of literals.

Example 2.1. Figure 2.3 demonstrates several literal types, where one proceeds from the simplest literal in Fig. 2.3a (i.e., RPL) to the most complex literal in Fig. 2.3e. For RPL in Fig. 2.3a, a value (K) is produced by the literal when the value of the variable is equal to a specific state, and in this particular example a value of K = 1 is generated by the 1-RPL when the value of variable x is equal to certain state (here this state is equal to one). Figure 2.3b shows PL where the value generated by the literal at a specific state is equal to the maximum value (i.e., radix) of that logic. WL in Fig. 2.3c generates a value equal to the radix for a "window" of specific states. GPL in Fig. 2.3d produces a value of radix for a set of distinct states. One notes that, in contrast to the other literals, universal literal (UL) in Fig. 2.3e can have any value of the logic system at distinct states, and thus universal literals have the highest complexity among the five different types of literals.

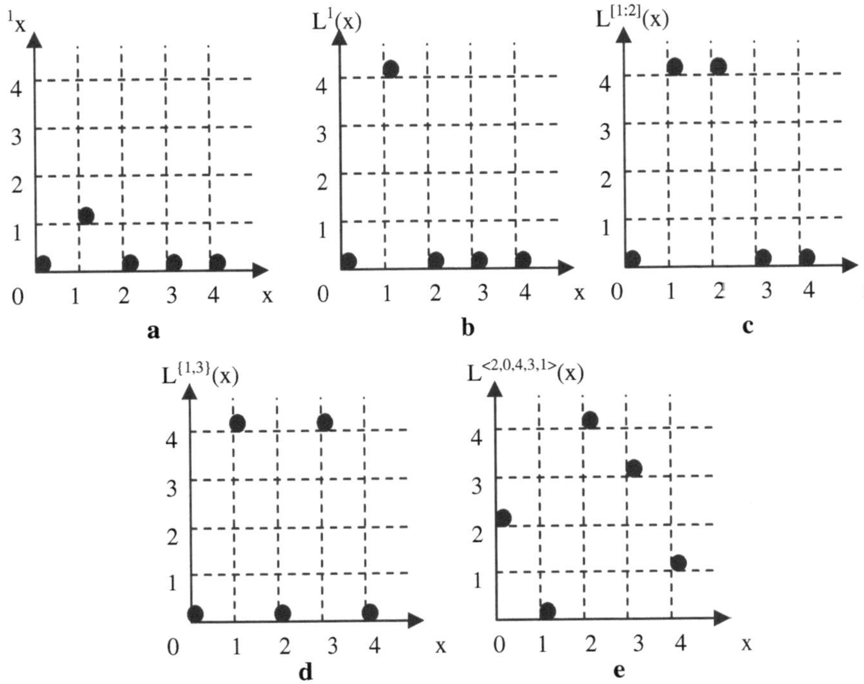

Fig. 2.3. Illustrating the different types of literals over an arbitrary five-radix logic: **a** 1-Reduced Post literal (RPL), **b** Post literal (PL), **c** Window literal (WL), **d** Generalized (Post) literal (GL), and **e** Universal literal (UL).

Since K-RPL-GFSOP is as simple as PL and it is simpler from implementation point of view than WL, GL or UL, we will perform all the GFSOP expansions utilizing 1-RPL-GFSOP. Let us define the 1-Reduced Post Literal as:

$$^ix = 1 \text{ iff } x = i \text{ else } ^ix = 0. \tag{2.5}$$

For example ^0x, ^1x, ^2x are the zero, first, and second polarities of the 1-Reduced Post Literal, respectively. Also, let us define the ternary shifts (over variable x) as x, x', x" as the zero, first, and second shifts of the variable x respectively (i.e., x = x + 0, x' = x +1, and x" = x + 2, respectively), and x can take any value in the set {0,1,2}. We chose to represent the 1-Reduced Post Literals in terms of shifts and powers, among others, because of the ease of the implementation of powers of shifted variables in hardware (for the production of RPL, see the Universal Logic Modules (ULMs) in

Appendix D). The fundamental Shannon expansion over GF(3) for a ternary function with a single variable is shown in the following theorem.

Theorem 2.1. Shannon expansion over GF(3) for a function with single variable is:

$$f = {}^{0}x f_0 + {}^{1}x f_1 + {}^{2}x f_2, \tag{2.6}$$

where f_0 is cofactor of f with respect to variable x of value 0, f_1 is cofactor of f with respect to variable x of value 1, and f_2 is cofactor of f with respect to variable x of value 2.

Proof. From Eq. (2.5), if we substitute the values of the 1-Reduced Post Literal in Eq. (2.6), we obtain the following Eqs.:

For $x = 0 \Rightarrow f_{x=0} = f_0$.

For $x = 1 \Rightarrow f_{x=1} = f_1$.

For $x = 2 \Rightarrow f_{x=2} = f_2$.

which are the cofactors of variable x of value 0, of value 1, and of value 2, respectively. **Q.E.D.**

Example 2.2. Let $f(x_1,x_2) = x_1'x_2 + x_2''x_1$.

Then the ternary truth vector of the function f is: $F = [0,2,1,1,2,0,2,2,2]^T$. Using Eq. (2.6), we obtain the following ternary Shannon expansion over *GF(3)* of the above function f (x_1,x_2):

$$f = {}^{0}x_1\,{}^{1}x_2 + 2\cdot{}^{0}x_1\,{}^{2}x_2 + 2\cdot{}^{1}x_1\,{}^{0}x_2 + 2\cdot{}^{1}x_1\,{}^{1}x_2 + 2\cdot{}^{1}x_1\,{}^{2}x_2 + {}^{2}x_1\,{}^{0}x_2 + 2\cdot{}^{2}x_1\,{}^{2}x_2.$$

Using the addition and multiplication over GF(3), and the axioms of GF(3), it can be shown that the 1-Reduced Post Literals defined in Eq. (2.5), are related to the shifts of variables over GF(3) in terms of powers as follows:

$${}^{0}x = 2(x)^2 + 1, \tag{2.7}$$
$${}^{0}x = 2(x')^2 + 2(x'), \tag{2.8}$$
$${}^{0}x = 2(x'')^2 + x'', \tag{2.9}$$
$${}^{1}x = 2(x)^2 + 2(x), \tag{2.10}$$
$${}^{1}x = 2(x')^2 + x', \tag{2.11}$$
$${}^{1}x = 2(x'')^2 + 1, \tag{2.12}$$
$${}^{2}x = 2(x)^2 + x, \tag{2.13}$$
$${}^{2}x = 2(x')^2 + 1, \tag{2.14}$$

$$^{2}x = 2(x'')^2 + 2(x''), \tag{2.15}$$

where ^{0}x, ^{1}x, ^{2}x are the zero, first, and second polarities of the 1-Reduced Post Literal, respectively, and x, x', x" are the zero, first, and second shifts of the variable x respectively. The variable x can take any value of the set {0,1,2}. After the substitution of Eqs. (2.7) through (2.15) in Eq. (2.6), and after the minimization of the terms according to the axioms of Galois field, one obtains the following Eqs.:

$$f = 1\cdot f_0 + x\cdot(2f_1+f_2) + 2(x)^2(f_0+f_1+f_2), \tag{2.16}$$
$$f = 1\cdot f_2 + x'\cdot(2f_0+f_1) + 2(x')^2(f_0+f_1+f_2), \tag{2.17}$$
$$f = 1\cdot f_1 + x''\cdot(2f_2+f_0) + 2(x'')^2(f_0+f_1+f_2). \tag{2.18}$$

Equations (2.6) and (2.16) through (2.18) are the ternary fundamental Shannon and Davio expansions for single variable, respectively. These Eqs. can be rewritten in the following matrix-based forms:

$$f = [\,{}^0x \ \ {}^1x \ \ {}^2x\,] \begin{bmatrix} 1 & 0 & 0 \\ 0 & 1 & 0 \\ 0 & 0 & 1 \end{bmatrix} \begin{bmatrix} f_0 \\ f_1 \\ f_2 \end{bmatrix}, \tag{2.19}$$

$$f = [1 \ \ x \ \ x^2] \begin{bmatrix} 1 & 0 & 0 \\ 0 & 2 & 1 \\ 2 & 2 & 2 \end{bmatrix} \begin{bmatrix} f_0 \\ f_1 \\ f_2 \end{bmatrix}, \tag{2.20}$$

$$f = [1 \ \ x' \ \ (x')^2] \begin{bmatrix} 0 & 0 & 1 \\ 2 & 1 & 0 \\ 2 & 2 & 2 \end{bmatrix} \begin{bmatrix} f_0 \\ f_1 \\ f_2 \end{bmatrix}, \tag{2.21}$$

$$f = [1 \ \ x'' \ \ (x'')^2] \begin{bmatrix} 0 & 1 & 0 \\ 1 & 0 & 2 \\ 2 & 2 & 2 \end{bmatrix} \begin{bmatrix} f_0 \\ f_1 \\ f_2 \end{bmatrix}. \tag{2.22}$$

We observe that Eqs. (2.19) - (2.22) are expansions for a single variable. Yet, these expansions can be recursively generated for arbitrary number of variables (N) using the Kronecker (tensor) product ($\otimes$), analogous to the binary case [4,9,88,238]. This can be expressed formally as in the following forms for ternary Shannon (S), and Davio (D_0, D_1, and D_2) expansions, respectively:

$$f = \bigotimes_{i=1}^{N} [{}^0x_i \ {}^1x_i \ {}^2x_i] \bigotimes_{i=1}^{N} [S][\vec{F}], \tag{2.23}$$

$$f = \bigotimes_{i=1}^{N} [1 \ x_i \ x_i^2] \bigotimes_{i=1}^{N} [D_0][\vec{F}], \tag{2.24}$$

$$f = \bigotimes_{i=1}^{N} [1 \ x_i' \ (x_i')^2] \bigotimes_{i=1}^{N} [D_1][\vec{F}], \tag{2.25}$$

$$f = \bigotimes_{i=1}^{N} [1 \ x_i'' \ (x_i'')^2] \bigotimes_{i=1}^{N} [D_2][\vec{F}]. \tag{2.26}$$

(The name "tensor product" is due to the fact that the growth of a transform matrix is in the form of a matrix of matrix elements.)

Analogously to the binary case, we can have expansions that are mixed of Shannon (S) for certain variables and Davio (D_0, D_1, and D_2) for the other variables. This will lead, analogously to the binary case, to the Kronecker Ternary Decision Trees (TDTs). Moreover, the mixed expansions can be extended to include Pseudo Kronecker TDT. (Full discussion of these TDTs that correspond to various expansions, as well as their hierarchy will be included in Chapt. 3).

Analogously to the ternary case, quaternary Shannon expansion over GF(4) for a function with single variable is [9]:

$$f = {}^0x\,f_0 + {}^1x\,f_1 + {}^2x\,f_2 + {}^3x\,f_3, \tag{2.27}$$

where f_0 is the cofactor of f with respect to variable x of value 0, f_1 is the cofactor of f with respect to variable x of value 1, f_2 is the cofactor of f with respect to variable x of value 2, and f_3 is the cofactor of f with respect to variable x of value 3.

Example 2.3. Let f $(x_1,x_2) = x_1''x_2 + x_2'''x_1$. The quaternary truth vector of this function f is F = $[0,3,1,2,2,1,3,0,3,0,2,1,1,2,0,3]^T$. Utilizing Eq. (2.27), we obtain the following quaternary Shannon expansion over GF(4) of the function f:

$f = 2 \cdot {}^0x_1\,{}^1x_2 + 3 \cdot {}^0x_1\,{}^2x_2 + {}^0x_1\,{}^3x_2 + 3 \cdot {}^1x_1\,{}^0x_2 + \cdot\,{}^1x_1\,{}^1x_2 + 2 \cdot {}^1x_1\,{}^3x_2 + {}^2x_1\,{}^0x_2 + 3 \cdot {}^2x_1\,{}^1x_2 + 2 \cdot {}^2x_1\,{}^2x_2 + 2 \cdot {}^3x_1\,{}^0x_2 + {}^3x_1\,{}^2x_2 + 3 \cdot {}^3x_1\,{}^3x_2$.

Using the axioms of GF(4), it can be derived that the 1-RPL defined in Eq. (2.5) are related to the shifts of variables over GF(4) in terms of powers as follows:

$$^{0}x = x^{3} + 1, \tag{2.28}$$
$$^{0}x = x' + (x')^{2} + (x')^{3}, \tag{2.29}$$
$$^{0}x = 3(x'') + 2(x'')^{2} + (x'')^{3}, \tag{2.30}$$
$$^{0}x = 2(x''') + 3(x''')^{2} + (x''')^{3}, \tag{2.31}$$
$$^{1}x = x + (x)^{2} + (x)^{3}, \tag{2.32}$$
$$^{1}x = (x')^{3} + 1, \tag{2.33}$$
$$^{1}x = 2(x'') + 3(x'')^{2} + (x'')^{3}, \tag{2.34}$$
$$^{1}x = 3(x''') + 2(x''')^{2} + (x''')^{3}, \tag{2.35}$$
$$^{2}x = 3(x) + 2(x)^{2} + (x)^{3}, \tag{2.36}$$
$$^{2}x = 2(x') + 3(x')^{2} + (x')^{3}, \tag{2.37}$$
$$^{2}x = (x'')^{3} + 1, \tag{2.38}$$
$$^{2}x = x''' + (x''')^{2} + (x''')^{3}, \tag{2.39}$$
$$^{3}x = 2(x) + 3(x)^{2} + (x)^{3}, \tag{2.40}$$
$$^{3}x = 3(x') + 2(x')^{2} + (x')^{3}, \tag{2.41}$$
$$^{3}x = x'' + (x'')^{2} + (x'')^{3}, \tag{2.42}$$
$$^{3}x = (x''')^{3} + 1, \tag{2.43}$$

where ^{0}x, ^{1}x, ^{2}x, ^{3}x are the zero, first, second, and third polarities of the 1-Reduced Post literal, respectively. Also, x, x', x'', x''' are the zero, first, second, and third shifts (inversions) of the variable x respectively, and variable x can take any value of the set {0, 1, 2, 3}. Analogous to the ternary case, we chose to represent the 1-Reduced Post literal in terms of shifts and powers, among others, because of the ease of the implementation of powers of shifted variables in hardware. After the substitution of Eqs. (2.28) through (2.43) in Eq. (2.27), and after the rearrangement and reduction of the terms according to the axioms of GF(4), we obtain the following Eqs.:

$$f = 1\cdot f_0 + x\,(f_1+3f_2+2f_3) + (x)^{2}\,(f_1+2f_2+3f_3)+(x)^{3}(f_0+f_1+f_2+f_3), \tag{2.44}$$
$$f = 1\cdot f_1 + (x')(f_0+2f_2+3f_3) + (x')^{2}\,(f_0+3f_2+2f_3)+(x')^{3}\,(f_0+f_1+f_2+f_3), \tag{2.45}$$
$$f = 1\cdot f_2 + (x'')(3f_0+2f_1+f_3) + (x'')^{2}\,(2f_0+3f_1+f_3)+(x'')^{3}\,(f_0+f_1+f_2+f_3), \tag{2.46}$$
$$f = 1\cdot f_3 + (x''')(f_2+3f_1+2f_0) + (x''')^{2}\,(f_2+2f_1+3f_0)+(x''')^{3}\,(f_0+f_1+f_2+f_3). \tag{2.47}$$

Equations (2.27) and (2.44) through (2.47) are the 1-Reduced Post literal quaternary Shannon and Davio expansions for single variable, respectively. These Eqs. can be rewritten in the following matrix-based convolution-like forms, respectively:

$$f = [{}^{0}x \;\; {}^{1}x \;\; {}^{2}x \;\; {}^{3}x] \begin{bmatrix} 1 & 0 & 0 & 0 \\ 0 & 1 & 0 & 0 \\ 0 & 0 & 1 & 0 \\ 0 & 0 & 0 & 1 \end{bmatrix} \begin{bmatrix} f_0 \\ f_1 \\ f_2 \\ f_3 \end{bmatrix}, \tag{2.48}$$

$$f = [1 \;\; x \;\; x^2 \;\; x^3] \begin{bmatrix} 1 & 0 & 0 & 0 \\ 0 & 1 & 3 & 2 \\ 0 & 1 & 2 & 3 \\ 1 & 1 & 1 & 1 \end{bmatrix} \begin{bmatrix} f_0 \\ f_1 \\ f_2 \\ f_3 \end{bmatrix}, \tag{2.49}$$

$$f = [1 \;\; x' \;\; (x')^2 \;\; (x')^3] \begin{bmatrix} 0 & 1 & 0 & 0 \\ 1 & 0 & 2 & 3 \\ 1 & 0 & 3 & 2 \\ 1 & 1 & 1 & 1 \end{bmatrix} \begin{bmatrix} f_0 \\ f_1 \\ f_2 \\ f_3 \end{bmatrix}, \tag{2.50}$$

$$f = [1 \;\; x'' \;\; (x'')^2 \;\; (x'')^3] \begin{bmatrix} 0 & 0 & 1 & 0 \\ 3 & 2 & 0 & 1 \\ 2 & 3 & 0 & 1 \\ 1 & 1 & 1 & 1 \end{bmatrix} \begin{bmatrix} f_0 \\ f_1 \\ f_2 \\ f_3 \end{bmatrix}, \tag{2.51}$$

$$f = [1 \;\; x''' \;\; (x''')^2 \;\; (x''')^3] \begin{bmatrix} 0 & 0 & 0 & 1 \\ 2 & 3 & 1 & 0 \\ 3 & 2 & 1 & 0 \\ 1 & 1 & 1 & 1 \end{bmatrix} \begin{bmatrix} f_0 \\ f_1 \\ f_2 \\ f_3 \end{bmatrix}. \tag{2.52}$$

One can observe, that Eqs. (2.48) through (2.52) are expansions for single variable. Yet, these canonical expressions can be generated for arbitrary number of variables (N) using the Kronecker (tensor) product. This can be expressed formally as in the following discrete convolution-like forms for Shannon (S), and Davio (D_0, D_1, D_2, and D_3) expressions, respectively:

$$f \;=\; \bigotimes_{i=1}^{N} [{}^{0}x_i \;\; {}^{1}x_i \;\; {}^{2}x_i \;\; {}^{3}x_i] \bigotimes_{i=1}^{N} [S][\vec{F}], \tag{2.53}$$

$$f = \bigotimes_{i=1}^{N} [\,1 \;\; x_i \;\; x_i^{2} \;\; x_i^{3}\,] \bigotimes_{i=1}^{N} [D_0][\vec{F}\,], \tag{2.54}$$

$$f = \bigotimes_{i=1}^{N} [\,1 \;\; x_i' \;\; (x_i')^{2} \;\; (x_i')^{3}\,] \bigotimes_{i=1}^{N} [D_1][\vec{F}\,], \tag{2.55}$$

$$f = \bigotimes_{i=1}^{N} [\,1 \;\; x_i'' \;\; (x_i'')^{2} \;\; (x_i'')^{3}\,] \bigotimes_{i=1}^{N} [D_2][\vec{F}\,], \tag{2.56}$$

$$f = \bigotimes_{i=1}^{N} [\,1 \;\; x_i''' \;\; (x_i''')^{2} \;\; (x_i''')^{3}\,] \bigotimes_{i=1}^{N} [D_3][\vec{F}\,]. \tag{2.57}$$

The following Sect. introduces generalizations of the multiple-valued fundamental expansions introduced in this Sect. These generalizations will be used in Chapt. 4 for building and minimizing the size of 2-D and 3-D classical lattice structures, Chapt. 6 for constructing reversible lattice structures, and in Chapt. 10 for constructing the quantum counterparts of reversible lattice structures.

2.2 Invariant Multi-Valued Families of Generalized Spectral Transforms

In this Sect. we present the invariant multiple-valued Galois field based spectral transforms, and their generalized notation. The new scaled expansions can be used to produce minimal size circuits for the three-dimensional lattice structures which will be presented in Chapt. 4. Also, the new scaled expansions will be used for the construction of a new type of logic primitives (as will be shown in Fig. 2.4) that implement "weights" into their inputs. Such new primitives can be useful in technological implementations where weighted inputs are used to realize logic functionalities.

2.2.1 General Notation for Operations on Transform Matrices

The following notation describes the operations on a transform matrix M over GF(K) [5,12]:

$$M_D \rightarrow M_O^{\,p_0 \quad \cdots \; p_{K-1} \;|\; q_0 \quad \cdots \; q_{K-1}}, \tag{2.58}$$

$$\rightarrow M_O^{\,\alpha \;\; \beta \quad \cdots \; \gamma \;|\; 1 \;\; 1 \quad \cdots \; 1}, \tag{2.59}$$

where M_D is the Derived (Modified) Matrix, M_O is the Original Matrix. The symbols p_0, p_1, …, p_{K-1} are row multiplication numbers $\in$ GF(K), {0,1,…,K-1} are indices referring to row_0, row_1,…, and row_{K-1}. The symbols q_0,q_1,…, q_{K-1} are column multiplication numbers $\in$ GF(K), and {0,1,…,K-1} are indices referring to $column_0$, $column_1$,…, and $column_{K-1}$. The operations performed utilizing the upper notation are done through the multiplication of all the elements of row_i of the matrix M_o by p_i and then multiplying each resulting element of ($row_i \cdot p_i$) by q_j, where i, j = 0, 1, …, k-1 (i.e., $\prod_{i,\forall j} (p_i \cdot q_j \cdot row_i)$). The mathematical interpretation of this notation, in terms of matrices, is as follows: if D is a diagonal matrix $\Rightarrow$ D = Diag (α, β, …, γ), then $M_D = D \cdot M_o \Rightarrow M_D^{-1} = (D \cdot M_o)^{-1} = M_o^{-1} D^{-1}$. The following Eq. can be applied to obtain the functional expansions for any modified transform matrix:

$$f \;=\; M_s^{-1} \;\; M \;\; \vec{F}\,, \tag{2.60}$$

where M is the transform matrix, and $\vec{F}$ is the truth vector of the function f.

Example 2.4. Let us produce some modified matrices from their unmodified counterparts utilizing the proposed notation for Galois field of radix four.

$$\text{GF(4): } T = \begin{bmatrix} 3 & 2 & 2 & 1 \\ 2 & 1 & 0 & 2 \\ 0 & 2 & 3 & 1 \\ 2 & 0 & 3 & 2 \end{bmatrix} \Rightarrow T \rightarrow T^{\,1231|3312} = \begin{array}{c} \begin{bmatrix} 3 & 2 & 2 & 1 \\ 2 & 1 & 0 & 2 \\ 0 & 2 & 3 & 1 \\ 2 & 0 & 3 & 2 \end{bmatrix} \begin{array}{l} \bullet\mathbf{1} \\ \bullet\mathbf{2} \\ \bullet\mathbf{3} \\ \bullet\mathbf{1} \end{array} \\ \bullet\mathbf{3} \;\; \bullet\mathbf{3} \;\; \bullet\mathbf{1} \;\; \bullet\mathbf{2} \end{array},$$

$$= \begin{bmatrix} 2 & 1 & 2 & 2 \\ 2 & 1 & 0 & 1 \\ 0 & 3 & 2 & 1 \\ 1 & 0 & 3 & 3 \end{bmatrix}.$$

$$\text{GF(4):} \quad T = \begin{bmatrix} 3 & 2 & 2 & 1 \\ 2 & 1 & 0 & 2 \\ 0 & 2 & 3 & 1 \\ 2 & 0 & 3 & 2 \end{bmatrix} \Rightarrow T \rightarrow T^{2332|3333} = \begin{bmatrix} 3 & 2 & 2 & 1 \\ 2 & 1 & 0 & 2 \\ 0 & 2 & 3 & 1 \\ 2 & 0 & 3 & 2 \end{bmatrix},$$

$$= \begin{bmatrix} 3 & 2 & 2 & 1 \\ 3 & 2 & 0 & 3 \\ 0 & 3 & 1 & 2 \\ 2 & 0 & 3 & 2 \end{bmatrix}.$$

2.2.2 Invariant Families of Multi-Valued Spectral Transforms

To introduce Theorems 2.1, 2.2, and 2.3, the following definitions are presented (where p is a prime number and k is a natural number of value $k \geq 1$).

Definition 2.1. The transform matrix that is generated by multiplying the rows of GF(p^k) Shannon matrix by the numbers $\{\alpha, \beta, \ldots, \gamma\} \in$ GF(p^k) respectively is called $\alpha\beta\ldots\gamma$ IS (invariant Shannon) matrix.

Definition 2.2. The transform matrix that is generated by multiplying the rows of GF(p^k) Davio of type t (denoted by D_t) matrix by the numbers $\{\alpha, \beta, \ldots, \gamma\} \in$ GF(p^k) respectively is called $\alpha\beta\ldots\gamma$ ID_t (invariant Davio of type t) matrix, where $t \in$ GF(p^k).

Definition 2.3. The transform matrix that is generated by multiplying the rows of GF(p^k) flipped Shannon matrix by the numbers $\{\alpha, \beta, \ldots, \gamma\} \in$ GF(p^k) respectively is called $\alpha\beta\ldots\gamma$ IfS (invariant flipped Shannon) matrix, where $t \in$ GF(p^k).

Definition 2.4. For GF(n) where $n = p^k$, a family of transforms is defined as a set that contains one Shannon transform and the corresponding n Davio transforms.

Definition 2.5. The Family (set) of spectral transforms that has the members (elements) of $\{\alpha\beta\ldots\gamma\text{IS, D}\}$ is called $\alpha\beta\ldots\gamma$ IS/D (invariant Shannon/Davio).

Definition 2.6. The Family (set) of spectral transforms that has the members (elements) of $\{\alpha\beta\ldots\gamma\text{ID, S}\}$ is called $\alpha\beta\ldots\gamma$ ID/S (invariant Davio/Shannon).

Definition 2.7. The total Family (set) of spectral transforms that has the members (elements) of $\{\alpha_1\beta_1\ldots\gamma_1\text{IS}, \alpha_2\beta_2\ldots\gamma_2\text{ID}\}$ is called $\alpha_1\beta_1\ldots\gamma_1\text{IS}/\alpha_2\beta_2\ldots\gamma_2\text{ID}$ (invariant Shannon/invariant Davio).

The following theorems are valid for arbitrary GF(n) fields for $n = p^k$, where p is a prime number and k is a natural number of value $k \geq 1$. Full details and proofs of all counts are presented in Appendix A.

Theorem 2.2. For $\{\alpha, \beta, \ldots, \gamma\} \in \text{GF}(n)$, there exist $(n-1)^n$ of $\alpha\beta\ldots\gamma$ IS nonsingular spectral transforms.

Proof. In general, Shannon matrix for GF(n) is the identity matrix I:

$$S = I = \begin{bmatrix} 1 & 0 & \ldots & 0 \\ 0 & 1 & \ldots & 0 \\ \ldots & \ldots & \ldots & \ldots \\ 0 & 0 & \ldots & 1 \end{bmatrix}.$$

If the rows of the matrix S are multiplied by $\{\alpha, \beta, \ldots, \gamma\}$ respectively, we obtain:

$$S^{\alpha\beta\ldots\gamma|11\ldots1} = \begin{bmatrix} \alpha & 0 & 0\ldots & 0 \\ 0 & \beta & 0\ldots & 0 \\ \ldots & \ldots & \ldots & \ldots \\ 0 & 0 & 0\ldots & \gamma \end{bmatrix}.$$

The inverse of such matrix is:

$$\left[{}_{S}\alpha\beta...\gamma|11...1\right]^{-1} = \begin{bmatrix} \hat{\alpha} & 0 & 0... & 0 \\ 0 & \hat{\beta} & 0... & 0 \\ ... & ... & ... & ... \\ 0 & 0 & 0... & \hat{\gamma} \end{bmatrix},$$

where: $\alpha\hat{\alpha} = 1, \beta\hat{\beta} = 1,..., \gamma\hat{\gamma} = 1$. Utilizing Eq. (2.60), we get the following invariant (p-adic) Shannon expansion:

$$f = [\hat{\alpha}\,{}^{0}x \;\; \hat{\beta}\,{}^{1}x \ldots \hat{\gamma}\,{}^{k-1}x] \begin{bmatrix} \alpha & 0 & 0... & 0 \\ 0 & \beta & 0... & 0 \\ ... & ... & ... & ... \\ 0 & 0 & 0... & \gamma \end{bmatrix} \vec{F}, \tag{2.61}$$

where ${}^{k-1}x$ is the 1-Reduced Post Literal defined previously. Note that Eq. (2.61) is an expansion of the function, f, that always preserves the values of the function (i.e., the cofactors) and does not transform the truth vector into a different domain. Therefore, the same set of Davio expansions (that correspond to the various invariant Shannon expansions) will always be produced. Consequently, the number of transform families is equal to the number of different invariant Shannon transforms that can be obtained. **Q.E.D.**

Theorem 2.3. For {α, β, ..., γ} ∈ GF(n), there exist $(n\text{-}1)^n$ of αβ...γ ID nonsingular spectral transforms per Davio type, and $n(n\text{-}1)^n$ of αβ...γ ID nonsingular spectral transforms for all Davio types of expansions.

Proof. Let us produce the proof for a single type of Davio (D) expansion in a third radix GF. Although this proof is for one type of Davio expansion and for the special case of GF(3), similar and straightforward proofs can be provided systematically for other Davio expansions of an arbitrary radix of GF (p^k), where p is a prime number and k is a natural number of value k ≥ 1. D_0 matrix for GF(3) is:

$$D_0 = \begin{bmatrix} 1 & 0 & 0 \\ 0 & 2 & 1 \\ 2 & 2 & 2 \end{bmatrix}.$$

If the rows of the matrix D_0 are multiplied by {α, β, γ} respectively, we obtain:

$$D_0^{\alpha\beta\gamma|111} = \begin{bmatrix} \alpha & 0 & 0 \\ 0 & 2\beta & \beta \\ 2\gamma & 2\gamma & 2\gamma \end{bmatrix}.$$

Utilizing Eq. (2.60), we get the following invariant (p-adic) D_0 functional expansion (for a single type of Shannon expansions I_{3x3}):

$$f = [\ \hat{\alpha} \quad \hat{\beta}\, x \quad \hat{\gamma}\, x^2] \begin{bmatrix} \alpha & 0 & 0 \\ 0 & 2\beta & \beta \\ 2\gamma & 2\gamma & 2\gamma \end{bmatrix} \vec{F}, \tag{2.62}$$

where $\alpha\hat{\alpha} = 1, \beta\hat{\beta} = 1, \ldots, \gamma\hat{\gamma} = 1$. Note that Eq. (2.62) is an expansion of the function, f, that always preserves the value of the function (i.e., the cofactors) and does not transform the truth vector into a different domain. Therefore, the same Shannon expansion will always be produced. **Q.E.D.**

As a consequence of Theorem 2.2, we observe that there exist for a certain radix of Galois field a fixed number of Davio expansions (i.e., same forms of Davio expansions), which is equal to the radix of GF, and many invariant Shannon expansions (i.e., different forms of Shannon expansions). Also, as a consequence of Theorem 2.3, we observe that there exist for a certain radix of Galois field and a certain type of Davio expansions a single Shannon expansion (i.e., one form) and many invariant Davio expansions (i.e., different forms of Davio expansions).

Theorem 2.4. For {α, β, …, γ} ∈ GF(n), there exist $(n-1)^n$ of αβ…γ IfS nonsingular spectral transforms.

Proof. In general, the flipped Shannon matrix for GF(n) is the matrix:

$$fS = \begin{bmatrix} 0 & 0 & ... & 1 \\ 0 & 0 & ..1 & 0 \\ ... & ... & ... & ... \\ 1 & 0 & ... & 0 \end{bmatrix}.$$

If the rows of the matrix of fS are multiplied by {α, β, …, γ} respectively, we obtain:

$$fS^{\alpha\beta...\gamma|11...1} = \begin{bmatrix} 0 & 0 & 0... & \alpha \\ 0 & 0 & ...\beta & 0 \\ ... & ... & ... & ... \\ \gamma & 0 & 0... & 0 \end{bmatrix}.$$

The inverse of such matrix is:

$$\left[fS^{\alpha\beta...\gamma|11...1}\right]^{-1} = \begin{bmatrix} 0 & 0 & 0... & \hat{\gamma} \\ ... & ... & ... & ... \\ 0 & \hat{\beta} & 0... & 0 \\ \hat{\alpha} & 0 & 0... & 0 \end{bmatrix},$$

where: $\alpha\hat{\alpha} = 1, \beta\hat{\beta} = 1, ..., \gamma\hat{\gamma} = 1$. Utilizing Eq. (2.60), we get the following invariant (p-adic) flipped Shannon expansion

$$f = [\hat{\alpha}\,{}^{k-1}x \;\; \hat{\beta}\,{}^{k-2}x ... \hat{\gamma}\,{}^{0}x] \begin{bmatrix} 0 & 0 & ...0 & \alpha \\ 0 & 0 & ...\beta & 0 \\ ... & ... & ... & ... \\ \gamma & 0 & 0... & 0 \end{bmatrix} \vec{F}, \qquad (2.63)$$

where ${}^{k-1}x$ is the 1-Reduced Post Literal defined previously. Eq. (2.63) is an expansion of function f, that always preserves the values of the function (i.e., the cofactors) and does not transform the truth vector into a different domain. **Q.E.D.**

Utilizing Eq. (2.61), all permutations of the invariant Shannon expansion can be obtained to produce the Invariant Permuted Shannon Expansions (IPSE). The invariant flipped Shannon in Eq. (2.63) represents one special case of such permutations. Similar permutations can be also obtained for the invariant Davio expansions in Eq. (2.62) and the other types of Davio expansions to produce the Invariant Permuted Davio Expansions (IPDE). One potential important utilization of such permutations is in the reduction of the size of the corresponding lattice structures analogous to the results for the binary case in [50,51].

Multi-valued spectral transforms that are generated by Theorems 2.2, 2.3, and 2.4 can be produced for an arbitrary number of variables utilizing the Kronecker-based *recursion.* From Theorems 2.2, 2.3, and 2.4, one can note the following interesting property of the new transforms: the basis functions of the new sets of multi-valued transforms are exactly the same as the basis functions of the fundamental Shannon, Davio, and flipped Shannon expansions but scaled by constants (i.e., $\hat{\alpha}, \hat{\beta},..., \hat{\gamma}$). Moreover, these constants are not generated arbitrarily; they are the multiplicative inverses of the corresponding constants that scale the rows of the corresponding basic multi-valued Shannon, Davio, and flipped Shannon transform matrices (i.e., the constants $\{ \alpha, \beta,..., \gamma \}$ in Eqs. (2.61), (2.62), and (2.63) respectively), and can be directly calculated according to the axioms of the Galois field which is operated upon. For illustration, Example 2.5 illustrates the use of Theorem 2.2.

Example 2.5. Utilizing Definition 2.1 and Theorem 2.2, The following is one of the invariant Shannon transform matrices that can be produced in GF(5):

$$S^{34222|21142} = \begin{bmatrix} 4 & 0 & 0 & 0 & 0 \\ 0 & 4 & 0 & 0 & 0 \\ 0 & 0 & 2 & 0 & 0 \\ 0 & 0 & 0 & 1 & 0 \\ 0 & 0 & 0 & 0 & 3 \end{bmatrix},$$

$$\Rightarrow [S^{34222|21142}]^{-1} = \begin{bmatrix} 2 & 0 & 0 & 0 & 0 \\ 0 & 2 & 0 & 0 & 0 \\ 0 & 0 & 4 & 0 & 0 \\ 0 & 0 & 0 & 1 & 0 \\ 0 & 0 & 0 & 0 & 3 \end{bmatrix},$$

$$f = [\,2\cdot{}^{0}x \quad 2\cdot{}^{1}x \quad 4\cdot{}^{2}x \quad 1\cdot{}^{3}x \quad 3\cdot{}^{4}x\,]\begin{bmatrix}4&0&0&0&0\\0&4&0&0&0\\0&0&2&0&0\\0&0&0&1&0\\0&0&0&0&3\end{bmatrix}\vec{F}\,.$$

Example 2.6. Utilizing Definition 2.2, and Theorem 2.3, The following is one of the invariant D_2 transform matrices that can be produced in GF(3):

$$D_2^{\,212|111} = \begin{bmatrix}0&2&0\\1&0&2\\1&1&1\end{bmatrix} \Rightarrow f = [\,2 \quad x \quad 2\cdot x^{2}\,]\begin{bmatrix}0&2&0\\1&0&2\\1&1&1\end{bmatrix}\vec{F}\,.$$

Example 2.7. The following logic circuits represent a comparison between the logic primitives of the fundamental ternary Shannon decomposition versus the invariant ternary Shannon decomposition, respectively.

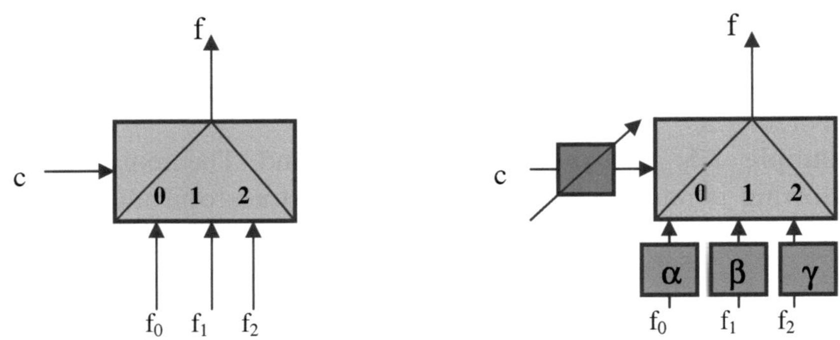

$$f = \left[{}^{0}c \quad {}^{1}c \quad {}^{2}c\right]\begin{bmatrix}1&0&0\\0&1&0\\0&0&1\end{bmatrix}\begin{bmatrix}f_0\\f_1\\f_2\end{bmatrix}$$

a

$$f = \left[\hat{\alpha}\,{}^{0}c \quad \hat{\beta}\,{}^{1}c \quad \hat{\gamma}\,{}^{2}c\right]\begin{bmatrix}\alpha&0&0\\0&\beta&0\\0&0&\gamma\end{bmatrix}\begin{bmatrix}f_0\\f_1\\f_2\end{bmatrix}$$

b

Fig. 2.4. Fundamental and invariant Shannon decompositions: **a** logic primitive for the ternary fundamental Shannon decomposition, and **b** logic primitive for the ternary invariant Shannon decomposition.

The resulting non-singular transforms in Theorems 2.2, 2.3, and 2.4 are a subset within the set of linearly independent (LI) transforms which is a subset of the whole space of singular transforms. By linearly independent spectral transforms we mean all possible transforms that have transform matrices for which no single column is a linear combination of the other columns, and no single row is a linear combination of the other rows. By singular transforms we mean all possible transforms that have transform matrices for which at least a single column is a linear combination of the other columns, and a single row is a linear combination of the other rows. Figure 2.5 illustrates a set-theoretic relationship between the non-permuted new set of spectral transforms and other sets of spectral transforms, where the shaded area represents the new sets of multiple-valued invariant Shannon and invariant Davio spectral transforms. Appendix A provides full counts of the new families over an arbitrary GF(n) fields for $n = p^k$, where p is a prime number and k is a natural number of value $k \geq 1$.

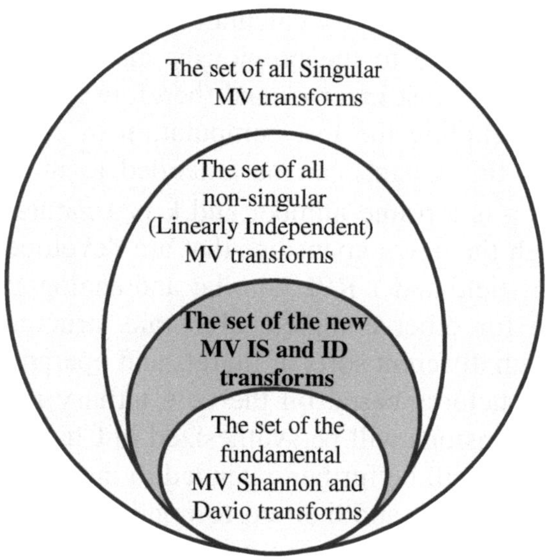

Fig. 2.5. A set-theoretic relationship between families of non-permuted multiple-valued spectral transforms.

Galois field forms that are introduced throughout this Chapt. can be realized in the corresponding GFSOP logic circuits. Some of the multiple-valued logic circuits that are used for the realization of GFSOP expressions are illustrated in Appendix B.

2.3 Summary

In this Chapt. we introduced a systematic method to create and classify new multiple-valued invariant non-singular spectral transforms based on multi-valued fundamental Shannon expansions and Davio expansions over an arbitrary radix of Galois field.

The new spectral transforms will have an application in the construction of regular layout in three-dimensions as will be shown in Chapt. 4. The new spectral transforms have an important property: their basis functions are exactly the same as the basis functions of the fundamental Shannon and Davio expansions but scaled by constants (i.e., $\hat{\alpha}, \hat{\beta}, \ldots, \hat{\gamma}$). Moreover, these constants are not generated arbitrarily; they are the multiplicative inverses of the corresponding constants that scale the rows of the corresponding basic multi-valued Shannon, Davio, and flipped Shannon transform matrices, and can be directly calculated according to the axioms of the Galois field. Due to the previously mentioned property, these transforms possess fast inverses and therefore are suitable for many applications including the fast computation of spectral transforms. All results in this Chapt. can be extended to an arbitrary GF(p^k) fields, where p is a prime number and k is a natural number $k \geq 1$. Also, although the new expansions that are developed in this Chapt. are for Galois field and 1-RPL, similar and analogous developments can be done for other complete algebraic structures and spectral transforms with different sorts of literals and operations.

Lattice structures based on the new ternary invariant Shannon and Davio expansions will be synthesized in Chapt. 4. The new 3-D lattice structures will be further extended to include reversible lattice structures in Chapt. 6, and their corresponding quantum circuits will be introduced in Chapt. 10. The new primitive in Fig. 2.4b will be extended to reversible logic in Chapt. 5 and will be used in Chapt. 6 to build reversible binary Shannon lattice structures. Also, the new families of multiple-valued invariant Shannon and Davio expansions that were introduced in this Chapt. will be fully generalized to include the reversible counterparts of such new expansions in Chapt. 5, from which new reversible primitives and structures will be constructed in the following Chapts.

3 New Multiple-Valued S/D Trees and their Canonical Galois Field Sum-Of-Product Forms

Economical and highly testable implementations of Boolean functions [99,198,199,204,217], based on Reed-Muller (AND-EXOR) logic, play an important role in logic synthesis and circuit design. AND-EXOR circuits include canonical forms (i.e., expansions that are unique representations of a Boolean function). Several large families of canonical forms: Fixed Polarity Reed-Muller (FPRM) forms, Generalized Reed-Muller (GRM) forms, Kronecker (KRO) forms, and Pseudo-Kronecker (PSDKRO) forms, referred to as the Green/Sasao hierarchy, have been described [4,9]. Because canonical families have higher testability and some other properties desirable for efficient synthesis, especially of some classes of functions, they are widely investigated. A similar ternary version of the binary Green/Sasao hierarchy was developed in [4]. This new hierarchy will find applications in minimizing Galois field Sum-Of-Product (GFSOP) expressions (i.e., expressions that are in the sum-of-product form which uses the additions and multiplications of arbitrary radix Galois field that was introduced in Chapt. 2), creation of new forms, decision diagrams, and regular structures (Such new structures will be discussed in details in Appendix D.)

The state-of-the-art minimizers of Exclusive Sum-Of-Product (ESOP) expressions [80,85,114,157,214,234,235,242] (i.e., expressions that are in the sum-of-product form which uses the addition and multiplication of Galois field of radix two that was introduced in Figs. 2.1a and 2.1b, respectively) are based on heuristics and give the exact solution only for functions with a small number of variables. The formulation for finding the exact ESOP was given in [52], but all known exact algorithms can deliver solutions for not all but only certain functions of more than five variables. Because GFSOP minimization is even more difficult, it is important to investigate structural properties and the counts of their canonical subfamilies.

Recently, two families of binary canonical Reed-Muller forms, called Inclusive Forms (IFs) and Generalized Inclusive Forms (GIFs) have been proposed [52]. The second family was the first to include all minimum ESOPs (binary GFSOPs). In this Chapt., we propose, as analogous to the binary case, two general families of canonical ternary Reed-Muller forms, called Ternary Inclusive Forms (TIFs), and their generalization, Ternary Generalized Inclusive Forms (TGIFs). The second family includes minimum GFSOPs over ternary Galois field GF(3). One of the basic motivations in this work is the application of these TIFs and TGIFs to find the minimum GFSOP for multiple-valued inputs multiple-valued outputs for reversible logic synthesis using, for instance, reversible cascades in Chapt. 8, a problem that has not yet been solved.

An ESOP minimizer for completely specified functions has been developed [157]. This minimizer does not work for functions with don't cares. The ESOP minimizer from [235] works for functions with few percent of don't cares, yet this minimizer does not work for functions with a high number of don't cares (like in machine learning where don't cares comprise more than 99% of the values of the functions). The best minimizer for functions with a high number of don't cares is based on the use of genetic algorithms from [80]. Yet, this type of minimizer is for binary input binary output functions and is restricted to GRM polarities only. The multiple-valued S/D tree developed in this Chapt. provides more general polarity of Inclusive Form (IF) polarity, which contains the GRM as a special case. A GFSOP minimizer based on IF polarity will be used to minimize the multiple-valued ESOP (GFSOP) expression for a given function, as will be shown in Sect. 3.7, to realize the logic function using reversible structures such as the reversible Cascades that are presented in Chapt. 8. GFSOP evolutionary algorithm for minimization using S/D trees will be presented in this Chapt. The main contributions of this Chapt. are:

- Multiple-valued Galois Shannon/Davio (S/D) trees.
- The generation and count of two new families of canonical multiple-valued Reed-Muller forms, called multiple-valued Inclusive Forms (IFs), and their generalization, multiple-valued

Generalized Inclusive Forms (GIFs). A new extended Green/Sasao hierarchy of families and forms with a new sub-family for multiple-valued Reed-Muller logic is also introduced.

- An evolutionary algorithm that implements the IF polarity from S/D trees to find the minimum GFSOP.

The remainder of this Chapt. is organized as follows: Green/Sasao hierarchy of binary canonical forms is presented in Sect. 3.1. The concept of S/D trees and Inclusive Forms is presented in Sect. 3.2. The ternary S/D trees and their corresponding Inclusive Forms and Generalized Inclusive Forms are presented in Sect. 3.3. Properties of the ternary Inclusive Forms and their Ternary Generalized Inclusive Forms are presented in Sect. 3.4. The new extended Green/Sasao hierarchy is presented in Sect. 3.5. Quaternary S/D trees are presented in Sect. 3.6. An evolutionary algorithm for the minimization of GFSOP expressions using the Inclusive Forms polarity for the corresponding S/D trees will be presented in Sect. 3.7. A Summary of the Chapt. is presented in Sect. 3.8. Although we discuss the ternary and quaternary cases, all results can be extended to an arbitrary $GF(p^k)$ fields, where p is a prime number and k is a natural number of value $k \geq 1$.

3.1 Green/Sasao Hierarchy of Binary Canonical Forms

The Green/Sasao hierarchy of families of canonical forms and corresponding decision diagrams is based on three generic expansions, Shannon, positive Davio, and negative Davio expansions. This includes [217]: Shannon Decision Trees and Diagrams, Positive Davio Decision Trees and Diagrams, Negative Davio Decision Trees and Diagarms, Fixed Polarity Reed-Muller Decision Trees and Diagrams, Kronecker Decision Trees and Diagrams, Pseudo Reed-Muller Decision Trees and Diagrams, pseudo Kronecker Decision Trees and Diagrams, and Linearly-Independent Decision Trees and Diagrams. A set-theoretic relationship between families of canonical forms over GF(2) was proposed and extended in [52] by introducing binary IF, GIF, and FGIF forms. Figure 3.1 illustrates the set-theoretic relationship between families of canonical forms over GF(2).

Analogously to the Green/Sasao hierarchy of binary Reed-Muller families of spectral transforms over GF(2) that is shown in Fig. 3.1, we will introduce the extended Green/Sasao hierarchy of spectral transforms, with a new sub-family, for ternary Reed-Muller logic over GF(3) in Sect. 3.5.

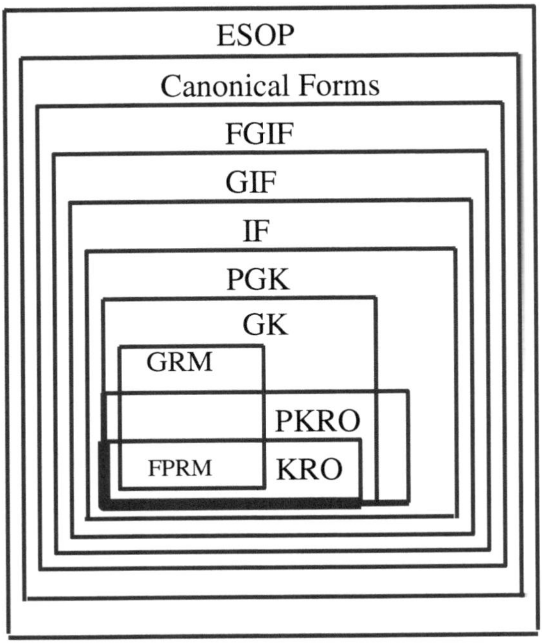

Fig. 3.1. Set-theoretic relationship between families of canonical forms over GF(2).

3.2 Binary S/D Trees and their Inclusive Forms

Two general families of DDs were introduced in [52]. These families are based on the Shannon expansion and the Generalized Davio expansion, and are produced using the S/D Trees. These families are called the Inclusive Forms (IFs) and the Generalized Inclusive Forms (GIFs), respectively. It was proven [52] that these forms include a minimum ESOP. The expansions over GF(2) are shown in Fig. 3.2, where Fig. 3.2d shows the new expansion, which is based on binary Davio expansions, called generalized Davio (D) expansion that generates the negative and positive Davio expansions as special cases.

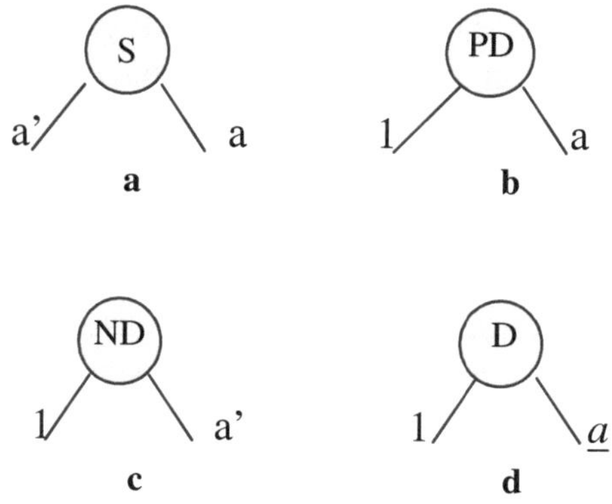

Fig. 3.2. Two-valued expansions: **a** Shannon, **b** positive Davio, **c** negative Davio, and **d** generalized Davio expansions.

The S/D trees for IFs of two variables of order {a,b}, and the S/D trees for IFs of two variables of order {b,a} were fully illustrated [52]. The set of Generalized Inclusive Forms (GIFs) for two variables is the union of the two sets of Inclusive Forms (IFs). The total number of the GIFs is equal to:

$$\# GIF = 2\cdot(\# IF_{a,b}) - \# (IF_{a,b} \cap IF_{b,a}). \tag{3.1}$$

Thus for two variables:
$\# IF_{a,b} = 1 + 2 + 2 + 4 + 4 + 8 + 8 + 16 = 45$,
$\# IF_{b,a} = 1 + 2 + 2 + 4 + 4 + 8 + 8 + 16 = 45$,
$\# GIF = 2\cdot(45) - (1 + 4 + 4 + 16) = 65$.

Properties and experimental results of the binary Inclusive Forms and the binary Generalized Inclusive Forms were investigated [52], where it was proven that GIFs include a minimum ESOP.

3.3 Ternary S/D Trees and their Inclusive Forms and Generalized Inclusive Forms

The following Sect. defines the ternary Shannon and ternary Davio decision trees over GF(3). As analogous to the binary case, we can have expansions that are mixed of Shannon (S) for certain variables and Davio (D_0, D_1, and D_2) for the other variables. This will lead, analogously to the binary case, to the Kronecker TDT. Moreover, the mixed expansions can be extended to include Pseudo Kronecker TDT. (Full discussion of these TDTs that correspond to various expansions, as well as their hierarchy will be included in Sect. 3.5). The basic S, D_0, D_1, and D_2 ternary expansions (i.e., flattened forms) over GF(3) can be represented in Ternary DTs (TDTs) and the corresponding varieties of Ternary DDs (TDDs) (according to the corresponding reduction rules that are used). For one variable (one level), Fig. 3.3 represents the expansion nodes for S, D_0, D_1, and D_2, respectively.

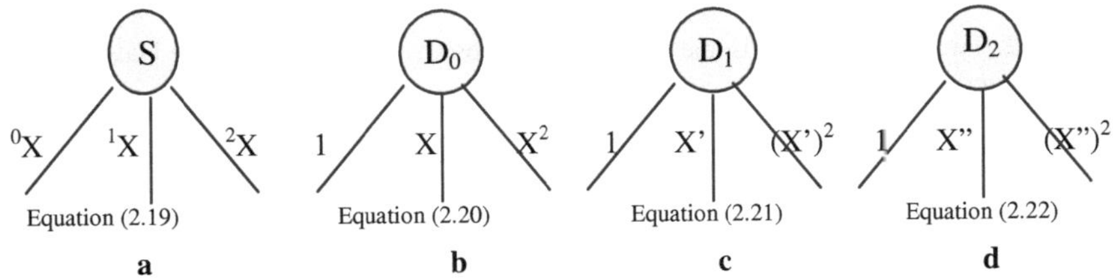

Fig. 3.3. Ternary expansion nodes for ternary DTs: **a** Shannon, **b** $Davio_0$, **c** $Davio_1$, and **d** $Davio_2$.

Utilizing Fig. 3.3, the following Sect. defines the Ternary S/D trees, and Ternary Inclusive Forms (TIFs), respectively.

3.3.1 Ternary S/D trees and Inclusive Forms

In correspondence to the binary S/D trees, we can produce the Ternary S/D Trees. To define the Ternary S/D Trees we will define the Generalized Davio expansion over GF(3) as shown in Fig. 3.4:

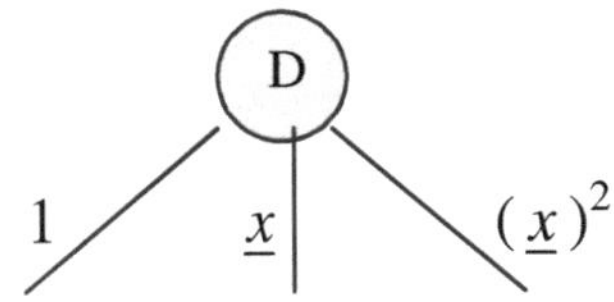

Fig. 3.4. Generalized ternary Davio expansion.

Our notation here is that ($\underline{x}$) corresponds to the three possible shifts of the variable x as follows:

$$\underline{x} \in \{x, x', x''\} \text{ over } GF(3). \tag{3.2}$$

Definition 3.1. The ternary tree with ternary Shannon and ternary Generalized Davio expansion nodes, that generates other ternary trees, is called the Ternary Shannon/Davio (S/D) tree.

Utilizing the definition of ternary Shannon (Fig. 3.3a) and ternary generalized Davio (Fig. 3.4), we obtain the ternary Shannon/Davio trees (ternary S/D trees) for two variables as shown in Fig. 3.5. From the ternary S/D DTs shown in Fig. 3.5, if we take any S/D tree and multiply the second-level cofactors (which are in the TDT leaves) each by the corresponding path in that TDT, and sum all the resulting cubes (terms) over GF(3), we obtain the flattened form of the function f, as a certain GFSOP expression. For each TDT in Fig. 3.5, there are as many forms obtained for the function f as the number of possible permutations of the polarities of the variables in the second-level branches of each TDT.

Definition 3.2. The family of all possible forms obtained per ternary S/D tree are called Ternary Inclusive Forms (TIFs).

The numbers of these TIFs per TDT for two variables are shown on top of each S/D TDT in Fig. 3.5. (General formalisms to obtain the exact number of TIFs for any number of variables over GF(3) are presented in Appendix C.)

By observing Fig. 3.5, we can generate the flattened forms by two methodologies. A classical methodology, per analogy with well-known binary forms, would be to create every transform matrix for every TIF S/D tree, and then expand using that transform matrix. A better methodology is to create one flattened form (expansion over certain transform matrix, i.e., certain TIF), and then transform systematically from one form to another form, without the need to create all transform matrices from the corresponding S/D trees. This general approach can lead to several algorithms of various complexity that generalize the binary algorithms to obtain FPRM, KRM, GRM, and IF forms, including the butterfly methods [89,57].

Example 3.1. Using the result of Example 2.2 for the expansion of $f(x_1,x_2)$ in terms of ternary Shannon expansion (that resembles the S/D tree for Shannon expansions in both levels as seen in Fig. 3.5):

$$f = {}^{0}x_1\,{}^{1}x_2 + 2\cdot{}^{0}x_1\,{}^{2}x_2 + 2\cdot{}^{1}x_1\,{}^{0}x_2 + 2\cdot{}^{1}x_1\,{}^{1}x_2 + 2\cdot{}^{1}x_1\,{}^{2}x_2 + {}^{2}x_1\,{}^{0}x_2 + 2\cdot{}^{2}x_1\,{}^{2}x_2. \qquad (3.3)$$

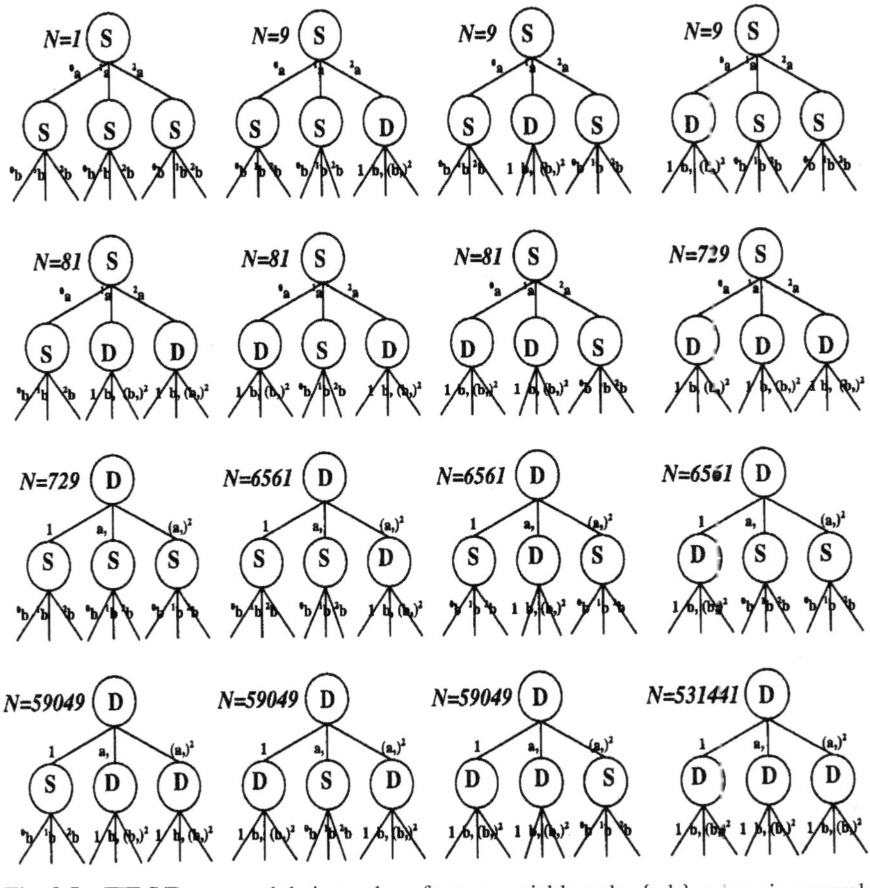

Fig. 3.5a. TIF S/D trees and their numbers for two-variable order {a,b}, where in general a, and b, are defined as in Eq. (3.2).

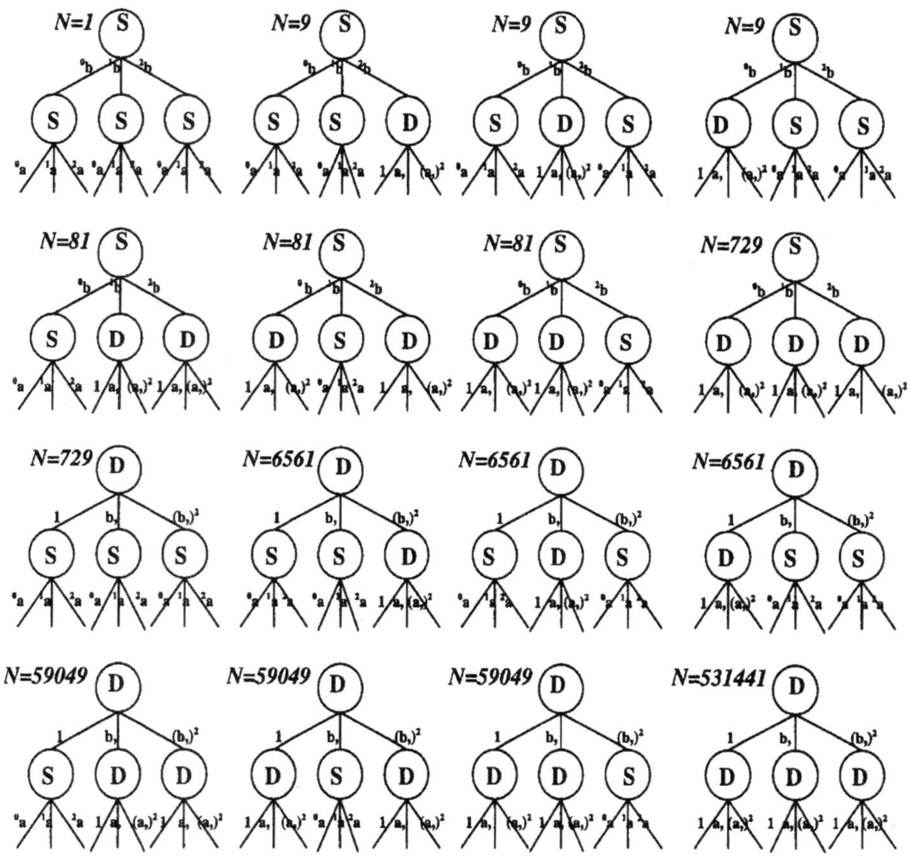

Fig. 3.5b. TIF S/D trees and their numbers for two-variable order {b,a}, where in general a, and b, are defined as in Eq. (3.2).

We can substitute any of Eqs. (2.7) through (2.15), or a mix of these Eqs., to transform one flattened form to another. For example, if we substitute Eq. (2.7) and Eq. (2.11), we obtain:

$$f = (2(x_1)^2 + 1)(2(x_2')^2 + x_2') + 2(2(x_1)^2 + 1)\,{}^2x_2 + 2(2(x_1')^2 + x_1')(2(x_2)^2 + 1) + 2(2(x_1')^2 + x_1')(2(x_2')^2 + x_2') + 2(2(x_1')^2 + x_1')\cdot{}^2x_2 + {}^2x_1(2(x_2)^2 + 1) + 2 \cdot {}^2x_1\,{}^2x_2. \qquad (3.4)$$

By utilizing the axioms of Galois field, Eq. (3.4) is transformed to:

$$
\begin{aligned}
f = {} & (x_1)^2(x_2')^2 + 2(x_1)^2(x_2') + 2(x_2')^2 + x_2' + (x_1)^2\,({}^2x_2) + \\
& 2({}^2x_2) + 2(x_1')^2(x_2)^2 + (x_1')^2 + (x_1')(x_2)^2 + 2x_1 + \\
& 2(x_1')^2(x_2')^2 + (x_1')^2(x_2') + (x_1')(x_2')^2 + 2x_1'x_2' + (x_1')^2\,{}^2x_2 + \\
& 2(x_1')^2x_2 + 2({}^2x_1)(x_2)^2 + {}^2x_1 + 2({}^2x_1)({}^2x_2).
\end{aligned}
\tag{3.5}
$$

Let us define, as one of possible definitions, the cost of the flattened form (expression) to be:

$$
Cost = \#\ Cubes. \tag{3.6}
$$

We observe that Eq. (3.3) has the cost of seven, while Eq. (3.5) has the cost of 19. Thus, the inverse transformations applied to Eq. (3.5) would lead to Eq. (3.3) and a reduction of cost from 19 to seven. Using the same approach, we can generate a subset of possible GFSOP expressions (flattened forms). Note that all these GFSOP expressions are equivalent (since they produce the same function in different forms). Yet, as can be observed from Eq. (3.5), by further transformations of Eq. (3.3) from one form to another, some transformations produce flattened forms with a smaller number of cubes than the others. From this observation rises the idea of a possible application of evolutionary computing [80] using the S/D trees and related transformations to produce the minimum GFSOPs.

3.3.2 Enumeration of Ternary Inclusive Forms

Each of the S/D trees shown in Fig. 3.5 is a generator of a set of flattened forms (TIFs). Each one of these TIFs is merely a Kronecker-based transform as can be obtained from Eqs. (2.23) through (2.26). The numbers of these TIFs generated by the corresponding S/D trees are shown on the top of each S/D tree for two variables in Fig. 3.5.

Example 3.2.

3.2a. For the S/D trees in Fig. 3.5a, and by utilizing the notation from Eq. (3.2), we obtain for Figs. 3.6a and 3.7a, the ternary trees in Figs. 3.6b - 3.6d and Figs. 3.7b - 3.7d, respectively.

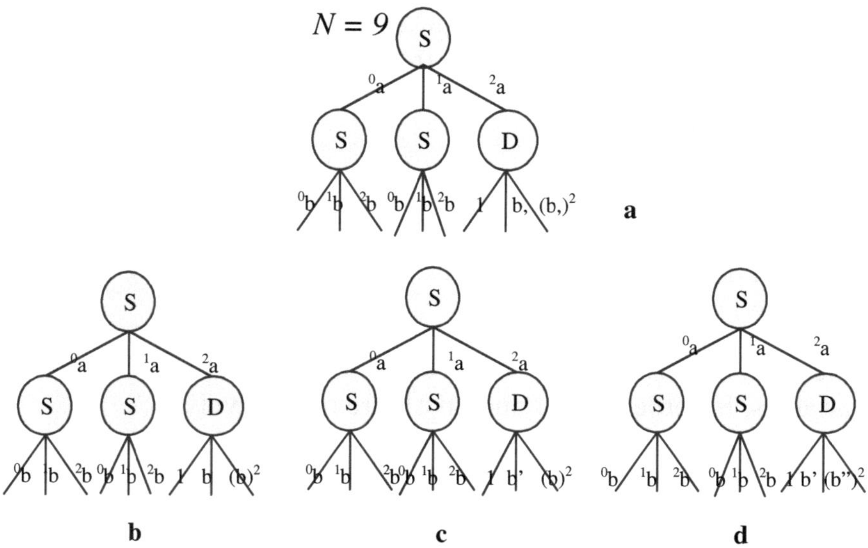

Fig. 3.6. a An S/D tree with three Shannon nodes and one generalized Davio node, and **(b, c, d)** some of the ternary trees that it generates.

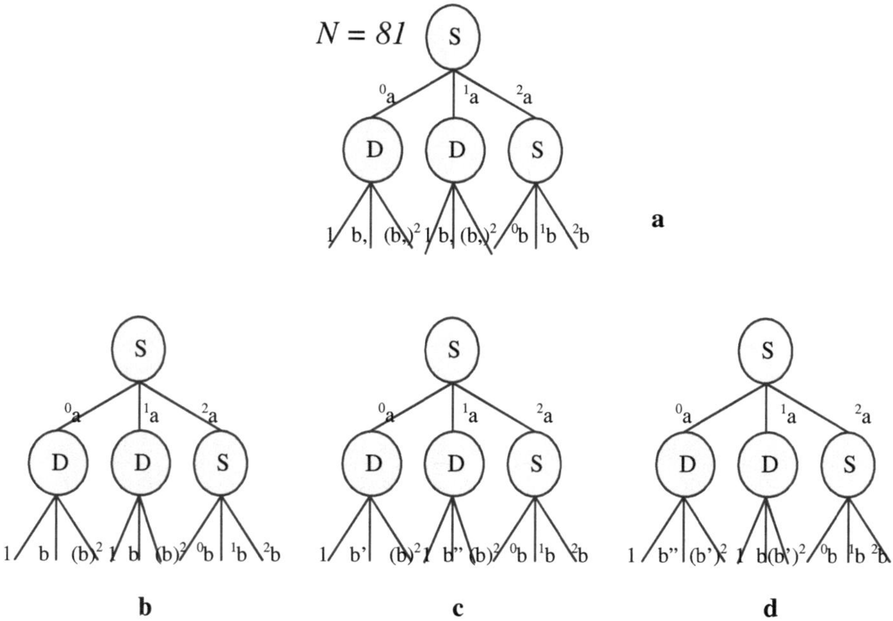

Fig. 3.7. a An S/D tree with two Shannon nodes and two generalized Davio node, and **(b, c, d)** some of the ternary trees that it generates.

3.2b. Let us produce some of the ternary trees for the S/D tree in Fig. 3.5b. Utilizing the notation from Eq. (3.2), we obtain, for the S/D tree in Fig. 3.8a, the ternary trees in Figs. 3.8b, 3.8c, and 3.8d, respectively.
The generalized IFs (GIFs) can be defined as the union of both IFs.

Definition 3.3. The family of forms, which is created as a union of sets of TIFs for all variable orders, is called Ternary Generalized Inclusive Forms (TGIFs).

Theorem 3.1. The total number of the ternary IFs (#TIFs), for two variables, for orders {a,b} and {b,a}, are respectively:

$$\# TIF_{a,b} = 1\cdot(3)^0 + 3\cdot(3)^2 + 3\cdot(3)^4 + 2\cdot(3)^6 + 3\cdot(3)^8 + 3\cdot(3)^{10} + 1\cdot(3)^{12} = 730{,}000, \tag{3.7}$$

$$\# TIF_{b,a} = 1\cdot(3)^0 + 3\cdot(3)^2 + 3\cdot(3)^4 + 2\cdot(3)^6 + 3\cdot(3)^8 + 3\cdot(3)^{10} + 1\cdot(3)^{12} = 730{,}000. \tag{3.8}$$

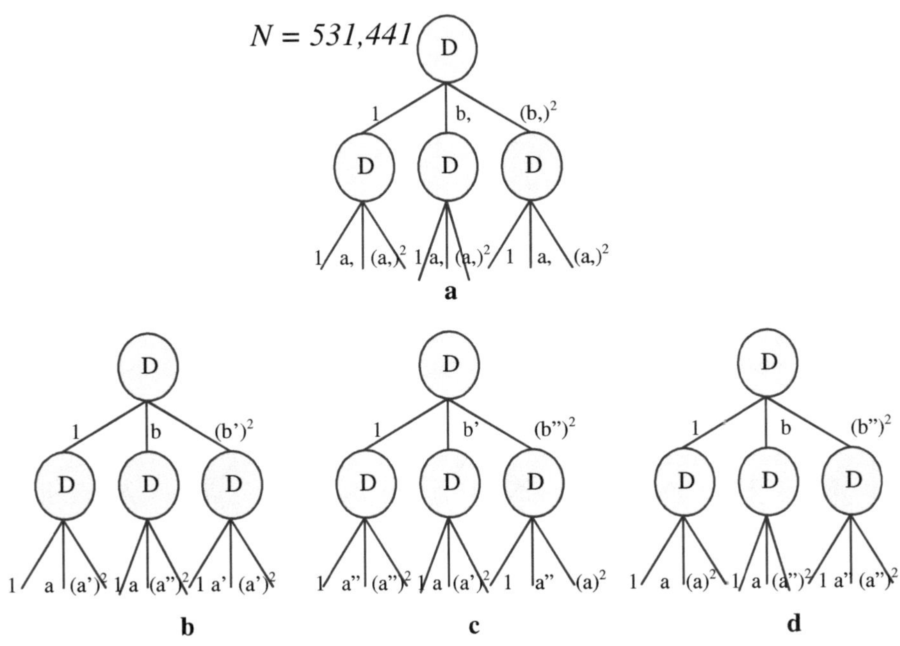

Fig. 3.8. a An S/D tree with four generalized Davio nodes, and **(b, c, d)** some of the ternary trees that it generates.

Proof. By observing Figs. 3.5a and 3.5b, we note that the total number of TIFs for orders {a,b} and {b,a} is the sum of the numbers on the top of S/D trees, that leads to Eqs. (3.7) and (3.8). **Q.E.D.**

The total number of the ternary Generalized IFs (#TGIFs), for two variables, is:

$$\begin{aligned} \#\,TGIF &= \#\,TIF_{a,b} + \#\,TIF_{b,a} - \#\,(\,TIF_{a,b} \cap TIF_{b,a}) \\ &= 2 \cdot \#\,TIF - \#\,(\,TIF_{a,b} \cap TIF_{b,a}), \\ &= 2 \cdot (730{,}000) - (1 \cdot (3)^{0} + 2 \cdot (3)^{6} + 1 \cdot (3)^{12}) = 927{,}100. \end{aligned} \tag{3.9}$$

3.4 Properties of TIFs and TGIFs

The following Sects. present are basic properties of TIFs and TGIFs.

3.4.1 Properties of TIFs

In this Sect., we prove that all TIFs for the given variable ordering are canonical and unique.

Theorem 3.2. Each TIF $\{t_i\}$, $1 \leq i \leq n$, is canonical, i.e., for any function F of the same number of variables, there exists one and only one set of coefficients $\{a_i\}$, such that $F = a_1 t_1 +_{GF(3)} \ldots +_{GF(3)} a_n t_n$.

Proof. In [52] (and references therein), it was shown that an expansion is canonical iff its terms are linearly independent, that is, none of the terms is equal to a linear combination of other terms (over the algebraic field used). Using this fact, it was proven that IFs over GF(2) are canonical. Using an approach which is analogous to the approach presented in [52], one can therefore prove, by induction on the number of variables, that terms in TIFs over ternary Galois field are linearly independent and thus canonical. **Q.E.D.**

3.4.2 Properties of TGIFs

It is easy to see that, for different variable orderings, some forms are *not* repeated while other forms are. For example, Kronecker forms and GRMs over GF(3) are repeated. Therefore the union of sets of TIFs for all variable orders contains more forms than any of the TIF sets taken separately and less forms than the total sum of all of these TIFs.

Theorem 3.3. Ternary Generalized Inclusive Forms (TGIFs) are canonical with respect to the given variable order.

Proof. The proof is analogous to the one in Theorem 3.2. **Q.E.D.**

Generalized Inclusive Forms include GRMs and PKROs over GF(3) as can be shown by considering all possible combinations of literals for all possible orders of variables. If we relax the requirement of fixed variable ordering, and allow any ordering of variables in the branches of the tree but do not allow repetitions of variables in the branches, we generate more general family of forms over GF(3).

Definition 3.4. The family of forms, generated by the S/D tree with no fixed ordering of variables, provided that variables are not repeated along the same branches, is called Ternary Free Generalized Inclusive Forms (TFGIFs).

The studies show that it is difficult to trace the relationship between the number of forms that are repeated for N > 2 and the number of forms that are not.

3.5 An Extended Green/Sasao Hierarchy with a New Sub-Family for Ternary Reed-Muller Logic

Here we introduce the extended Green/Sasao hierarchy with a new sub-family for ternary Reed-Muller logic over GF(3). Definitions 3.2, 3.3, and 3.4 defined the Ternary Inclusive Forms (TIFs), Ternary Generalized Inclusive Forms (TGIFs), and Ternary Free Generalized Inclusive Forms (TFGIFs), respectively. Analogously to the binary Reed-Muller case, we introduce the following definitions over GF(3).

Definition 3.5. The Decision Tree (DT) that results from applying the Ternary Shannon Expansion (Eq. (2.23)) recursively to a ternary input-ternary output logic function (i.e., all levels in a DT) is called Ternary Shannon Decision Tree (TSDT). The result expression (flattened form) from the TSDT is called Ternary Shannon Expression, which is a canonical expression.

Definition 3.6. The Decision Trees (DTs) that result from applying the Ternary Davio expansions (Eqs. (2.24), (2.25), and (2.26)) recursively to a ternary-input ternary -output logic function (i.e., all levels in a DT) are called: Ternary Zero-Polarity Davio Decision Tree (TD_0DT), Ternary First-Polarity Davio Decision Tree (TD_1DT), and Ternary Second-Polarity Davio Decision Tree (TD_2DT), respectively. The resulting expressions (flattened forms) from TD_0DT, TD_1DT, and TD_2DT are called: TD_0, TD_1, and TD_2 expressions, respectively. These expressions are canonical.

Definition 3.7. The Decision Tree (DT) that results from applying any of the Ternary Davio expansions (nodes) for all nodes in each level (variable) in the DT is called Ternary Reed-Muller Decision Tree (TRMDT). The corresponding expression is called Ternary Fixed Polarity Reed-Muller (TFPRM) Expression. This expression is canonical for a given set of polarities.

Definition 3.8. The Decision Tree (DT) that results from using any of the Ternary Shannon (S) or Davio (D_0, D_1, or D_2) expansions (Nodes) for all nodes in each level (variable) in the DT (that has fixed order of variables), is called Ternary Kronecker Decision Tree (TKRODT). The resulting expression is called Ternary Kronecker Expression. This expression is canonical.

Definition 3.9. The Decision Tree (DT) that results from using any of the Ternary Davio expansions (Nodes) for each node (per level) of the DT is called Ternary Pseudo-Reed-Muller Decision Tree (TPRMDT). The resulting expression is called Ternary Pseudo-Reed-Muller Expression.

Definition 3.10. The Decision Tree (DT) that results from using any of the Ternary Shannon Expansion or Ternary Davio expansions (Nodes) for each node (per level) of the DT is called Ternary

Pseudo-Kronecker Decision Tree (TPKRODT). The resulting expression is called Ternary Pseudo-Kronecker Expression.

Definition 3.11. The Decision Tree (DT) that results from using any of the Ternary Shannon Expansion or Ternary Davio expansions (Nodes) for each node (per level) of the DT , disregarding order of variables, provided that variables are not repeated along the same branches, is called Ternary Free Kronecker Decision Tree (TFKRODT). The result is called Ternary Free-Kronecker Expression.

Definition 3.12. The Ternary Kronecker DT that has at least one Ternary Generalized Reed-Muller expansion node is called Ternary Generalized Kronecker Decision Tree (TGKDT). The result is called Ternary Generalized Kronecker Expression.

Definition 3.13. The Ternary Kronecker DT that has at least one TGIF node is called Ternary Generalized Inclusive Forms Kronecker Decision Tree (TGIGKDT). The result is called Ternary Generalized Inclusive Form Kronecker Expression.

Figure 3.9 illustrates this extended Green/Sasao hierarchy with a new sub-family (TGIFK) for ternary Reed-Muller logic over GF(3). TGIF nodes can be realized with Universal Logic Modules (ULMs) for pairs of variables, as shown in Appendix D, analogously as done for binary. Although the S/D trees that have been developed so far are for the ternary radix, interesting properties emerge when applying S/D trees to higher radices, like radix four for example. One important property is that the upper bound counts for S/D trees using the Inclusive Forms Traingle (IF Triangle), which is presented in Appendix C. Thus, next Sect. will introduce the quaternary S/D trees as a generalization for ternary and binary cases.

3.6 Quaternary S/D Trees

The basic S, D_0, D_1, D_2, and D_3 quaternary expansions (i.e., flattened forms) over GF(4) introduced previously in Eqs. (2.53) through (2.57) can be represented in quaternary DTs (QuDTs) and the corresponding varieties of reduced quaternary DDs (RQuDDs) (i.e., according to the corresponding reduction rules that are used). For one variable (i.e., one level of the DT), Fig. 3.10 represents the expansion nodes for S, D_0, D_1, D_2, and D_3, respectively.

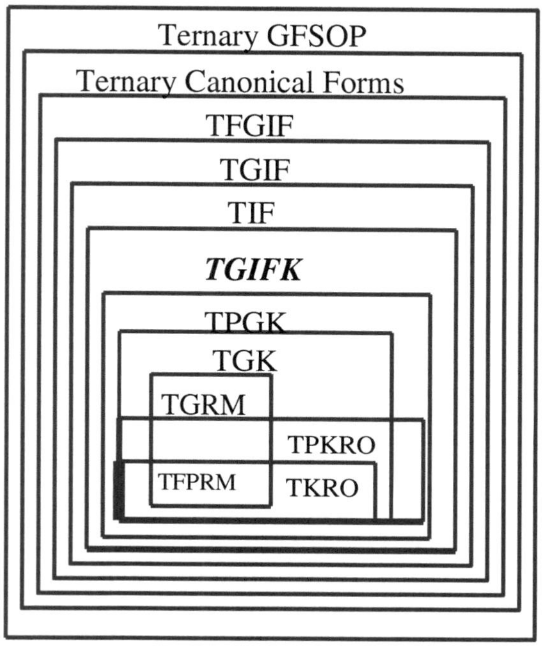

Fig. 3.9. An extended Green/Sasao hierarchy with a new sub-family (TGIFK) for ternary Reed-Muller logic over GF(3).

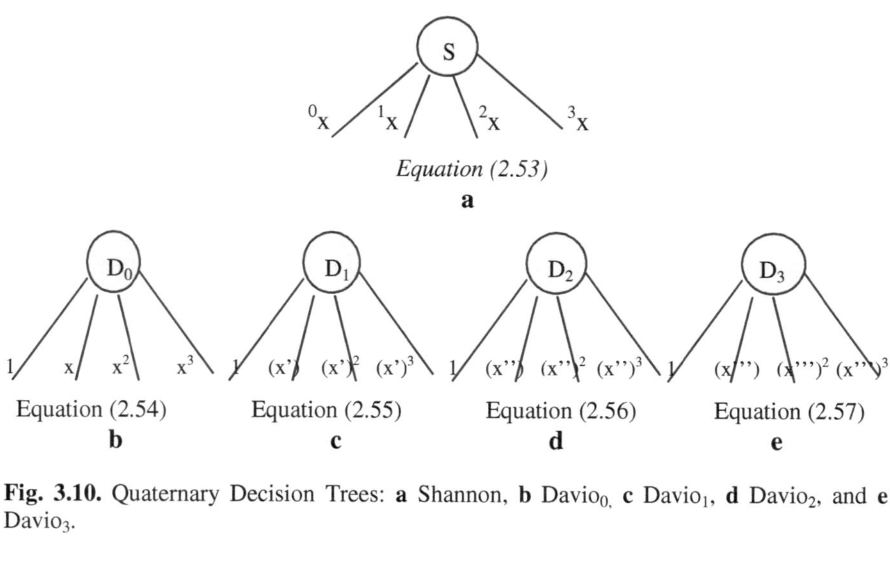

Fig. 3.10. Quaternary Decision Trees: **a** Shannon, **b** Davio_0, **c** Davio_1, **d** Davio_2, and **e** Davio_3.

In correspondence to the binary S/D trees, and ternary S/D trees, the concept of the quaternary S/D trees can be introduced. The quaternary S/D trees are generated through the definition of the Generalized Quaternary Davio (GQD) expansion over GF(4) as in Fig. 3.11.

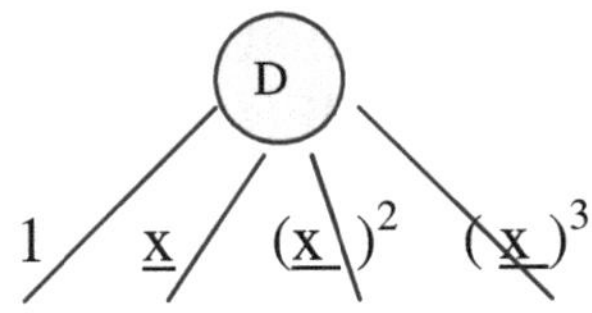

Fig. 3.11. Generalized quaternary Davio (GQD) expansion.

The notation in Fig. 3.11 is that $(\underline{x})$ corresponds to the four possible shifts of the variable x as follows:

$$\underline{x} \in \{x, x', x'', x'''\} \text{ over } GF(4). \qquad (3.10)$$

Utilizing the definition of quaternary Shannon (Fig. 3.10.a) and quaternary Generalized Davio (Fig. 3.11), and analogously to the work done for the binary and ternary cases, one can obtain the quaternary Shannon/Davio trees (QS/DT) for two variables. The number of these S/D trees per variable order is $2^{(4+1)} = 32$. The number of QIFs per S/D tree will be later derived in two different ways: (1) the first method is by using the general formula for an arbitrary number of variables over GF(4) developed in Appendix C, and (2) the second method is using the very general formula developed in Appendix C as well for any radix. The Count of the number of all possible forms is important because it can be used as an upper-bound parameter in a search heuristic that searches for a minimum GFSOP expression using the S/D trees. Example 3.3 illustrates some of the quaternary S/D trees and some of the quaternary trees they produce. The numbers on top of S/D trees in Figs. 3.12a and 3.13a are the numbers of the total QIFs (i.e., total number of quaternary trees) that are generated (As stated previously, derivation of these numbers is shown later in Appendix C).

Example 3.3. Let us produce some of the quaternary trees for the following quaternary S/D trees. Utilizing the notation in Eq. (3.10), we obtain, for the S/D trees in Figs. 3.12a and 3.13a, the S/D trees in Figs. 3.12b and 3.12c and Figs. 3.13b and 3.13c, respectively.

From the quaternary S/D DTs shown in Figs. 3.12 and 3.13, by taking any S/D tree, multiplying the two-level cofactors (which are in the QuDT leafs) each by the corresponding path in that QuDT, and next summing all the resulting cubes (terms; products) over GF(4), one obtains the flattened form for the function f, as a certain GFSOP expression (expansion). For each QuDT in Figs. 3.12a and 3.13a, there are as many forms obtained for the function f as the number of all possible permutations of the polarities of the variables in the second level branches of each QuDT. Properties for the quaternary S/D trees can be developed similar to the binary and ternary cases. The following Sect. presents a minimization algorithm that utilizes the S/D trees that are developed in this Chapt.

3.7 An Evolutionary Algorithm for the Minimization of GFSOP Expressions Using IF Polarity from Multiple-Valued S/D Trees

This Sect. presents an evolutionary algorithm (See Appendix E) that uses the multiple-valued IF polarity from S/D trees to minimize GFSOP functions. Evolutionary algorithms [80,102,137] have proven superiority to heuristic algorithms [157,233] in the minimization of incompletely specified functions with a high number of don't cares. Consequently, this process of minimization is very important to realize a minimum GFSOP expression using reversible structures such as the reversible Cascades that will be presented in Chapt. 8.

An ESOP minimizer for completely specified functions has been developed [157]. This minimizer does not work for functions with don't cares. The ESOP minimizer from [235] works for functions with few percent of don't cares, yet this minimizer does not work for functions with a high number of don't cares. The best minimizer for functions with a high number of don't cares is based on the use of genetic algorithms from [80]. Yet, as stated previously, this type of minimizer is for two-valued input two-valued output functions and is restricted to GRM polarities only.

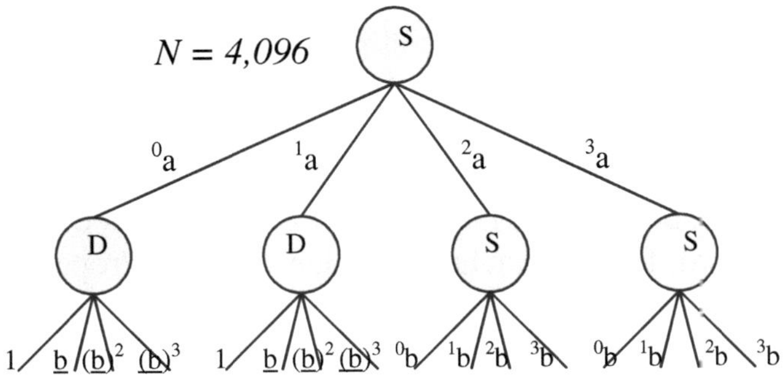

a

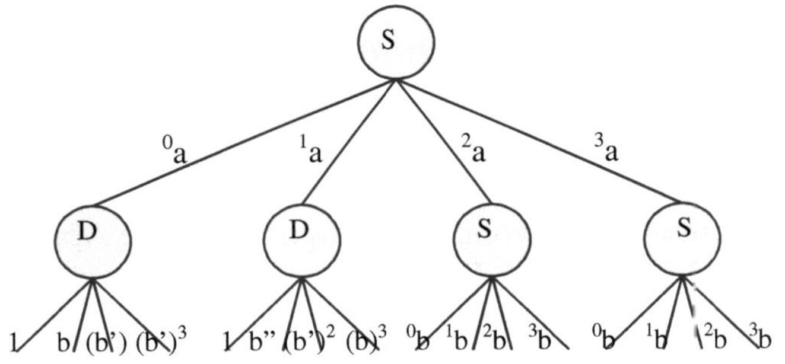

b

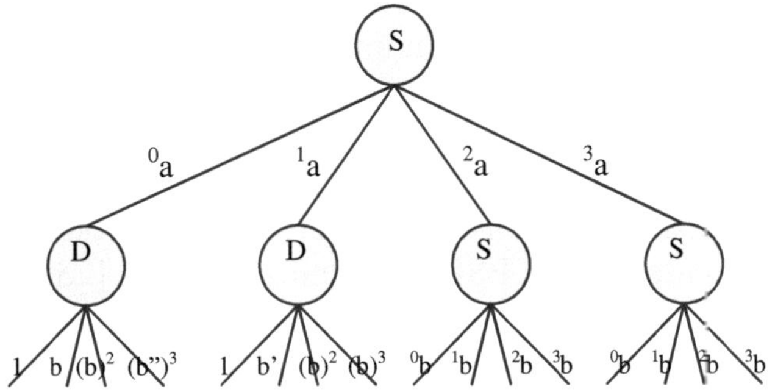

c

Fig. 3.12. a A quaternary S/D tree for two variables of order {a,b} with three Shannon nodes and two generalized Davio nodes, and **(b, c)** some of the quaternary trees that it generates.

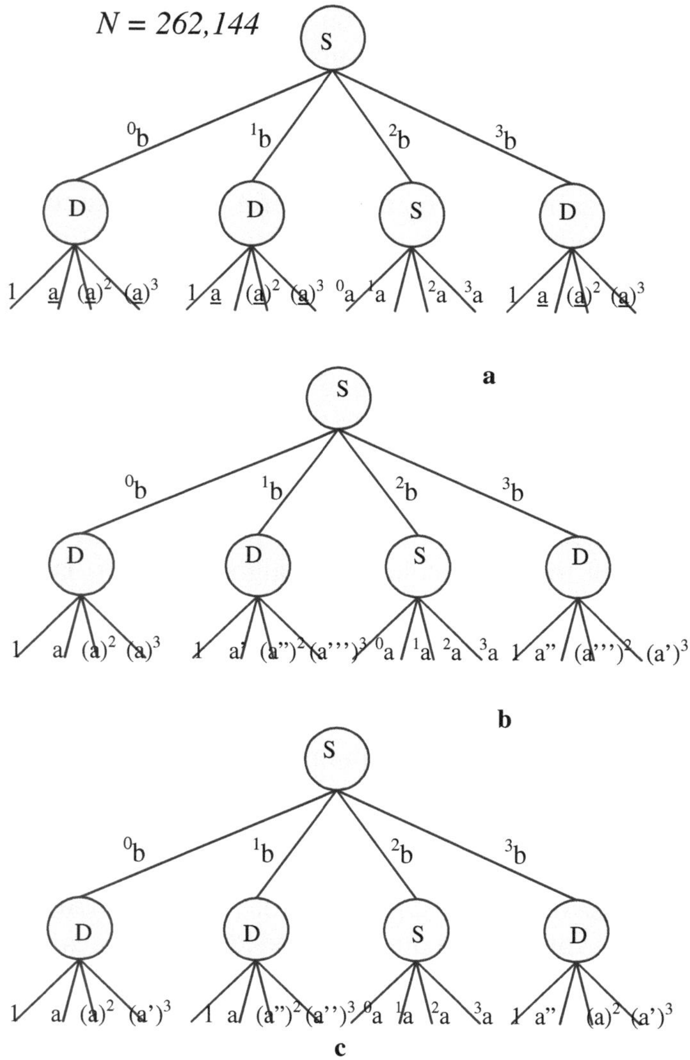

Fig. 3.13. a A quaternary S/D tree for two variables of order {b,a} with two Shannon nodes and three generalized Davio nodes, and **(b, c)** some of the quaternary trees that it generates.

Multiple-valued S/D trees developed previously provide more general polarity of Inclusive Form (IF) polarity, which contains the GRM as a special case. In this Sect., a proposition for GFSOP evolutionary minimizer based on IF polarity is provided. This is important in order to realize smallest functional forms using the reversible structures that will be presented in Chapt. 8.

In addition to the reason that evolutionary minimizers are valid for functions with a high percentage of don't cares [80], evolutionary minimizers (and in general heuritic-based minimizers) are far more efficient than exact (exhaustive) formal minimizers for problems of high dimensions. Numerical results from [80] prove that evolutionary methods can be a good fit for the minimization of the general case of incompletely specified functions. Such results are illustrated in Table 3.1, where (*) indicates a population of size 25, otherwise the population is of size 50.

The minimization using this evolutionary algorithm utilizes a heuristic inside it, which uses functional properties to ensure the convergence of the algorithm in each iteration [80]. This functional property, that the new evolutionary algorithm uses, is basically the selection of "good" products of literals in a greedy algorithm which when incorporated into the genetic algorithm (GA) will ensure convergence of the result (i.e., the resulting function from the GA minimizer is always correct). This internal heuristic was missing from previous attempts to minimize functions using GA, which made previous evolutionary algorithms for functional minimization to be non-convergent (i.e., does not produce the correct functionality after an iteration in the GA). The new GA minimizer utilizes the Darwinian evolution (Appendix E), as well Lamarckian and Baldwinian evolutions to minimize the logic functions.

By referring to Appendix E, one observes that Darwinian evolution evolves the polarity chromosome using the evolutionary operations of mutation and crossover [102] to produce a modified polarity chromosome upon which the fitness function compares the cost and functional correctness (convergence) of the new polarity chromosome, and iterations occur until the optimization criteria in the fitness function are met.

Table 3.1. iGRMIN minimization results over a number of benchmarks.

Benchmark	# Inputs	# Outputs	Terms	Generation	Run-time
con175	7	2	10	g1	00:01:09.15*
con195	7	2	4	g1	00:00:04.97
rd7375	7	3	20	g2	00:01:50.52*
rd7395	7	3	4	g1	00:00:06.59*
5xp175	7	10	56	g3	00:05:01.52*
5xp195	7	10	11	g1	00:00:26.75
rd8475	8	4	79	g2	00:20:30.71
rd8495	8	4	9	g1	00:00:36.16
log8mod75	8	5	88	g1	00:30:12.22
log8mod95	8	5	13	g1	00:01:08.73
misex195	8	7	15	g1	00:04:08.00
dc295	8	7	9	g2	00:00:24.55*
clip95	9	5	19	g1	00:08:40.06
rd84275	8	1	19	g2	00:06:23.65
rd84295	8	1	3	g1	00:00:04.67*
rd84475	8	1	9	g1	00:03:08.79*
rd84495	8	1	2	g1	00:00:18.79
9sym95	9	1	2	g1	00:02:11.47*
sao2175	10	1	7	g2	01:53:55.23
sao2195	10	1	1	g1	00:15:39.01
misex6475	10	1	1	g1	00:53:55.89*
misex6495	10	1	1	g1	00:07:56.95

This type of evolution produces the modified polarity chromosome through genetic operations only. The idea of Darwinian evolution is modified to Baldwinian evolution by generating the phenotype (function) from the polarity chromosome and then evaluating the phenotype rather than the polarity chromosome itself. The Baldwinian evolution can be further generalized into the Lamarckian evolution by performing operations on the phenotype and then evaluate the phenotype, which will lead to the production of a new polarity chromosome. Thus, in the Lamarckian model, a local heuristic search is used to modify and improve the chromosomes, and thus Lamarckian evolution requires an inverse mapping from phenotype and environment to genotype. In contrast, Baldwinian learning uses the local search to improve the fitnesses of the chromosomes but the chromosomes themselves remain the same without modification. The functional minimization using the new Baldwinian evolutionary algorithm combined with a greedy heuristic for Generalized Reed-Muller polarity results in functional minimization of incompletely specified functions which is much more optimal than previously reported results [80]. The new suggested algorithm for multiple-valued S/D-based minimizer generalizes the algorithm from [80] as shown in Fig. 3.14.

Old Method → New Method

Parameters	Old	New
Radix	GF(2)	GF(p^k)
Evolutionary Learning Method	Darwinian Baldwinian	Darwinian Baldwinian Lamarckian
Polarity	GRM	IF

Fig. 3.14. A comparison between previous evolutionary ESOP minimizer and the newly proposed evolutionary S/D-based GFSOP minimizer.

The idea of the Darwinian, Baldwinian, and Lamarckian learning methodologies [80] for the minimization of GFSOP expressions is illustrated in Fig. 3.15. The chromosome that is used in the evolutionary minimization represents the polarity of the corresponding canonical S/D based expansions. The phenotype in Fig. 3.15 represents the GFSOP expression. Figure 3.16 demonstrates such polarity chromosome for the binary expansions that are included in the Green/Sasao hierarchy from Fig. 3.1. The new S/D–based evolutionary minimizer generalizes the value of the polarity chromosome in Fig. 3.16 into multiple-valued logic. The following is a generic algorithm for Lamarckian evolutionary minimizer for the new multiple-valued IF polarity (variations from this algorithm can be implemented as well.)

```
{run:= 0
Maximum run:= i
cost = # literals
initial genotype:= G_0
 1.     if run:= 0
          then (G_n = G_0)
          else G_n = G_z+run
 2.  perform random genetic operations on G_n
 3.  using a greedy search heuristic obtain the initial phenotype P_x+run
     using logic transformation produce a modified phenotype P_y+run
               if cost(P_x+run) < cost( P_y+run)
                 then (P_z+run = P_x+run)
                 else P_z+run = P_y+run
               if cost(P_z+run) < cost(P_z+(run-1))
                 then P_z+run = P_z+run
                 else P_z+run = P_z+(run-1)
               produce genotype G_z+run for P_z+run
                 run = run ++
                   if run = i
                     then go to 4
                     else go to 1
                 4. print P_z+run
             print G_z+run
   end}
```

Example 3.4. Let us demonstrate the idea of modifying the genotype in Lamarckian evolution in contrast to Baldwinian evolution using the K-map in Fig. 3.17.

It can be observed that two possible ESOP-based expressions (phenotypes) are possible for the function shown in Fig. 3.17a. The first one is $F1 = ab \oplus cd$, and the second one is $F1 = ab \oplus c$. Both phenotypes have the same genotype (polarity vector) of {a = 1, b = 1, c = 1, d = 1} or more compactly the polarity vector [1111]. Consequently, the Baldwinian evolutionary minimizer will use a greedy search heuristic to search for the least cost (e.g., minimum number of literals) phenotype for a specific non-changing polarity vector. This is obviously the second phenotype $F1 = ab \oplus c$ which has the cost of 3. On the other hand, Lamarckian evolutionary minimizer will search for a phenotype without necessarily maintaining the same chromosome (polarity vector). This can potentially lead to less costly phenotypes, but on the expense of having more run times of the evolutionary minimizer algorithm.

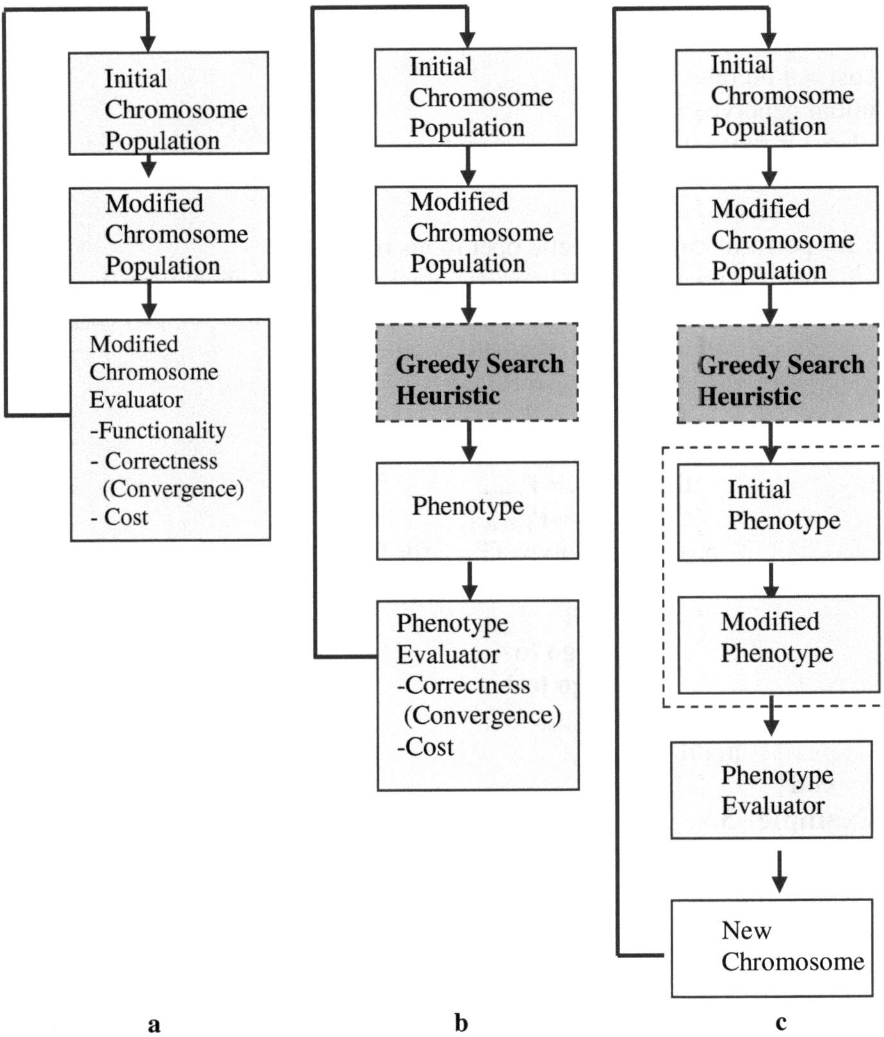

Fig. 3.15. Evolutionary Algorithms: **a** Darwinian, **b** Baldwinian, and **c** Lamarckian, where the dashed box in (c) indicates the effect of the environment as knowledge and optimization. (Greedy heuristic is used for a "good" selection of products of literals.)

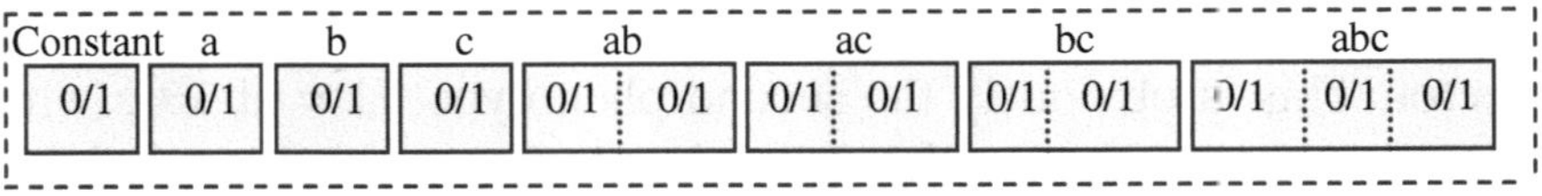

Fig. 3.16. Polarity chromosome (string; polarity vector) for GRM.

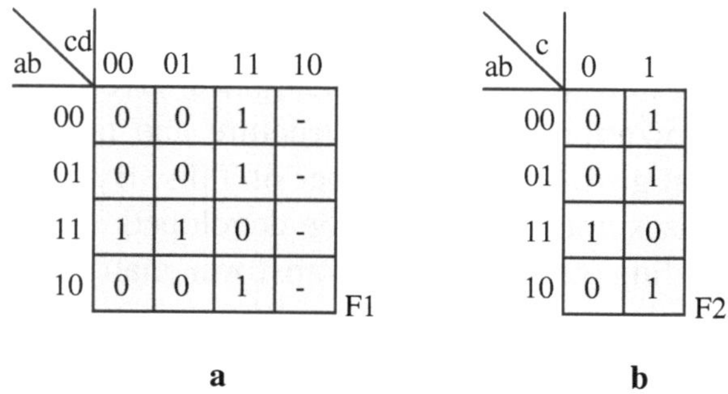

Fig. 3.17. K-maps for **a** Boolean function F1, and **b** Boolean function F2.

This can be observed utilizing the K-map in Fig. 3.17b, where one possible phynotype is $F2_{Initial}$ = ab ⊕ abc. Yet, it can be observed that due to the Boolean law $x \oplus 1 = \bar{x}$, then this F2 phenotype can be transformed into $F2_{Modified}$ = ab(1⊕c) = $ab\bar{c}$. It can be observed that while the intial phenotype with polarity vector [111] has the cost of 5, the modified phenotype with the polarity vector [110] has the cost of 3. Thus, a more compact (minimized) expression (phenotype) has been obtained by changing the chromosome (polarity vector) using the Lamarckian evolutionary algorithm.

3.8 Summary

In this Chapt., we introduced a new family of Ternary S/D Trees, defined new families of Ternary Inclusive Forms (TIFs) and Ternary Generalized Inclusive Forms (TGIFs), and calculated their numbers for two variables. We introduced a new subfamily of Trees called Ternary Generalized Inclusive Forms Kronecker (TGIFK) Decision Tree, and showed its hierarchical position with respect to the Extended Green/Sasao hierarchy over GF(3). It can be observed that the number of S/D trees for ternary logic is very large, and it grows exponentially with the increase of the number of variables. Therefore, searching for efficient algorithms to find the minimum GFSOP flattened form expression of ternary logic over GF(3) would be a real challenge, and much more difficult than for the well-researched binary case, where it is already practically intractable.

Therefore, new strategies and heuristics have to be invented for GFSOP minimization, using TIF structure and taking into account the exponential growth of the number of TIFs. By analogy to binary logic, in all cases, methods should be developed to find high quality GFSOPs avoiding generating all transform matrices, by searching efficiently in the space of TGIFs. We propose an evolutionary algorithm to solve the problem of searching for a minimum GFSOP using IF polarity from the corresponding multiple-valued S/D trees.

The proposed evolutionary minimizer will be used to minimize the GFSOP expression for a given function to realize the function using reversible structures that will be developed in Chapt. 8. Such GFSOP minimization for the realization of GFSOP expressions in multiple-valued reversible structures will have an effect on producing minimal size quantum circuits which will be shown in Chapt. 10, and consequently minimizing the total number of arithmetic calculations that are used in quantum computing as will be demonstrated in Chapt. 11.

4 Novel Methods for the Synthesis of Boolean and Multiple-Valued Logic Circuits Using Lattice Structures

This Chapt. presents a new type of regular structures that will be used to produce regular reversible lattice structures in Chapt. 6. With future logic realization in technologies that are scaled down rapidly in size, the emphasis will be increasingly on the mutually linked issues of regularity, predictable timing, high testability, and self-repair. For the current leading technologies with the active-device count reaching the hundreds of millions, and most of the circuit areas occupied by local and global interconnects, the delay of interconnects is responsible for about 40-50% or more of the total delay associated with a circuit [51,229]. In future technologies, interconnects will take an even higher percent of area and delay which creates interest in cellular (regular) structures [109,209], especially for nano technologies [109].

As it has been shown [229] that most of the circuit area is occupied by local and global interconnects, and the delay of interconnects is responsible for most of the total delay associated with a circuit, maintaining equal length of local inter-connects will minimize the total length of the used wires and consequently minimize the delay and power consumed. Also, it has been shown in [229] that the relative delay for global interconnects with or without repeaters over all process technologies are much larger than their counterparts of local interconnects. This suggests that using lengthy interconnects between the circuit elements will produce higher delays of the signal propagation throughout the interconnects and thus one wants to use shorter interconnects. This problem becomes even more serious for circuits that switch at very high speeds, where the power consumption increases with the increasing operation frequencies, and even the smallest capacitance or inductance that exists naturally within the wirings will be of extreme importance to maintain the electrical "signal integrity" as much as possible. Figure

4.1 illustrates the trends for electrical signal delays for global interconnects with repeaters and without repeaters versus the local interconnects [229].

Recently, regular layout fabrics are becoming more popular with the new hardware implementation technologies such as single-electron devices (SET) [111,112] and quantum dots [259]. Fabrication of two dimensional hexagonal regular structures has been reported in [112] (See Figs. 11.1 and 11.2). Other circuits of interest for regular structure approach use the Chemically Assembled Electronic Nanotechnology (CAEN) [91,103], which is expected to offer significantly denser devices than CMOS technology. For instance, a single RAM cell that requires roughly 100 nm^2, will occupy in CMOS technology an area of 100,000 nm^2 [103].

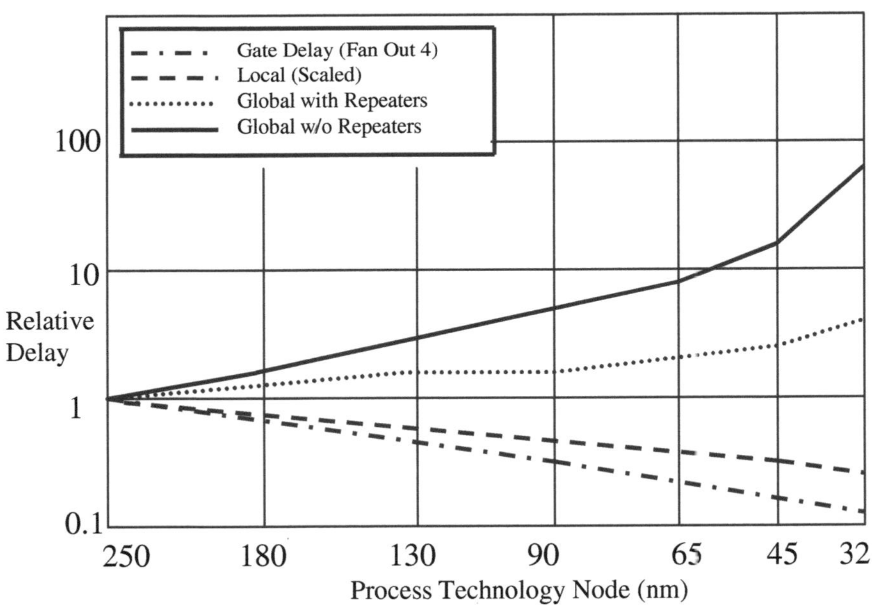

Fig. 4.1. Delay for local and global wiring versus feature size.

Lattice structures [5,13,18,51,144,161,177,178,179] generalize the ideas from the well-known regular structures: Fat trees, Generalized PLAs, Maitra cascades [148], and Akers Arrays [1], into a more systematic framework which is closely and naturally related to the symmetry of functions, and "symmetric networks" from [136]. In this Chapt., the concept of 2-D lattice structures is extended to the case of regular 3-D lattice structures and a new algorithm for the iterative decomposition of such 2-D and 3-D regular structures is provided. This subject is essential for the future nano technologies [103], as it shows the best way to place combinational logic functions in a three-dimensional space where all local connections are of the same length and global connections are only inputs on parallel oblique planes. Although the concept of symmetry for multiple-valued functions, which is closely related to lattice structures, have been previously investigated [46,47,247], the author is not aware of any previously published research about expansions into regular structures that have more than two dimensions. The main results of this Chapt. are:

- The application of expansions into regular three-dimensional lattice structures.
- The realization of non-symmetric ternary functions in 3-D lattice structures.
- Generic methodology of joining lattice nodes in three-dimensional space.
- The invariant (3,3) 3-D Shannon and (3,3) 3-D Davio lattice structures.
- Iterative Symmetry Indices Decomposition (ISID): A novel layout-driven method to decompose two-dimensional and three-dimensional lattice structures.

The remainder of this Chapt. is organized as follows. Basic background of symmetry indices is presented in Sect. 4.1. Basic definitions of the fundamental (2,2) two-dimensional binary lattice structures are given in Sect. 4.2. The concept of (3,3) two-dimensional lattice structures is presented in Sect. 4.3. Three-dimensional lattice structures based on the new generalized sets of Shannon and Davio canonical expansions (from Sect. 2.2) are presented in Sect. 4.4. An algorithm for the creation of three-dimensional lattice structures is presented in Sect. 4.5. Complete example of such three-dimensional lattice structures is presented in Sect. 4.6. Iterative Symmetry Indices Decomposition (ISID) is presented in Sect. 4.7. A Summary of the Chapt. is presented in Sect. 4.8.

4.1 Symmetry Indices

It is known in logic synthesis that certain classes of logic functions exhibit specific types of symmetries [83,118,136,213,244]. Such symmetries include symmetries between different functions under negation, symmetries within a logic function under the negation of its variables, and symmetries within a logic function under the permutation of its variables. Accordingly, the following is one possible classification of logic functions:
(1) P-Equivalence class: a family of identical functions obtained by the operation of permutation of variables.
(2) NP-Equivalence class: a family of identical functions obtained by the operations of negation or permutation of one or more variables.
(3) NPN-Equivalence class: a family of identical functions obtained by the operations of negation or permutation of one or more variables, and also negation of function (cf. Table G.1)

Example 4.1. The following represents symmetric function: F = ab ⊕ bc ⊕ ac.

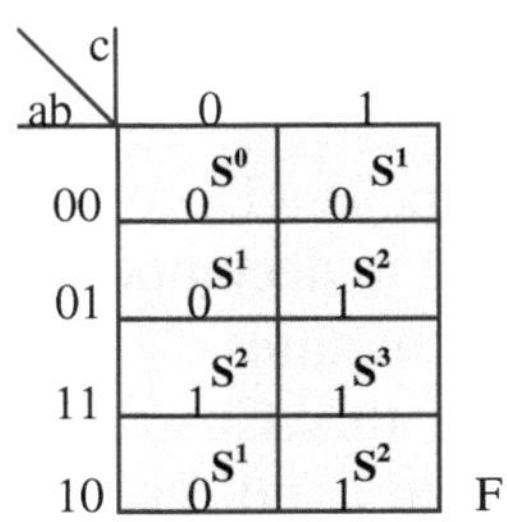

Fig. 4.2. K-map of three-variable symmetric function F = ab ⊕ bc ⊕ ac.

In Fig. 4.2, a symmetry index S^i specifies a K-map cell that counts value "1" in the specified minterm i number of times.

Definition 4.1. A single index symmetric function, denoted as $S^k(x_1,x_2,\ldots,x_n)$ has value 1 when exactly k of its n inputs are equal to 1, and exactly (n-k) of its remaining inputs are 0.

Definition 4.2. The elementary symmetric functions of n variables are:

$$S^0 = \bar{x}_1\bar{x}_2\ldots\bar{x}_n,$$

$S^1 = x_1\bar{x}_2...\bar{x}_n + \bar{x}_1 x_2\bar{x}_3...\bar{x}_n + ... + \bar{x}_1\bar{x}_2...\bar{x}_{n-1}x_n$,
..., and
$S^n = x_1 x_2 ... x_n$.

Thus, for a Boolean function of three variables one obtains the following sets of symmetry indices: $S^0 = \{\bar{a}\bar{b}\bar{c}\}$, $S^1 = \{\bar{a}\bar{b}c, \bar{a}b\bar{c}, a\bar{b}\bar{c}\}$, $S^2 = \{\bar{a}bc, ab\bar{c}, a\bar{b}c\}$, and $S^3 = \{abc\}$. It has been shown [213,219] that an arbitrary n-variable symmetric function f is uniquely represented by elementary symmetric functions S^0, S^1, ..., S^n as follows: f = $\sum_{i \in A} S^i = S^A$, where A $\in$ {0, 1, ..., n}. Also it can be shown that, for f = S^A and g = S^B, the following are obtained:

$$f.g = S^{A \cap B}, \tag{4.1}$$

$$f + g = S^{A \cup B}, \tag{4.2}$$

$$f \oplus g = S^{A \oplus B}, \tag{4.3}$$

$$\bar{f} = S^{\bar{A}}. \tag{4.4}$$

It has been shown in [1] that a non-symmetric function can be symmetrized by repeating its variables. This method of variable repetition transforms the values of K-map cells which make the function non-symmetric into don't cares which make the function symmetric.

Example 4.2. The following K-map demonstrates the symmetrization by repeating the variables of a non-symmetric Boolean function: F = a' + b.

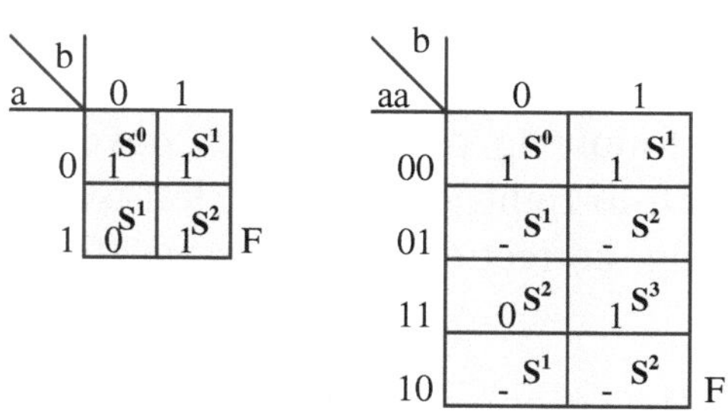

a **b**

Fig. 4.3. Symmetrization of a non-symmetric Boolean function by repeating its variables: **a** non-symmetric Boolean function, and **b** symmetric Boolean function obtained by repeating variable {a}.

One notes that while in Fig. 4.3a conflicting values occur for symmetry index S^1 in minterms $\overline{a}b$ and $a\overline{b}$, thus producing a non-symmetric function, non-conflicting values are produced for the same non-symmetric function in Fig. 4.3b by repeating variable {a} two times.

4.2 Fundamental (2,2) Two-Dimensional Lattice Structures

The concept of lattice structures for switching functions involves three components: (1) expansion of a function, that corresponds to the initial node (root) in the lattice, which creates several successor nodes of the expanded node, (2) joining (collapsing) of several nodes of a decision tree's level to a single node, which is the reverse operation of the expansion process, and (3) regular geometry to which the nodes are mapped that guides which nodes of the level are to be joined.

While the realization of non-symmetric functions in Akers arrays [1] requires an exponential growth of repetition of variables in the worst case, the realization of non-symmetric functions in lattice structures requires a linear growth of repetition of variables [50,51], and consequently one need not to repeat the variables of non-symmetric functions many times to realize such functions in lattice structures for most practical benchmarks. It has been shown [50,51] that one needs to repeat variables to realize benchmarks in lattice structures by 2.5 times on average. Figure 4.4 illustrates, as an example, the geometry of 4-neighbors and joining operations on the nodes where each cell has two inputs and two outputs (i.e., four neighbors). The construction of the lattice structure in Fig. 4.4 implements the following one possible convention: top-to-bottom expansion and left-to-right joining (i.e., left-to-right propagation of the corresponding correction functions in Figs. 4.4c and 4.4d, respectively).

Definition 4.3. The function that is generated by joining two nodes (sub-functions) in the lattice structure is called the joined function. The function that is generated in nodes other than the joining nodes, to preserve the functionality in the lattice structure, is called the correction function.

Note that the lattices presented in Fig. 4.4 preserve the functionality of the corresponding sub-functions f and g. This can be observed, for instance, in Fig. 4.4b as the negated variable {a'} will cancel the un-complemented variable {a}, when propagating the cofactors from the lower levels to the upper levels or vice versa, without the need for any correction functions to preserve the output functionality of the corresponding lattice structure. This simple observation cannot be seen directly in Figs. 4.4c and 4.4d, as the correction functions are involved to cancel the effect of the new joining nodes for the preservation of the functionality of the new lattice structures (these correction functions are shown in the extreme right of the second level in Figs. 4.4c and 4.4d, respectively).

It is shown in [1] that every function that is not symmetric can be symmetrized by repeating variables, and that a totally symmetric function can be obtained from an arbitrary non-symmetric function by the repetition of variables. Consequently, lattice structures and the symmetry of functions are very much related to each other. Example 4.3 will illustrate such close relationship.

Example 4.3. For the following non-symmetric function: F = ab + a'c. Utilizing the joining rule that was presented in Fig. 4.4b for a two-dimensional lattice structure with binary Shannon nodes, one obtains the lattice structure shown in Fig. 4.5.

One can observe that in order to represent the non-symmetric function in Example 4.3 in the 2-D lattice structure, variable {b} is repeated. The nodes in Fig. 4.5 are Shannon nodes, which are merely two-input one-output multiplexers, whose output goes in two directions, with the variables {a, b, c} operating as control signals.

The results from this Sect. will be generalized to ternary logic in a later Sect., and thus from two-dimensional space to three-dimensional space. It is important to prove that the repetition of variables will have an end in the process of the symmetrization of the non-symmetric functions. An intuitive proof is as follows: for totally symmetric functions the number of variables are equal to the number of levels of the lattice structure, as there is no need to repeat variables, and as it is known that by the repetition of variables every

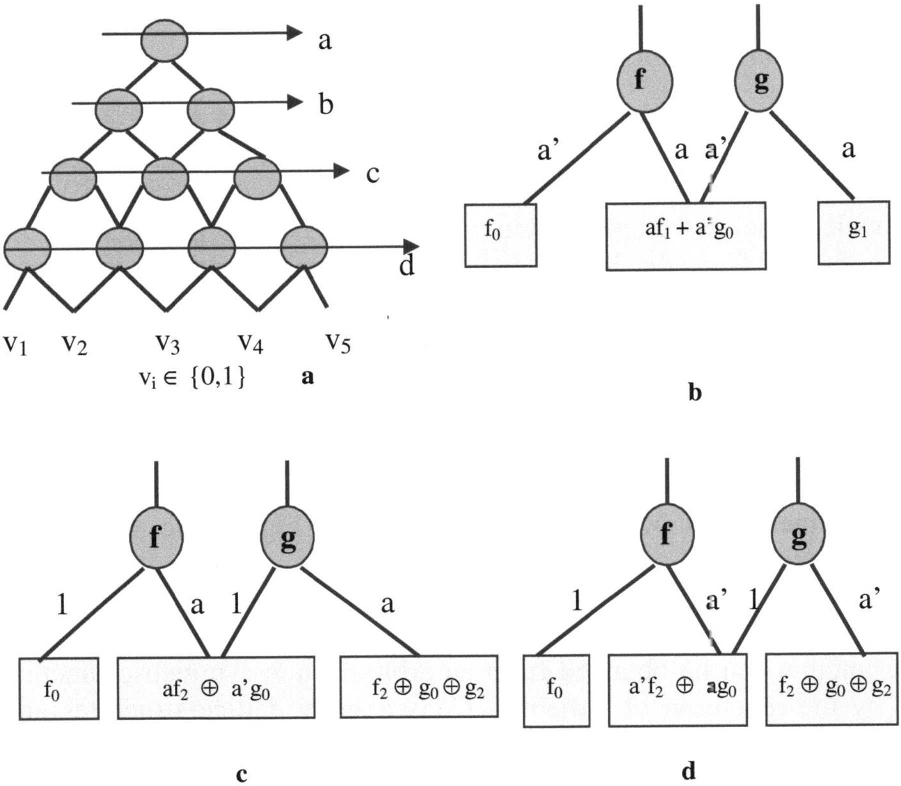

Fig. 4.4. **a** A two-dimensional 4-neighbor lattice structure, **b** joining rules for binary Shannon lattice structure, **c** binary positive Davio lattice structure, and **d** binary negative Davio lattice structure.

non-symmetric function is symmetrized, then this must result in definite number of levels in the corresponding lattice structure and as a consequence in certain number of total variables, repeated and non-repeated, that will result in the termination of the process of symmetrization. Lattice synthesis is typically performed from top to bottom when the levels of gates are synthesized one at a time until the level with constant cofactors is reached.

Example 4.4. Figure 4.6 illustrates the close relationship between the concept of the two-dimensional lattice structures and the symmetry of functions.

Note that in Fig. 4.6b the symmetry indices represent the sets of all possible paths from the leaves to the root through the internal nodes in Fig. 4.6a.

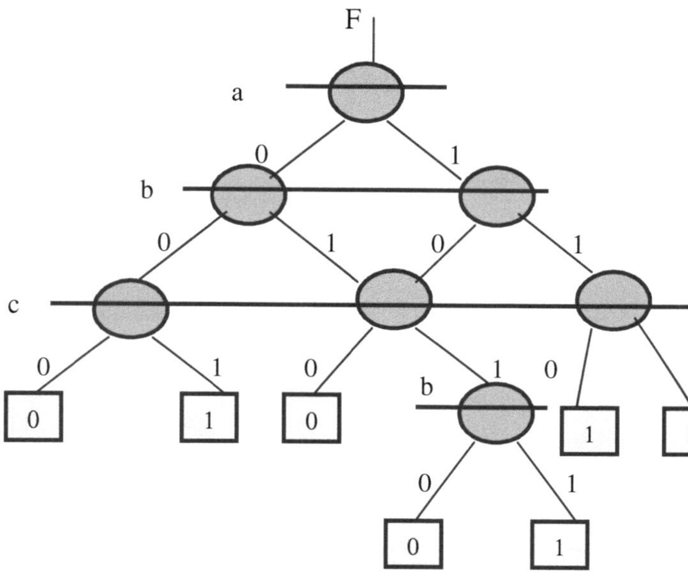

Fig. 4.5. Shannon lattice structure for the non-symmetric function: F = ab + a'c.

Figure 4.6c shows that the same symmetry indices, that are shown in Fig. 4.6b, are the counts of the number of ones in the Gray-encoded cell indices of the K-map. The concept of lattice structures is closely related to the concept of symmetric networks [136]. In symmetric networks switches are allocated on data paths that are controlled by control variables. The terminal nodes of such structure are the value of the corresponding symmetry index. This type of structures which is based on the concept of symmetry indices is illustrated in Fig. 4.7 for a Boolean function of three variables.

The realization of non-symmetric functions using lattice structures require the repetition of variables. To minimize the size of lattices, search heuristics for variable ordering for symmetric functions and search heuristics for variable ordering and repetition of non-symmetric functions are needed. This is similar to the case of Binary Decision Diagrams (BDDs) [2,45,142] where various variable ordering produces different sizes of the corresponding BDDs. The following example illustrates the idea of the realization of non-symmetric functions using lattice structures.

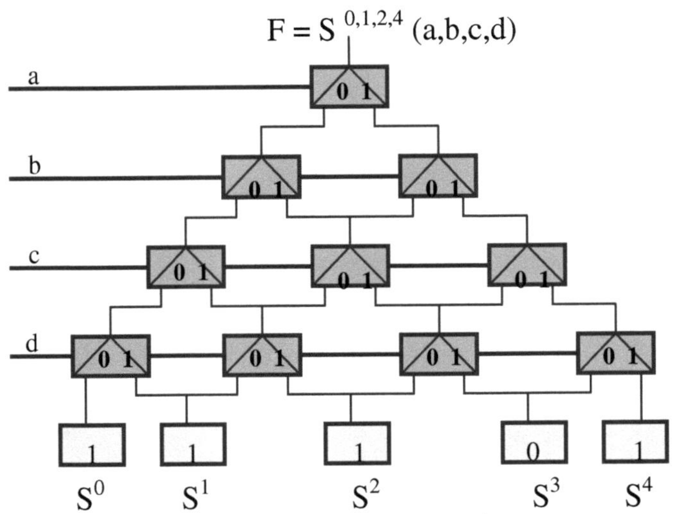

a

S^i is the symmetry indices, where i = # of ones in the Gray-encoded cell indices of the K-map:

S^0 = {a'b'c'd'}
S^1 = {a'b'c'd, a'b'cd', a'bc'd', ab'c'd'}
S^2 = {a'b'cd, a'bc'd, a'bcd', abc'd', ab'c'd, ab'cd'}
S^3 = {a'bcd, abc'd, abcd', ab'cd}
S^4 = {abcd}

b

ab \ cd	00	01	11	10
00	S^0	S^1	S^2	S^1
01	S^1	S^2	S^3	S^2
11	S^2	S^3	S^4	S^3
10	S^1	S^2	S^3	S^2

c

Fig. 4.6. **a** Two-dimensional lattice structure for function of four variables: lattice structure with inter-connected nodes which are 2-to-1 inter-connected multiplexers, **b** sets of binary symmetry indices, and **c** K-map interpretation of the binary symmetry indices.

ab \ c	0	1
00	s^0	s^1
01	s^1	s^2
11	s^2	s^3
10	s^1	s^2

a

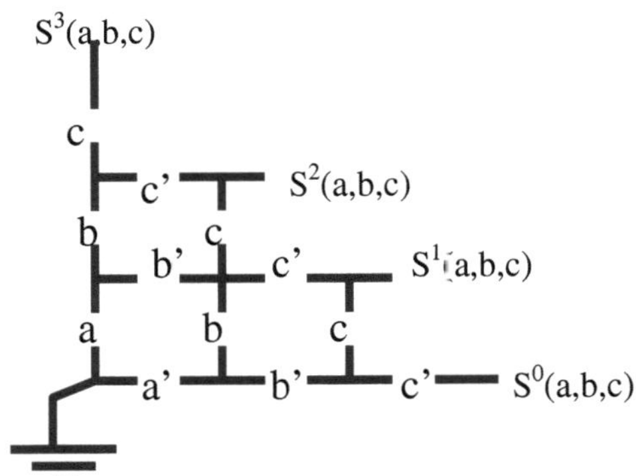

b

Fig. 4.7. Symmetric network: **a** K-map that illustrates the distribution of symmetry indices, and **b** the corresponding symmetric network which produces the symmetry indices as terminal nodes.

Example 4.5. For the binary non-symmetric implication function: F = a' + b, Fig. 4.8 illustrates the relationship between the k-map with non-conflicting symmetry indices and the two-dimensional lattice structure with non-conflicting leaves.

The concept of two-dimensional (2,2) lattice structures has been generalized to many types of two-dimensional (k,k) lattice structures. The following Sect. introduces one generalization of 2-D (2,2) lattice structures into 2-D (3,3) lattice structures.

4.3 (3,3) Two-Dimensional Lattice Structures

The idea of (2,2) lattice structures have been extended to the case of planar (3,3) lattice structures [195,188,221]. This new planar regular structure has some advantages over the (2,2) lattice structures especially for self-repair for Field Programmable Gate Arrays (FPGAs) [58,221].

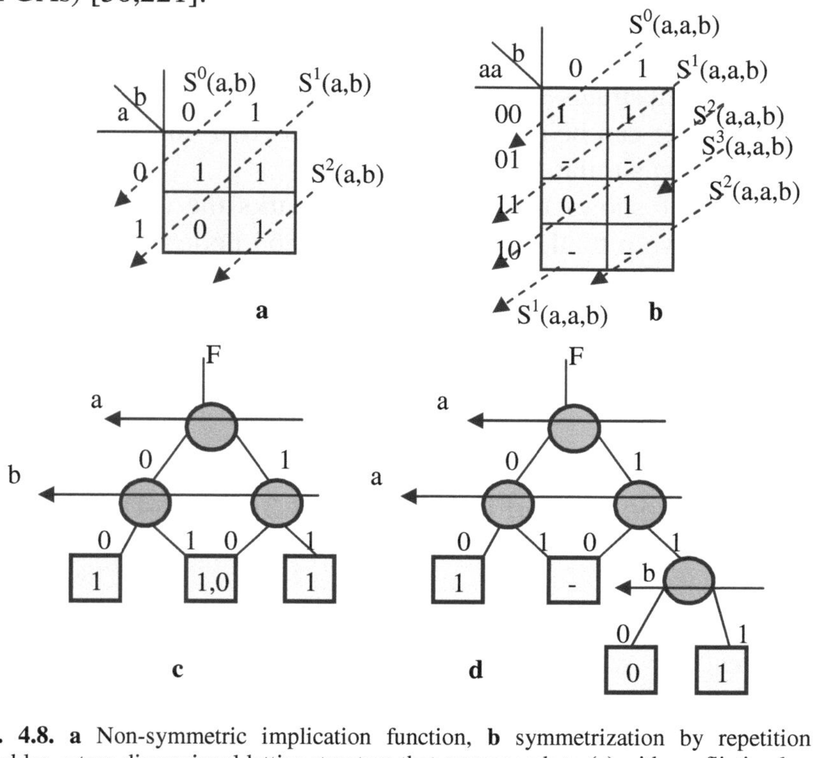

Fig. 4.8. **a** Non-symmetric implication function, **b** symmetrization by repetition of variables, **c** two-dimensional lattice structure that corresponds to (a) with conflicting leaves, and **d** two-dimensional lattice structure that corresponds to (b) with non-conflicting leaves.

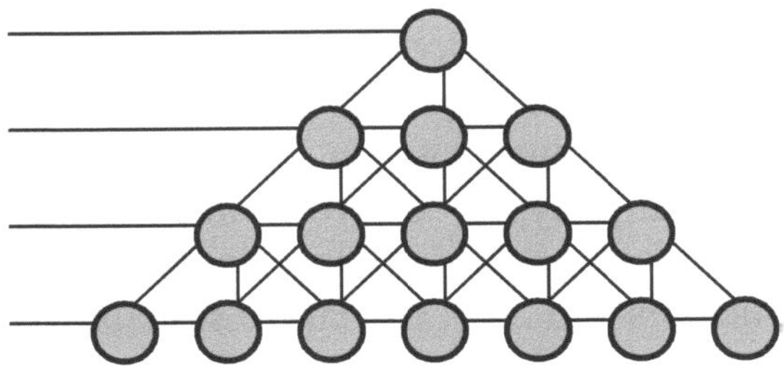

Fig. 4.9. (3,3) two-dimensional lattice structure.

The (3,3) 2-D lattice structures that are shown in Fig. 4.9 have the following additional features over their counterparts of (2,2) 2-D lattice structures: (1) the data inputs to a gate can be complemented, and (2) both data inputs of a gate can be connected to the same gate below. Numerical results have been obtained [188] that compared the size of benchmarks realized in (2,2) and (3,3) Shannon lattice structures, which showed a clear advantage of the (3,3) lattice structures over (2,2) lattice structures in terms of size. Table 4.1 from [188] illustrates such comparison. The (3,3) two-dimensional lattice stucture can be extended to any (k,k) two-dimensional lattice structure using similar procedures.

Table 4.1 shows that for 18 benchmarks the (3.3) 2-D lattice structures have a total of 291 levels and 863 nodes as compared to 383 levels and 2011 nodes in the case of (2,2) 2-D lattice structures. Thus for an initial evaluation, (3,3) 2-D lattice structures show a significant improvement in the minimization of the total size of lattice logic circuits over the (2,2) 2-D lattice structures.

4.4 New Three-Valued Families of (3,3) Three-Dimensional Shannon and Davio Lattice Structures

The concept of binary two-dimensional Shannon and Davio lattice structures that was presented in Sect. 4.2 can be generalized to include the case of three-dimensional Shannon and Davio lattice structures with function expansions that implement the fundamental

Table 4.1. Experimental results for the realization of MCNC benchmarks in (2,2) and (3,3) two-dimensional lattice structures.

Name	# Outputs	# Inputs	(2,2) Lattice		(3,3) Lattice	
			# Levels	# Nodes	# Levels	# Nodes
apex7	37(30)	49(17)	25	148	19	33
clip	5(1)	9(9)	18	103	9	30
	5(2)	9(9)	27	220	9	30
cm162a	5(3)	14(10)	11	24	10	18
cps	109(1)	24(22)	26	134	24	61
	109(2)	24(18)	26	164	21	66
	109(3)	24(22)	39	342	30	128
duke2	29(3)	22(15)	18	52	15	69
	29(6)	22(17)	18	47	17	36
	29(18)	22(15)	22	92	15	44
example2	66(23)	85(16)	21	52	16	45
	66(59)	85(14)	17	31	14	21
	66(63)	85(13)	15	37	13	26
frg2	139(99)	143(20)	22	189	20	28
	139(100)	143(19)	28	164	20	61
sao2	4(2)	10(10)	18	71	14	56
	4(3)	10(10)	16	73	14	65
	4(4)	10(10)	16	68	11	46
Total			383	2011	291	863
Ratio, %			100	100	**76**	**43**

multi-valued Shannon and Davio decompositions, as well as the new invariant set of multi-valued Shannon and Davio decompositions from Sect. 2.2. Since the most natural way to think about binary lattice structures is the two-dimensional 4-neighbor lattice structure that was shown in Fig. 4.4a, one can extend the same idea to utilize the full three-dimensional space in the case of ternary lattices. Such lattices represent three-dimensional 6-neighbor lattice structures. Although regular lattices can be realizable in the three-dimensional space for radix three while maintaining their full regularity, they are unrealizable for radices higher than three (i.e., 4, 5, etc). Higher dimensionality lattices can be implemented in 3-D space but at the expense of losing the full regularity. This is because the circuit realization for the ternary case produces a regular structure in three dimensions that is fully regular in terms of connections; all connections are of the same length. Realizing the higher dimensionality lattices in lower dimensionality space is possible but at the expense of regularity; the lattices will not be fully regular due to the uneven length of the inter-connections between nodes.

As a topological concept, and as stated previously, lattice structures can be created for two, three, four, and any higher radix. However, because our physical space is three-dimensional, lattice structures, as a geometrical concept, can be realized in solid material, with all the inter-connections between the cells of the same length, only for radix two (2-D space) or radix three (3-D space). It is thus interesting to observe that the characteristic geometric regularity of the lattice structure realization which is observed for binary and ternary symmetric functions will be no longer observable for quaternary functions. Thus, the ternary lattice structures have a unique position as structures that make the best use of three-dimensional space (we do not claim here that a regular structure that would use 3-D space better than 3-D lattice structures can not be invented, and the statement is restricted only to lattice-type structures). The following Sect. will introduce the proposed general three-dimensional logic circuit of ternary lattice structures. The new 3-D lattice structures that realize ternary functions, which will be presented in the next Sects., will be further extended to the reversible case in Chapt. 6, and then mapped into quantum circuits as will be illustrated in Chapt. 10.

4.4.1 Three-Dimensional Lattice Structures

In general, to reserve the fully regular realization of expansions over n^{th} radix, it is sufficient to join n nodes in n-dimensional space to obtain the corresponding lattice structures. For instance, as was shown in Fig. 4.4, it is sufficient in the binary case to join two nodes. Analogously, it is sufficient in the ternary case to join three nodes to form the corresponding 3-D lattice structures [5,13,18]. Analogously to the work presented previously, fully symmetric ternary functions do not need any joining operations to repeat variables in order to realize them in three-dimensional lattice structures. Because three-dimensional lattice structures exist in a three-dimensional space, a geometrical reference of coordinate systems is needed in order to be systematic in the realizations of the corresponding logic circuits. Consequently, the right-hand rule of the Cartesian coordinate system is adopted. Example 4.6 illustrates lattice realizations for such fully symmetric ternary functions.

Example 4.6. For the fully symmetric three-variable ternary input/ternary output function: F = ab + ac + bc. Adopting the right-hand rule of the Cartesian coordinate system, the following is the three-dimensional logic circuit realization of the symmetric function:

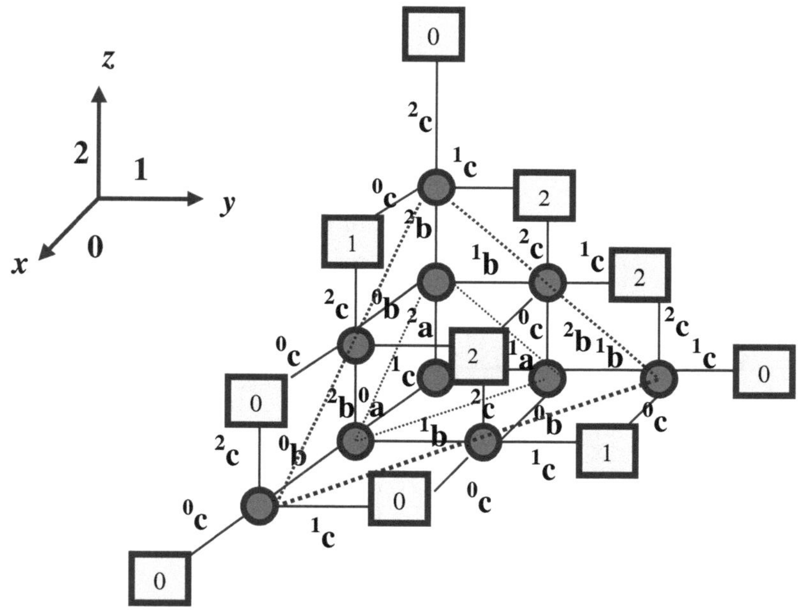

Fig. 4.10. Three-dimensional logic circuit realization for the ternary symmetric function: F = ab + ac + bc.

One can observe that Fig. 4.10 represents a *fully regular* lattice structure in three-dimensions. Each dimension corresponds to a value of the corresponding control variable; value zero of the control variable propagates along the x-axis, value one of the control variable propagates along the y-axis, and value two of the control variable propagates along the z-axis. Since the ternary function in Example 4.6 is fully symmetric [18], no variables need to be repeated in the corresponding lattice structure. In 3-D space, each control variable spreads in a plane to control the corresponding nodes (these parallel planes are represented using the dotted triangles in Fig. 4.10), in contrast to the binary case where each control variable spreads in a line to control the corresponding nodes (these control signals are in the solid bold lines in Fig. 4.4a). Each node in Fig. 4.10 is a three-input one-output multiplexer, whose output goes in three directions.

Example 4.7. The following is the ternary modsum addition.

+	0	1	2
0	0	1	2
1	1	2	0
2	2	0	1

Fig. 4.11. Ternary modsum addition.

The following is the logic circuit of the ternary 3-digit full adder:

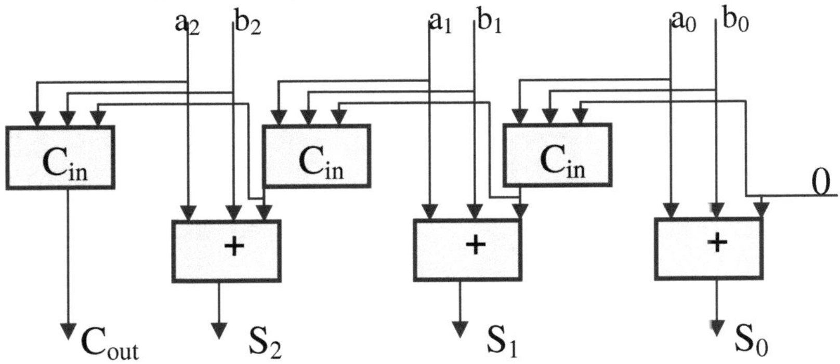

Fig. 4.12. Logic circuit of a ternary 3-digit full adder.

The following maps resemble such functions for the Sum (S) and the output carry (C_{out}) that appear in the logic circuit in Fig. 4.12.

a b \ C_{in}	0	1
0 0	0	1
0 1	1	2
0 2	2	0
1 0	1	2
1 1	2	0
1 2	0	1
2 0	2	0
2 1	0	1
2 2	1	2

a

a b \ C_{in}	0	1
0 0	0	0
0 1	0	0
0 2	0	1
1 0	0	0
1 1	0	1
1 2	1	1
2 0	0	1
2 1	1	1
2 2	1	1

b

Fig. 4.13. Ternary sum and ternary carry out maps.

The following are the 3-D lattice realizations of the functions in Fig. 4.13.

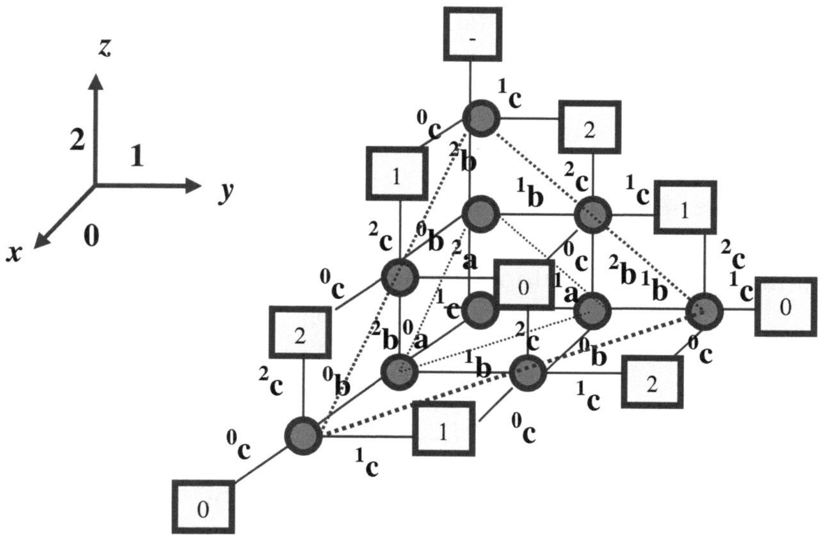

Fig. 4.14. The sum function of the ternary 3-digit full adder.

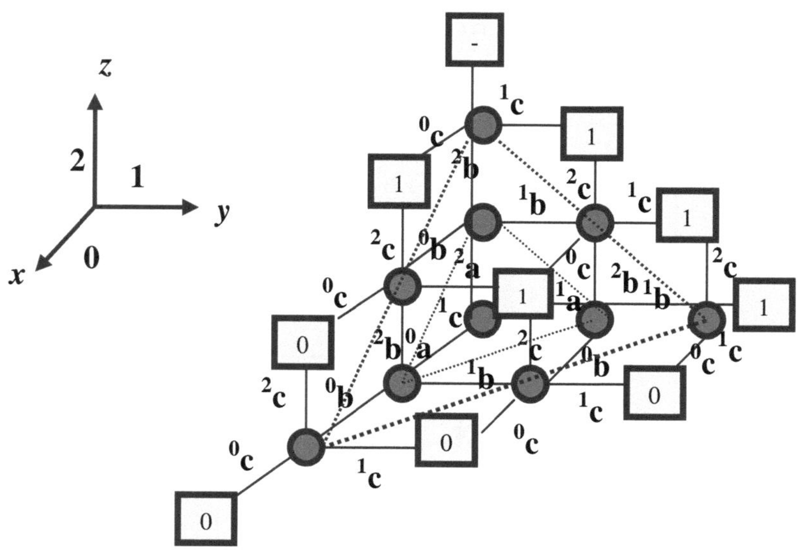

Fig. 4.15. The carry out function of the ternary 3-digit full adder.

From Figs. 4.14 and 4.15, one can observe that the sum function and the carry out function are both symmetric functions, and consequently there is no need to repeat variables to make the ternary functions realizable in the 3-D lattice structures. In the case of ternary non-symmetric functions, one needs to repeat variables, to symmetrize the corresponding non-symmetric functions, in order to represent such functions in the corresponding lattice structures, analogous to the binary case.

Example 4.8. Let us design a ternary 2-digit multiplier in a 3-D lattice structure. Two-digit multi-valued multiplication is performed utilizing the mod-multiplication operator as follows:

$$\begin{array}{ccccc} & & & B_1 & B_0 \\ & & & \underline{A_1} & \underline{A_0} \\ & & C_{out1} & m_{01} & m_{00} \\ & \underline{C_{out2}} & \underline{m_{11}} & \underline{m_{10}} & \underline{0} \\ C_{out} & S_3 & S_2 & S_1 & S_0 \end{array}$$

Figure 4.16 shows the logic circuit of the ternary 2-digit multiplier, and Fig. 4.17 shows the maps for the ternary multiplication and the output carry (C_{out}) that appear in the logic circuit (in Fig. 4.16). Figures 4.18 and 4.19 are the 3-D lattice realizations of the functions in Fig. 4.17.

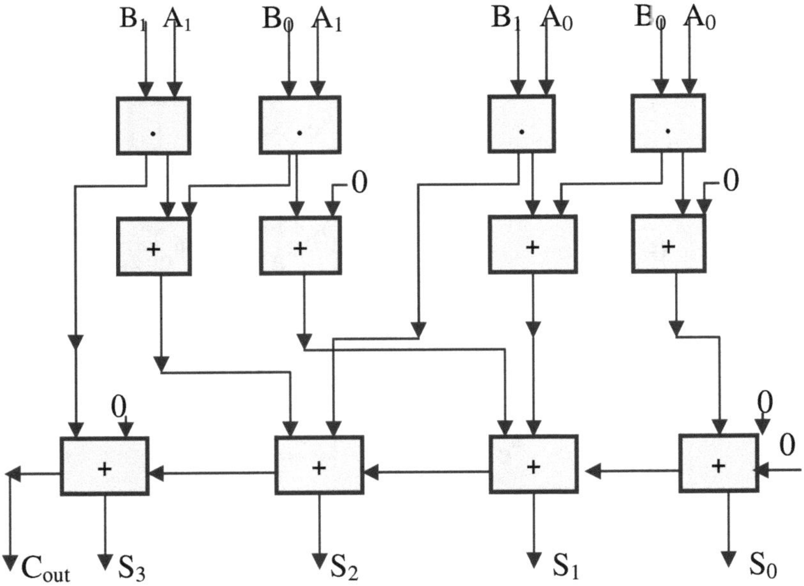

Fig. 4.16. Logic circuit of a ternary 2-digit multiplier.

b \ a	0	1	2
0	0	0	0
1	0	1	2
2	0	2	1

a

b \ a	0	1	2
0	0	0	0
1	0	0	0
2	0	0	1

b

Fig. 4.17. **a** Ternary multiply, and **b** ternary carry out for the ternary 2-digit multiplier.

The three input ternary adder that appears in the logic circuit in Fig. 4.16 can be implemented directly from the 3-D lattices shown in Figs. 4.14 and 4.15 for the sum and carry out, respectively. The two input ternary adder that appears in the logic circuit in Fig. 4.16 can also be implemented directly from the 3-D lattices similar to those shown in Figs. 4.14 and 4.15 for the sum and carry out, respectively, by using only two control variables, and setting the third digits (a_2 and b_2) in Fig. 4.12 to the value zero.

Example 4.9. For the non-symmetric two-variable ternary input/ternary output function: F = ab + a'b'', Fig. 4.20 illustrates the three-dimensional logic circuit for such non-symmetric functions. Figure 4.20 also indicates the relationship between the ternary natural-encoded map with non-conflicting symmetry indices and the three-dimensional lattice structure with non-conflicting leaves.

Example 4.10. Figure 4.21 illustrates the close relationship between the concept of the three-dimensional lattice structure and the symmetry of ternary functions.

Note that, in Fig. 4.21b, the ternary symmetry indices represent the sets of all possible paths from the leaves to the root through the internal nodes of the three-dimensional lattice structure which is shown in Fig. 4.21a. Figure 4.21c shows that the same ternary symmetry indices, that are shown in Fig. 4.21b, correspond to the counts of the number of ones and twos in the natural-encoded cell indices of the ternary map. In general, as a convention, let us denote the nodes in the lattice structure by their three-dimensional Cartesian coordinates; the tuple {x, y, z}. Also, let us denote the edge between two nodes $\{x_1, y_1, z_1\}$ and $\{x_2, y_2, z_2\}$ by $\{x_1, y_1, z_1\}$-$\{x_2, y_2, z_2\}$.

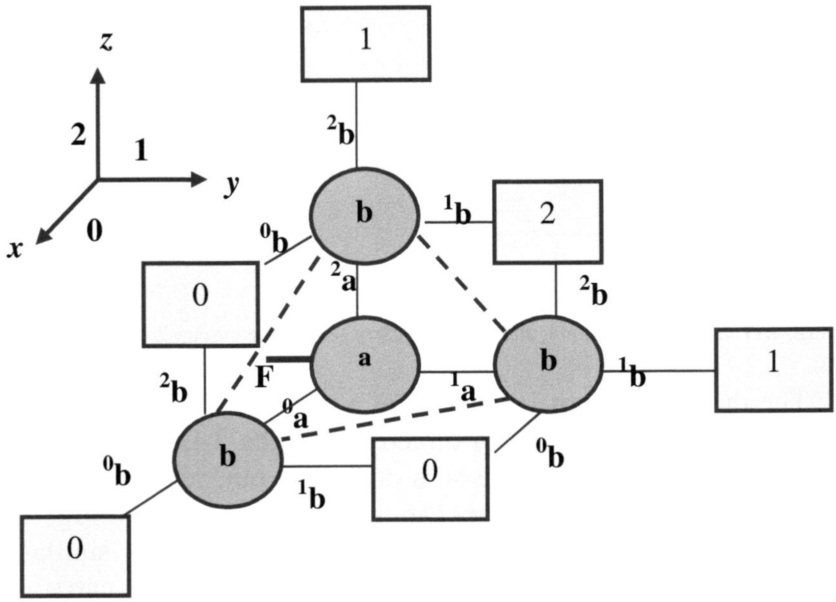

Fig. 4.18. The multiply function of ternary 2-digit multiplier.

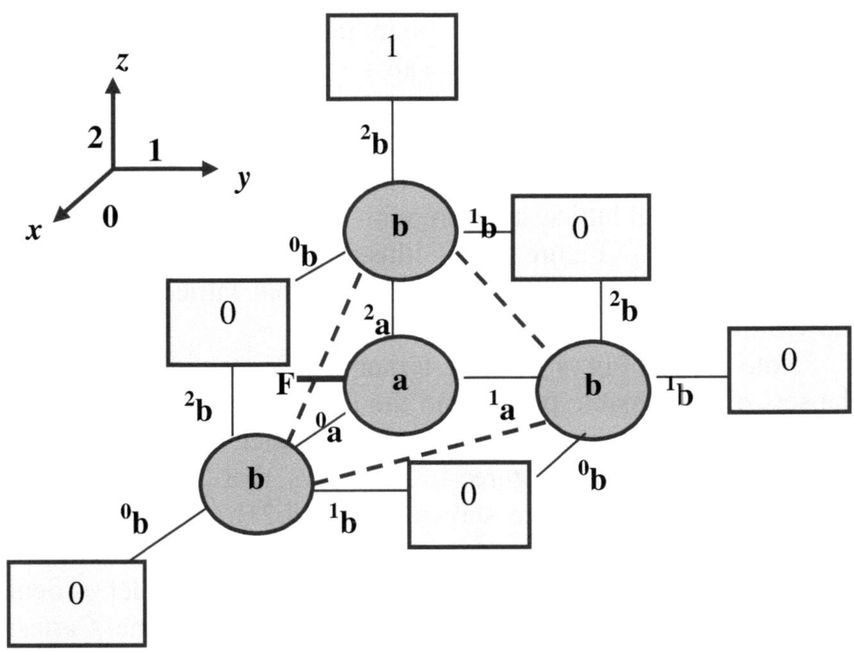

Fig. 4.19. The carry out function of ternary 2-digit multiplier.

a	b	F
0	0	2
0	1	0,1
0	2	1,0
1	0	0,1
1	1	1
1	2	2,1
2	0	1,0
2	1	2,1
2	2	1

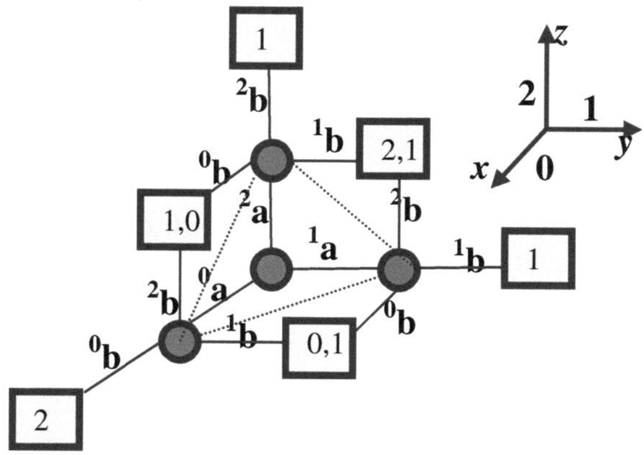

a

a	b	F
0	0	2
0	1	0
0	2	1
1	0	1
1	1	1
1	2	1
2	0	0
2	1	2
2	2	1

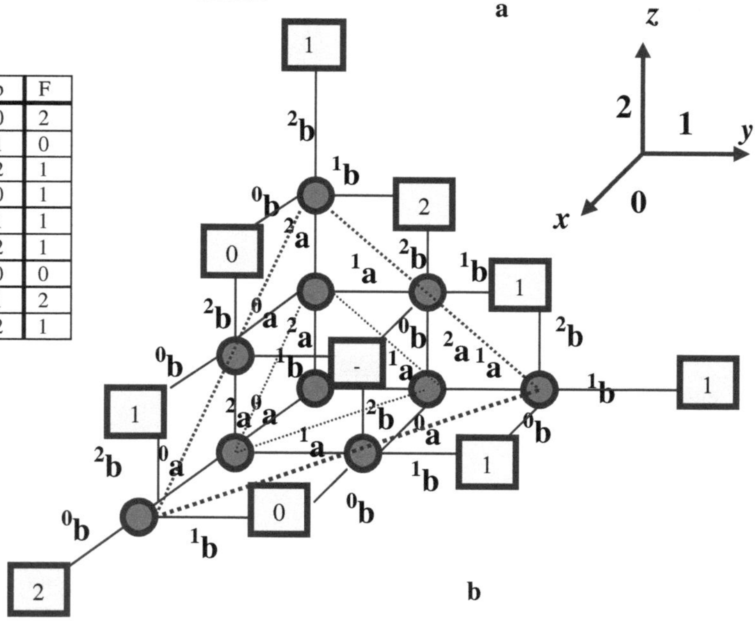

b

Fig. 4.20. a Three-dimensional lattice structure that corresponds to the non-symmetric ternary function F = ab + a'b" with conflicting leaves (shaded cells in the corresponding ternary map), and **b** three-dimensional lattice structure that corresponds to the symmetrized function F (by repeating variable a two times) with non-conflicting leaves.

As an example, the symbol (-) in the node {1, 1, 1} in Fig. 4.20b represents a complete three-valued don't care (i.e., 0, 1, or 2). This geometrical notation will be used in this Sect. and in the following Sect. To produce the repetition of variables for non-symmetric ternary functions, the joining operators are needed to join the corresponding nodes to produce the corresponding correction functions, in order to preserve the output function of the 3-D lattice.

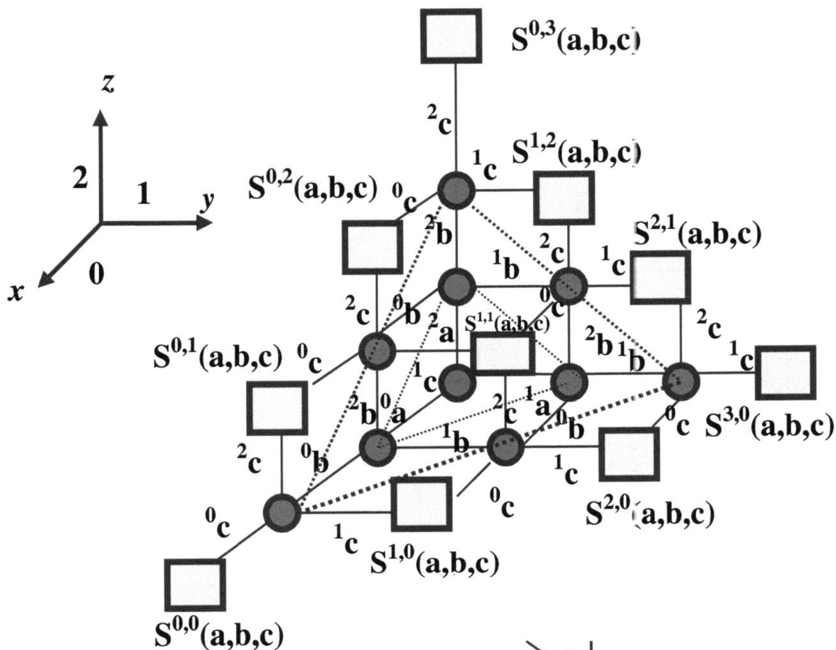

a

$S^{i,j}$ is the ternary symmetry indices, where i = # of ones in the natural-encoded cell indices of the ternary map, and j = # of twos in the natural-encoded cell indices of the ternary map :

$S^{0,0}(a,b,c) = \{{}^0a\,{}^0b\,{}^0c\}$
$S^{1,0}(a,b,c) = \{{}^1a\,{}^0b\,{}^0c,\ {}^0a\,{}^1b\,{}^0c,\ {}^0a\,{}^0b\,{}^1c\}$
$S^{2,0}(a,b,c) = \{{}^1a\,{}^1b\,{}^0c,\ {}^1a\,{}^0b\,{}^1c,\ {}^0a\,{}^1b\,{}^1c\}$
$S^{3,0}(a,b,c) = \{{}^1a\,{}^1b\,{}^1c\}$
$S^{0,1}(a,b,c) = \{{}^2a\,{}^0b\,{}^0c,\ {}^0a\,{}^2b\,{}^0c,\ {}^0a\,{}^0b\,{}^2c\}$
$S^{0,2}(a,b,c) = \{{}^2a\,{}^2b\,{}^0c,\ {}^2a\,{}^0b\,{}^2c,\ {}^0a\,{}^2b\,{}^2c\}$
$S^{0,3}(a,b,c) = \{{}^2a\,{}^2b\,{}^2c\}$
$S^{1,1}(a,b,c) = \{{}^1a\,{}^2b\,{}^0c,\ {}^1a\,{}^0b\,{}^2c,\ {}^2a\,{}^1b\,{}^0c,\ {}^2a\,{}^0b\,{}^1c,\ {}^0a\,{}^2b\,{}^1c,\ {}^0a\,{}^1b\,{}^2c\}$
$S^{1,2}(a,b,c) = \{{}^1a\,{}^2b\,{}^2c,\ {}^2a\,{}^1b\,{}^2c,\ {}^2a\,{}^2b\,{}^1c\}$
$S^{2,1}(a,b,c) = \{{}^1a\,{}^1b\,{}^2c,\ {}^1a\,{}^2b\,{}^1c,\ {}^2a\,{}^1b\,{}^1c\}$

b

ab \ c	0	1	2
00	$S^{0,0}$	$S^{1,0}$	$S^{0,1}$
01	$S^{1,0}$	$S^{2,0}$	$S^{1,1}$
02	$S^{0,1}$	$S^{1,1}$	$S^{0,2}$
10	$S^{1,0}$	$S^{2,0}$	$S^{1,1}$
11	$S^{2,0}$	$S^{3,0}$	$S^{2,1}$
12	$S^{1,1}$	$S^{2,1}$	$S^{1,2}$
20	$S^{0,1}$	$S^{1,1}$	$S^{0,2}$
21	$S^{1,1}$	$S^{2,1}$	$S^{1,2}$
22	$S^{0,2}$	$S^{1,2}$	$S^{0,3}$

c

Fig. 4.21. Three-dimensional lattice structure for ternary functions of three 3-valued variables: **a** 3-D lattice structure made of 3-to-1 inter-connected multiplexers, **b** sets of ternary symmetry indices, and **c** the ternary natural-encoded map interpretation of the ternary symmetry indices.

Figure 4.22 represents such joining for three-dimensional lattice structure.

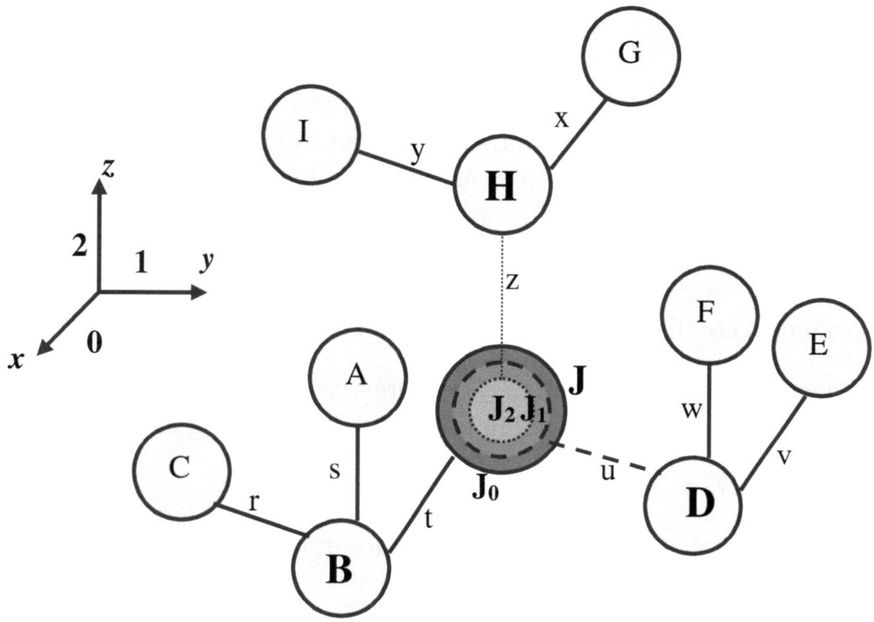

Fig. 4.22. General 3-D lattice structure for three joining nodes and a single joined node.

In Fig. 4.22, three nodes: B, D, and H are joining (superimposing) their nodes J_0, J_1, and J_2, respectively, to form the superimposed node J. The set of nodes {C, A, J_0} are the cofactors of the node B. The set of nodes {E, F, J_1} are the cofactors of the node D. The set of nodes {G, I, J_2} are the cofactors of the node H. The geometrical distribution of the nodes and edges in Fig. 4.22 are as follows:

Axis	*Nodes*	*Edges*	*Weights*
x-axis	$\{J_0, E, I\}$	$\{t, v, y\}$	α
y-axis	$\{C, J_1, G\}$	$\{r, u, x\}$	β
z-axis	$\{A, F, J_2\}$	$\{s, w, z\}$	γ

Section 4.4.2 will introduce the joining rules for the general structure in Fig. 4.22 by using the new sets of Shannon decompositions introduced in Sect. 2.2.2.

4.4.2 New (3,3) Three-Dimensional Invariant Shannon Lattice Structures

In the following derivation, two correction functions for the case of ternary logic are implemented. In general, for n^{th} radix Galois logic, no correction functions are needed for lattice structures with *n-valued* invariant Shannon nodes as will be shown in Theorem 4.1. So, for instance, for the case of binary Shannon, no correction functions are needed, due to the fact that all of the Shannon cofactors are disjoint, as was shown in Fig. 4.4b.

Theorem 4.1. For lattice structures with all invariant ternary Shannon nodes, the following is one possible joining rule:

$$J = {}^0aJ_0 + {}^1a\,J_1 + {}^2aJ_2. \tag{4.5}$$

Proof. Utilizing Eq. (2.61), and by joining in Fig. 4.22 the following invariant Shannon nodes:

$$\begin{bmatrix} \alpha_1 & 0 & 0 \\ 0 & \beta_1 & 0 \\ 0 & 0 & \gamma_1 \end{bmatrix}, \begin{bmatrix} \alpha_2 & 0 & 0 \\ 0 & \beta_2 & 0 \\ 0 & 0 & \gamma_2 \end{bmatrix}, \begin{bmatrix} \alpha_3 & 0 & 0 \\ 0 & \beta_3 & 0 \\ 0 & 0 & \gamma_3 \end{bmatrix}.$$

And by assigning the following values for the set of edges {r, s, t, u, v, w, x, y, z} in Fig. 4.22:

$$t = \hat{\alpha}_1\,{}^0a\;,\; v = \hat{\alpha}_2\,{}^0a\,,\; y = \hat{\alpha}_3\,{}^0a.$$

$$r = \hat{\beta}_1\,{}^1a\,,\; u = \hat{\beta}_2\,{}^1a\,,\; x = \hat{\beta}_3\,{}^1a.$$

$$s = \hat{\gamma}_1\,{}^2a\,,\; w = \hat{\gamma}_2\,{}^2a\,,\; z = \hat{\gamma}_3\,{}^2a.$$

One obtains the following set of Eqs. *before* and *after* joining the three nodes J_0, J_1, and J_2 in Fig. 4.22 (where: {A, C, J_0} are the set of functions for node B, {E, F, J_1} are the set of functions for node D, and {I, G, J_2} are the set of functions for node H, respectively):

Before joining the nodes:

$$B = \hat{\alpha}_1\,{}^0a\,J_0 + \hat{\beta}_1\,{}^1a\,C + \hat{\gamma}_1\,{}^2a\,A, \tag{4.6}$$

$$D = \hat{\alpha}_2\,{}^0a\,E + \hat{\beta}_2\,{}^1a\,J_1 + \hat{\gamma}_2\,{}^2a\,F, \tag{4.7}$$

$$H = \hat{\alpha}_3\,{}^0a\,I + \hat{\beta}_3\,{}^1a\,G + \hat{\gamma}_3\,{}^2a\,J_2. \tag{4.8}$$

After joining the nodes:

$$B = \hat{\alpha}_1\,{}^0a\,J + \hat{\beta}_1\,{}^1a\,C + \hat{\gamma}_1\,{}^2a\,A, \tag{4.9}$$

$$D = \hat{\alpha}_2 \, {}^0a\, N + \hat{\beta}_2 \, {}^1a\, J + \hat{\gamma}_2 \, {}^2a\, F, \tag{4.10}$$

$$H = \hat{\alpha}_3 \, {}^0a\, I + \hat{\beta}_3 \, {}^1a\, q + \hat{\gamma}_3 \, {}^2a\, J, \tag{4.11}$$

where N and q are the correction functions, and J is the superimposed node in Fig. 4.22. By equalizing Eq. (4.6) to Eq. (4.9), Eq. (4.7) to Eq. (4.10), and Eq. (4.8) to Eq. (4.11), and utilizing the axioms of GF(3), we obtain the following results:

$N = E,$

$q = G,$

$J = {}^0aJ_0 + {}^1a\, J_1 + {}^2aJ_2.$ **Q.E.D.**

From Eq. (4.5) one observes the fact that the joining rule of any corresponding invariant Shannon decomposition does not depend on the scaling numbers {α, β, γ} and does not need any correction function. The methodology that has been presented in this Sect. can be used for all possible permutations of the invariant Shannon decompositions. The following example illustrates Theorem 4.1.

Example 4.11. Let us produce the joining rule for the following invariant Shannon nodes:

$$\begin{bmatrix} 2 & 0 & 0 \\ 0 & 1 & 0 \\ 0 & 0 & 2 \end{bmatrix}, \begin{bmatrix} 2 & 0 & 0 \\ 0 & 2 & 0 \\ 0 & 0 & 2 \end{bmatrix}, \begin{bmatrix} 1 & 0 & 0 \\ 0 & 1 & 0 \\ 0 & 0 & 2 \end{bmatrix}.$$

Then according to Eq. (4.5), one obtains:

$J = {}^0aJ_0 + {}^1a\, J_1 + {}^2aJ_2.$

4.4.3 New (3,3) Three-Dimensional Invariant Davio Lattice Structures

In the following derivation, two correction functions are implemented for the case of ternary logic. In general, for n^{th} radix Galois logic, at least (n-1) correction functions are needed for lattice structures with n-valued invariant Davio nodes. So, for instance, one needs a single correction function in the case of binary Davio expansions (as was shown in the extreme right nodes of Figs. 4.4c and 4.4d, respectively). The following methodology, which is used in Theorems 4.2 through 4.4, can be used for all possible permutations of the invariant Davio decompositions as well.

Theorem 4.2. For lattice structures with all invariant ternary Davio$_0$ (D_0) nodes (that were presented in Theorem 2.3), the following is one possible set of joining rules, and correction functions, respectively:

$$J = J_0, \tag{4.12}$$
$$N = 2\,\alpha_2\,\hat{\beta}_2\,a\,J_0 + E + \alpha_2\,\hat{\beta}_2\,a\,J_1, \tag{4.13}$$
$$q = 2\,\hat{\gamma}_3\,\beta_3\,a\,J_0 + G + \hat{\gamma}_3\,\beta_3\,a\,J_2. \tag{4.14}$$

Proof. Utilizing Eq. (2.62), and by joining in Fig. 4.22 the following invariant D_0 nodes:

$$\begin{bmatrix} \alpha_1 & 0 & 0 \\ 0 & 2\beta_1 & \beta_1 \\ 2\gamma_1 & 2\gamma_1 & 2\gamma_1 \end{bmatrix}, \begin{bmatrix} \alpha_2 & 0 & 0 \\ 0 & 2\beta_2 & \beta_2 \\ 2\gamma_2 & 2\gamma_2 & 2\gamma_2 \end{bmatrix}, \begin{bmatrix} \alpha_3 & 0 & 0 \\ 0 & 2\beta_3 & \beta_3 \\ 2\gamma_3 & 2\gamma_3 & 2\gamma_3 \end{bmatrix}.$$

By assigning the following values for the edges {r, s, t, u, v, w, x, y, z} in Fig. 4.22:

$$t = \hat{\alpha}_1\,,\ v = \hat{\alpha}_2\,,\ y = \hat{\alpha}_3\,.$$
$$r = \hat{\beta}_1\,a\,,\ u = \hat{\beta}_2\,a\,,\ x = \hat{\beta}_3\,a.$$
$$s = \hat{\gamma}_1\,a^2\,,\ w = \hat{\gamma}_2\,a^2\,,\ z = \hat{\gamma}_3\,a^2.$$

And by following the same procedure that was used in Theorem 4.1, one obtains:

$$J = J_0,$$
$$N = 2\,\alpha_2\,\hat{\beta}_2\,a\,J_0 + E + \alpha_2\,\hat{\beta}_2\,a\,J_1,$$
$$q = 2\,\hat{\gamma}_3\,\beta_3\,a\,J_0 + G + \hat{\gamma}_3\,\beta_3\,a\,J_2.$$

Q.E.D.

Theorem 4.3. For lattice structures with all invariant ternary Davio$_1$ (D_1) nodes, the following is one possible set of joining rules, and correction functions, respectively:

$$J = J_0, \tag{4.15}$$
$$N = 2\,\alpha_2\,\hat{\beta}_2\,a'\,J_0 + E + \alpha_2\,\hat{\beta}_2\,a'\,J_1, \tag{4.16}$$
$$q = 2\,\hat{\gamma}_3\,\beta_3\,a'\,J_0 + G + \hat{\gamma}_3\,\beta_3\,a'\,J_2. \tag{4.17}$$

Proof. The proof of Theorem 4.3 follows the same methodology that is used to prove Theorem 4.1. **Q.E.D.**

Theorem 4.4. For lattice structures with all invariant ternary Davio$_2$ (D_2) nodes, the following is one possible set of joining rules, and correction functions, respectively:

$$J = J_0, \tag{4.18}$$
$$N = 2\ \alpha_2\ \hat{\beta}_2\, a'' J_0 + E + \alpha_2\ \hat{\beta}_2\, a''\ J_1, \tag{4.19}$$
$$q = 2\,\hat{\gamma}_3\ \beta_3\, a''\ J_0 + G + \hat{\gamma}_3\ \beta_3\, a''\ J_2. \tag{4.20}$$

Proof. The proof follows the method used for Theorem 4.1. **Q.E.D.**

Example 4.12. For following invariant Davio$_0$ (D_0) nodes over GF(3):

$$\begin{bmatrix} 2 & 0 & 0 \\ 0 & 1 & 2 \\ 1 & 1 & 1 \end{bmatrix}, \begin{bmatrix} 1 & 0 & 0 \\ 0 & 2 & 1 \\ 2 & 2 & 2 \end{bmatrix}, \begin{bmatrix} 2 & 0 & 0 \\ 0 & 2 & 1 \\ 2 & 2 & 2 \end{bmatrix}.$$

Then according to Eqs. (4.12), (4.13), and (4.14), one obtains:
$J = J_0$, $N = 2\,a\,J_0 + E + a\,J_1$, $q = 2\,a\,J_0 + G + a\,J_2$.

From the previous examples, it can be observed that the structural property of lattice structures depends on the functional property of the functions that are decomposed: if the ternary function is symmetric then there is no need to repeat variables to realize the function in three-dimensional lattice structure, otherwise there is a need to repeat variables (as was shown in Example 4.9 for instance). The following Sect. introduces an algorithm for the realization of three-dimensional logic circuits using Theorem 4.1.

4.5 An Algorithm for the Expansion of Ternary Functions into (3,3) Three-Dimensional Lattice Structures

This Sect. introduces as an example an algorithm for realizing multi-valued invariant Shannon expansion of ternary functions in 3-D lattice structures that were proposed in Theorem 4.1. Analogously to the convention that was established for the binary case, this algorithm is developed for the convention: in the octant that corresponds to the positive x-axis, positive y-axis, and positive z-axis, expand the nodes in-to-out and join the cofactors counter clock wise (CCW). Similar algorithms can be developed for other invariant expansions.

```
{for i = j = k = 0
   Utilizing Eq. (2.61) Expand n(i, j, k) into n(i+1, j, k), n(i, j+1, k), n(i, j, k+1)
         if all nodes are constants
            then go to 3
         else (
              // for the nodes that have common indices //
      1.  if there exist nodes with conflicting values
               then (
                           ( Utilizing Eq. (4.5)
                            Join nodes with common indices
                              if three nodes exist
                                        then (apply Eq. (4.5))
                              else (set the non-existing nodes to zero
                                       apply Eq. (4.5))
                           )
                       for all joined nodes
                            (
                             for each node ( Utilizing Eq. (2.61)
          // for l, m, and n as general positional indices that can be
             expressed in terms of i, j, and k, respectively //.
                 Expand n(l ,m, n) into n(l+1, m, n), n(l, m-1, n), n(l, m, n+1) )
                            )
                         go to 2
                     )
              else go to 2
          2. Utilizing Eq. (2.61)
          i = i ++: Expand n(i, j, k) into n(i+1, j, k), n(i, j+1, k), n(i, j, k+1)
          j = j ++: Expand n(i, j, k) into n(i+1, j, k), n(i, j+1, k), n(i, j, k+1)
          k = k ++: Expand n(i, j, k) into n(i+1, j, k), n(i, j+1, k), n(i, j, k+1)
           if all nodes are constants with no conflicting values in
             the same indices
                then go to 3
             else go to 1)
         3. end}
```

The following Sect. presents a complete example that illustrates the use of the invariant Shannon and Davio spectral transforms in the proposed three-dimensional (3,3) lattice structures.

4.6 Example of the Implementation of Ternary Functions Using the New Three-Dimensional Lattice Structures

The following example follows one convention for constructing three-dimensional lattice structures using the iterative process of expanding (decomposing) and collapsing (joining) in three-dimensional space: in the sub-space that corresponds to the positive x-axis, positive y-axis, and positive z-axis, expand the nodes in-to-out and join the cofactors counter clock wise (CCW). For the following ternary function:

a \ b	0	1	2
0	2	1	1
1	2	1	1
2	2	2	2

F

Fig. 4.23. Ternary input/ternary output map for the function: $F = 2\,{}^{0}a\,{}^{0}b + {}^{0}a\,{}^{1}b + {}^{0}a\,{}^{2}b + 2\,{}^{1}a\,{}^{0}b + {}^{1}a\,{}^{1}b + {}^{1}a\,{}^{2}b + 2\,{}^{2}a\,{}^{0}b + 2\,{}^{2}a\,{}^{1}b + 2\,{}^{2}a\,{}^{2}b$.

Utilizing Fig. 4.22, the joining operations presented in Eqs. (4.5) and (4.12) - (4.14) for the invariant Shannon and Davio_0 decompositions respectively, and the expansions from Eqs. (2.61) and (2.62), one obtains in Figs. 4.24 and 4.25 the 3-D lattice structures that realize the non-symmetric ternary function F from Fig. 4.23. In Fig. 4.24, one obtains the corresponding 3-D lattice, by utilizing the following:

<u>Step1</u>: Expanding nodes: Expand the non-symmetric function: $F = 2\,{}^{0}a\,{}^{0}b + {}^{0}a\,{}^{1}b + {}^{0}a\,{}^{2}b + 2\,{}^{1}a\,{}^{0}b + {}^{1}a\,{}^{1}b + {}^{1}a\,{}^{2}b + 2\,{}^{2}a\,{}^{0}b + 2\,{}^{2}a\,{}^{1}b + 2\,{}^{2}a\,{}^{2}b$ in node (0, 0, 0) according to Eq. (2.61) as:

$F_0 = F(a = 0) = 2\,{}^{0}b + {}^{1}b + {}^{2}b$ into node (1, 0, 0),

$F_1 = F(a = 1) = 2\,{}^{0}b + {}^{1}b + {}^{2}b$ into node (0, 1, 0),

$F_2 = F(a = 2) = 2\,{}^{0}b + 2\,{}^{1}b + 2\,{}^{2}b$ into node (0, 0, 1).

<u>Step 2</u>: Joining nodes: As a result from step 1, conflicting values occur in nodes (1, 1, 0), (0, 1, 1), and (1, 0, 1), then *Join* cofactors according to Eq. (4.5), as follows:

y-axis cofactor of node (1, 0, 0) and x-axis cofactor of node (0, 1, 0) into node (1, 1, 0) $\Rightarrow$ the joined node (1, 1, 0) is: $1\,{}^{1}b + 2\,{}^{0}b$,

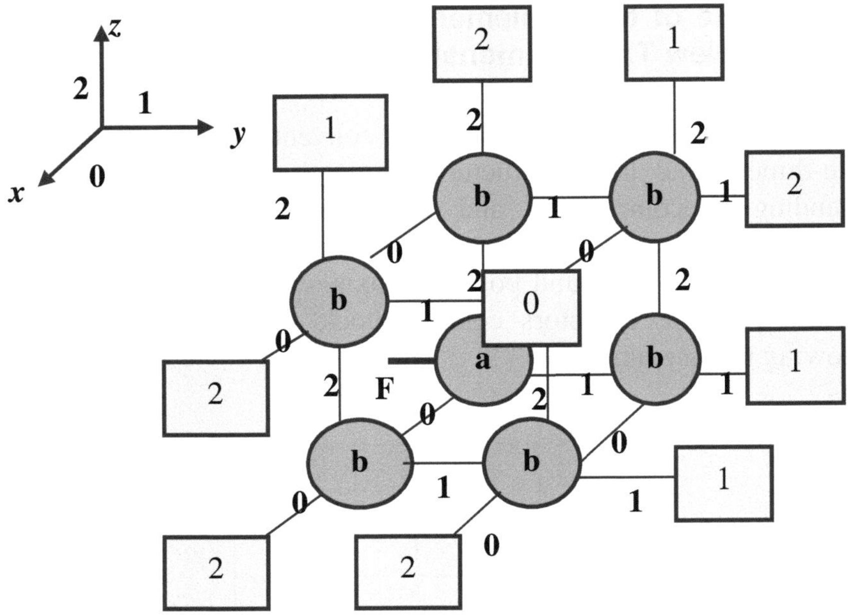

Fig. 4.24. First three-dimensional logic circuit of the non-symmetric function in Fig. 4.23: $F = 2\ {}^{0}a\ {}^{0}b + {}^{0}a\ {}^{1}b + 2\ {}^{1}a\ {}^{0}b + {}^{1}a\ {}^{1}b + {}^{1}a\ {}^{2}b + 2\ {}^{2}a\ {}^{2}b + 2\ {}^{2}a\ {}^{1}b + 2\ {}^{2}a\ {}^{0}b + {}^{)}a\ {}^{2}b$.

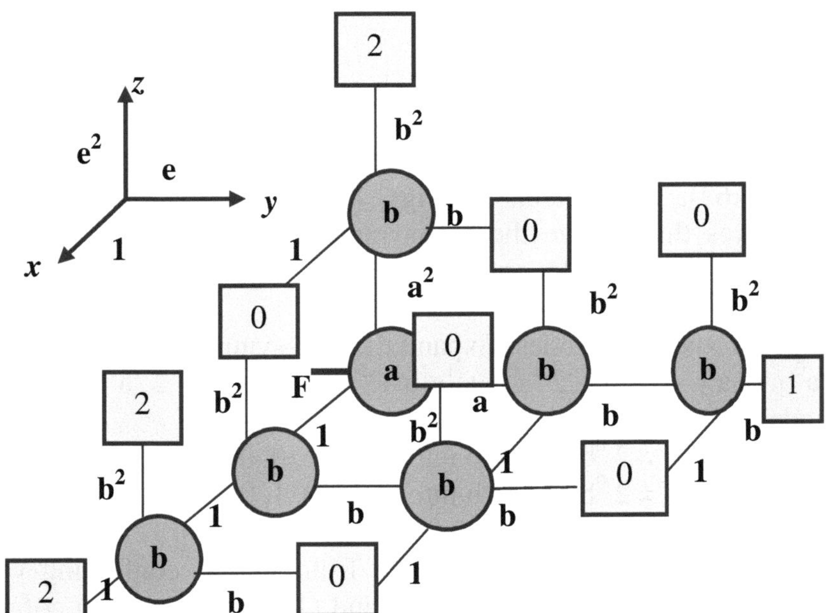

Fig. 4.25. Second three-dimensional logic circuit of the non-symmetric function in Fig. 4.23: $F = 2 + a{\cdot}b^2 + 2\ a^2{\cdot}b^2 + 2\ b^2$.

z-axis cofactor of node (0, 1, 0) and y-axis cofactor of node (0, 0, 1) into node (0, 1, 1) ⇒ the joined node (0, 1, 1) is: $1\,{}^{2}b + 2\,{}^{1}b$,
z-axis cofactor of node (1, 0, 0) and x-axis cofactor of node (0, 0, 1) into node (1, 0, 1) ⇒ the joined node (1, 0, 1) is: $1\,{}^{2}b + 2\,{}^{0}b$.
Step 3: Expanding nodes: Expand the lattice structure nodes, that result from step 2, as follows:
node (1, 0, 0) into node (2, 0, 0) of value 2,
node (1, 1, 0) into nodes: (2, 1, 0) of value 2, (1, 2, 0) of value 1, and (1, 1, 1) of value 0,
node (0, 1, 0) into node (0, 2, 0) of value 1,
node (0, 1, 1) into nodes: (0, 2, 1) of value 2, (0, 1, 2) of value 1, and (1, 1, 1) of value 0,
node (0, 0, 1) into node (0, 0, 2) of value 2,
node (1, 0, 1) into nodes: (2, 0, 1) of value 2, (1, 0, 2) of value 1, and (1, 1, 1) of value 0.
Note that by joining cofactors according to Eq. (4.5), the repetition of variable {b} is created, and therefore a new level in the 3-D lattice structure is created to create leaves with non-conflicting values.

By applying the same previous procedure of *expanding-joining* steps, and utilizing the expansion in Eq. (2.62) for expansion nodes of type $D_0 = \begin{bmatrix} 1 & 0 & 0 \\ 0 & 2 & 1 \\ 2 & 2 & 2 \end{bmatrix}$, and the joining operations in Eqs. (4.12) through (4.14), one obtains the three-dimensional logic circuit which is presented in Fig. 4.25. Note that by joining cofactors according to Eqs. (4.12) through (4.14) the repetition of variable {b} is created, and therefore a new level in the 3-D lattice structure is created to create leaves with non-conflicting values.

Observing Figs. 4.24 and 4.25, one obtains the size-based comparison results in Table 4.2. One can note that while the Shannon lattice structure in Fig. 4.24 has only one zero-valued leaf, Davio_0 lattice structure in Fig. 4.25 has six zero-valued leaves. This is important when considering power consumption in such lattices, since value "0" represents ground (Fig. 4.26a) and thus does not need to be supplied from a power supply in contrast to values "1"

(Fig. 4.26b) and “2” (Fig. 4.26c) that are obtained from a power supply and thus consume power.

For example, from Figs. 4.18 and 4.19, one can observe that the multiplication function and the carry function are both symmetric functions, as there is no need to repeat variables to make the ternary functions realizable in the 3-D lattice structures.

Table 4.2. Size-based comparison between the lattice realizations in Figs. 4.24 and 4.25, for the non-symmetric function $F = 2\,{}^{0}a^{0}b + {}^{0}a^{1}b + {}^{0}a^{2}b + 2\,{}^{1}a^{0}b + {}^{1}a^{1}b + {}^{1}a^{2}b + 2\,{}^{2}a^{0}b + 2\,{}^{2}a^{1}b + 2\,{}^{2}a^{2}b$.

Parameter/Type	Shannon (Fig. 4.24)	Davio_0 (Fig. 4.25)
Total # of Internal Nodes	7	7
Total # of Leaves	10	10
Total # of Zero-Valued Leaves	1	6

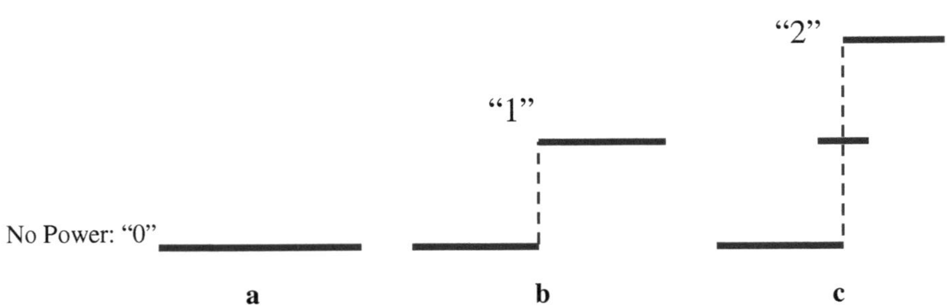

Fig. 4.26. Power levels for ternary logic.

Yet, as can be also observed in Example 4.8, by the substitution of the 3-D lattice structures of Figs. 4.18 and 4.19 into the logic circuit in Fig. 4.16, one can see that there is a need for placement of the individual 3-D lattices and the routing of connections between the placed lattices. Therefore, appropriate software tools should be created for such placement and routing that would start from the floor plan created with these 3-D regular lattice structures. One also can observe that although the individual lattices that are created in Figs. 4.18 and 4.19 are fully regular, the total structure of the logic circuit in Fig. 4.16, that is composed of all of the individual lattices, is only semi-regular because of the use of interconnects of different lengths. One possible solution for this, to obtain a fully regular total structure, is to produce a fully regular "macro" lattice from the corresponding "micro" lattices to produce a multi-input multi-output 3-D lattice, analogous to the binary case. Another solution is to combine the set of individual Eqs. into one Eq. to produce a single ternary function for the 2-digit ternary multiplier, that would produce all the individual multiply and C_{out} ternary functions from it, and implementing this single ternary function in one big 3-D lattice structure. All these variants should be combined in future EDA tools for a comprehensive (1) system, (2) logic, and (3) physical design levels into solid space. Other design issues to be considered by EDA tools is that for the previously designed ternary 3-digit full adder and ternary 2-digit multiplier one other choice of implementation would be the use of modulo-3 addition and multiplication in different design of logic circuits other than 3-D lattices. Another option, is to use Galois circuits whenever possible. Other multi-valued input multi-valued output designs are also possible for the ternary 2-digit multiplier, utilizing a mix of higher and lower radix logics in the same design. The following Sect. introduces a new algorithm called Iterative Symmetry Indices Decomposition (ISID) for the design of two-dimensional and three-dimensional lattice structures to fit specific layout boundaries.

4.7 ISID: Iterative Symmetry Indices Decomposition

As was illustrated in Example 4.5, as a simple case, realizing non-symmetric functions using lattice structures demands the repetition of variables. In many cases, one has to repeat variables so many times that will result in a big size lattice structure that does not fit specified area (or volume in case of three-dimensional lattice structures from Sects. 4.4.2 and 4.4.3). On the other hand, one can re-route the interconnects between the internal nodes of the lattice structure using optimization methods in a way such that the structure will ultimately fit into the specified layout area. Yet, this process will make the interconnects between cells of many different lengths, and consequently "strips" the lattice structure from one of its important features; all the inter-connects are of equal length. This idea of maintaining interconnects of equal length for a large size lattice structure that does not fit specific layout boundaries can be achieved using the new decomposition called Iterative Symmetry Indices Decomposition (ISID), as follows [16,25]: Suppose one has a k-map of a non-symmetric Boolean function. This means that conflicting values of "0" and "1" exist within some symmetry indices S^i. One way of removing such conflicting values is to repeat variables as was shown in previous Sects. Another way of removing such conflicting values is to decompose the non-symmetric function into a symmetric part superimposed with an error part [16,25]. The error part can be then iteratively decomposed into a "superposition" of two parts. The "superposition" of the decomposed parts to produce the total function can be done using the Exclusive-OR (⊕) operator or the Equivalence operator (⊗). The following algorithm demonstrates the ISID decomposition [16,25]:

(1) For a given area (or a given volume as in the case of 3-D lattice structures) specifications, synthesize a non-symmetric function using a symmetry-based structure like a lattice structure by repeating variables.

(2) If repeating variables will force the lattice structure to grow out of the layout boundaries, decompose the non-symmetric function into two super-imposed parts: a symmetric part and an error part (the error part can be alternatively named as the correction part). The original function is equal to the Exclusive-OR (⊕) or the Equivalence (⊗) of the two decomposed functions. This is denoted as ⊕-ISID and ⊗-ISID, respectively. Since there are many possible ways to obtain a symmetric function from the original non-symmetric function using ISID, one can choose a symmetric part by using the criterion of minimum number of changes of function values that are needed to transform the non-symmetric Boolean function into a symmetric one (e.g., minimum Hamming distance).

(3) Synthesize the symmetric part using lattice structure. If the synthesis fits layout boundaries then synthesize the error function.

(4) If the resulting synthesis does not fit layout boundaries go to step (2), and perform in serial-mode a single decomposition of the symmetric or error sub-functions, or perform in parallel-mode a multi-decomposition on all symmetric and error sub-functions.

(5) Repeat step (4) until the synthesis fits the layout boundaries.

Example 4.13. For the non-symmetric function in Fig. 4.27a, the corresponding 2-D lattice structure is realized in Fig. 4.27b by repeating the control variable {b} three times.

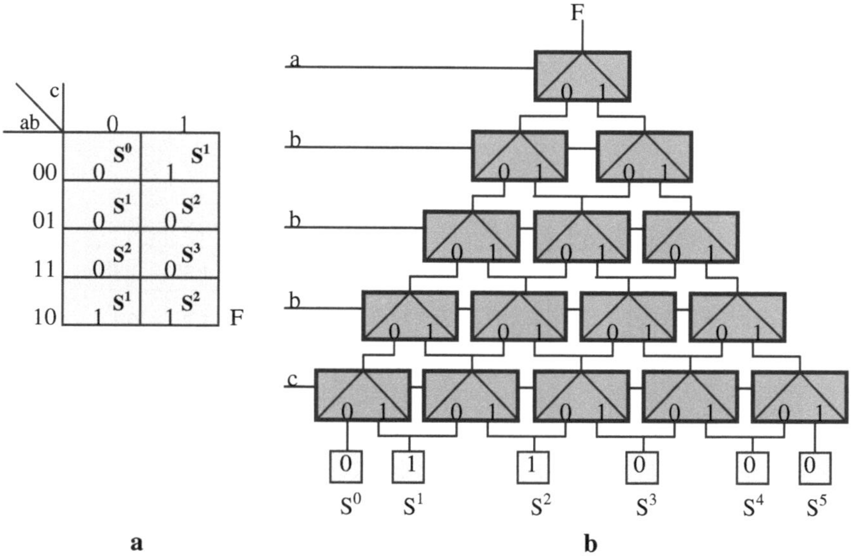

Fig. 4.27. Binary (2,2) 2-D Shannon lattice structure.

One notes that the lattice structure in Fig. 4.27 is made up of 15 2-to-1 multiplexers. Each 2-to-1 multiplexer consists of three logic gates. Thus Fig. 4.27 is made up of a total of 45 logic primitives. The same non-symmetric function can be synthesized using ⊕-ISID as in Fig. 4.28.

Consequently, one has the ⊕-ISID synthesis of the non-symmetric function where F1 realizes the symmetric part and F2 realizes the error part as shown in Fig. 4.28. Note that in Fig. 4.28 one has six 2-to-1 multiplexers and three primitives, namely AND, OR, and XOR. Thus one has a total of $6 \cdot 3 + 3 = 21$ logic primitives. Consequently, by comparing the total number of gates needed in Fig. 4.28 to those in Fig. 4.27, one observes that we economized a total of $45 - 21 = 24$ primitives.

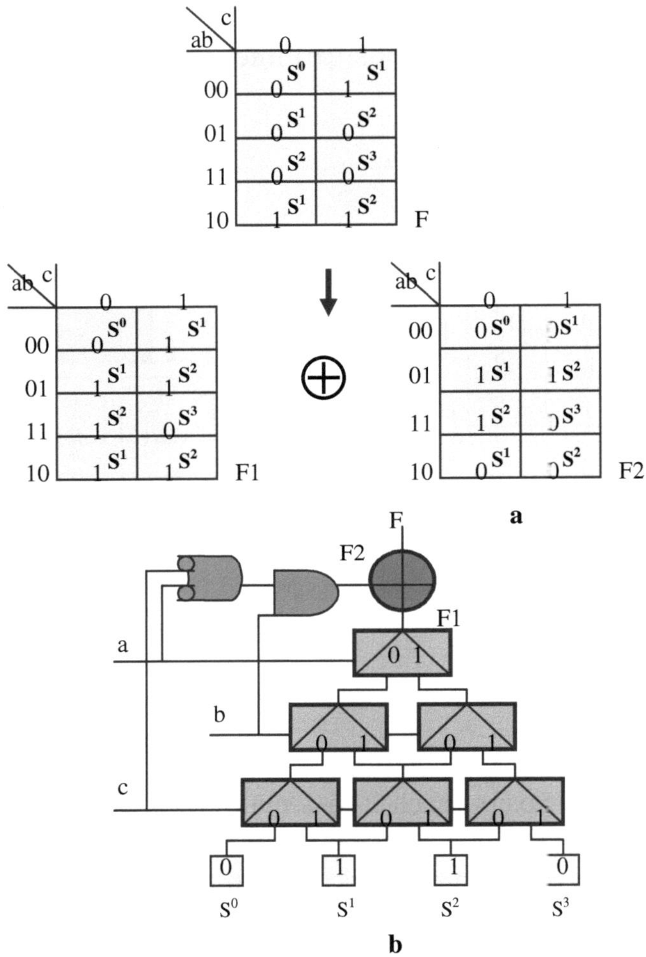

Fig. 4.28. ⊕-ISID Shannon lattice structure.

On the other hand, if one uses ⊗-ISID we obtain the decomposition in Fig. 4.29a and the synthesis of the non-symmetric function as shown in Fig. 4.29b. The cost of the lattice structure in Fig. 4.29b is 21 logic primitives which is the same as the cost from Fig. 4.28b.

One observes that 2-level Sum-Of-Product (SOP) structures have been used to synthesize the error functions in Figs. 4.28b and 4.29b, respectively. This choice has the advantage of the use of a minimal number of logic primitives, but has the disadvantage of transforming the lattice structure from a fully-regular structure where only one type of primitive has been used (namely 2-to-1 multiplexer) to "semi-regular" structure where many different logic primitives have been used, namely 2-to-1 multiplexer, AND, and OR gates. To retain full regularity in terms of using one type of logic primitives, one can synthesize the error function using a separate multiplexer-based logic structure like a Binary Decision Tree (BDT) for instance. This idea is demonstrated in Fig. 4.30 using the ⊗-ISID Shannon lattice structure from Fig. 4.29b.

One observes that the structure in Fig. 4.30 has 10 2-to-1 multilpexers and thus has 10·3 = 30 Boolean gates, which is still less than the total number of gates obtained in Fig. 4.27. Using the procedure for ISID decomposition, one can have a complicated decomposed structure that fits certain specifications. Figure 4.31a shows an iterative use of ISID using both operations of EXOR and EXNOR. This issue can be important by observing that in some certain cases the complement of a function can be much simpler than the function itself. Figure 4.31b demonstrates the use of serial-mode ISID decomposition versus parallel-mode ISID decomposition. This will "chop" the total rectangular area into smaller and smaller triangles of lattices as demonstrated in Fig. 4.31c.

The same idea of 2-D ISID can be used for 3-D ISID by using the algebraic identities over GF(3) to decompose the corresponding three-valued input three-valued output maps, such as the GF(3) Eqs. that are shown in Eqs. (4.21) and (4.22), respectively.

$$a \, *_{GF(3)} \, a \, *_{GF(3)} \, a = a, \tag{4.21}$$

$$a \, +_{GF(3)} \, a \, +_{GF(3)} \, a = 0. \tag{4.22}$$

Example 4.14. Lets us apply multiple-valued ISID to decompose the map in Fig. 4.32. The ternary non-symmetric map shown in Fig. 4.32 can be decomposed using ISID to a symmetric part and non-symmetric part. Yet, in contrast to the binary case where this can be done only in one specific EXOR expression, it can be done in the multiple-valued case in many different ways. This can be illustrated in Example 4.14 since the cell with {a = 0, b = 2, c = 2} has the value of "0". To produce a symmetric part the value of the cell {a = 0, b = 2, c = 2} has the value of "1".

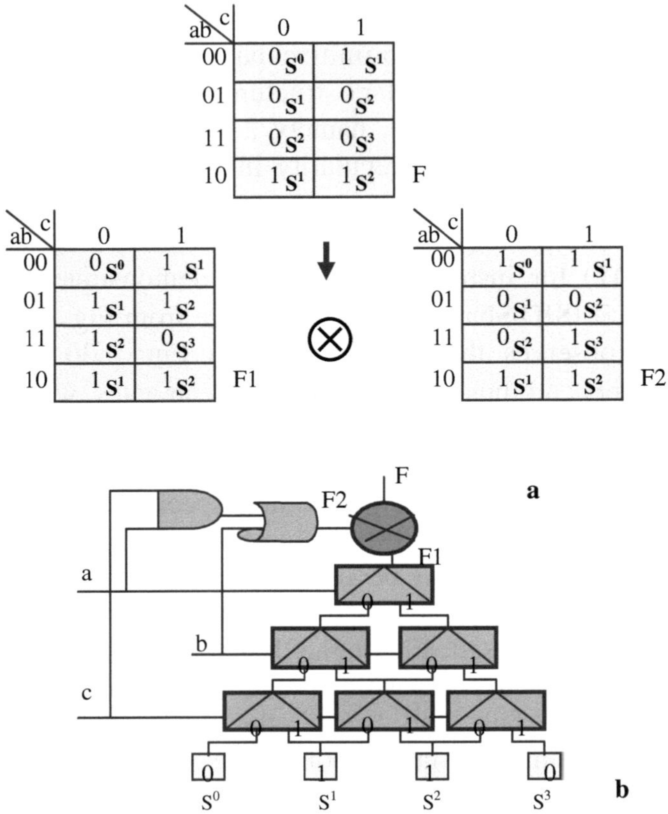

Fig. 4.29. ⊗-ISID Shannon lattice structure: **a** k-maps for ⊗-ISID, and **b** realization in lattice structure.

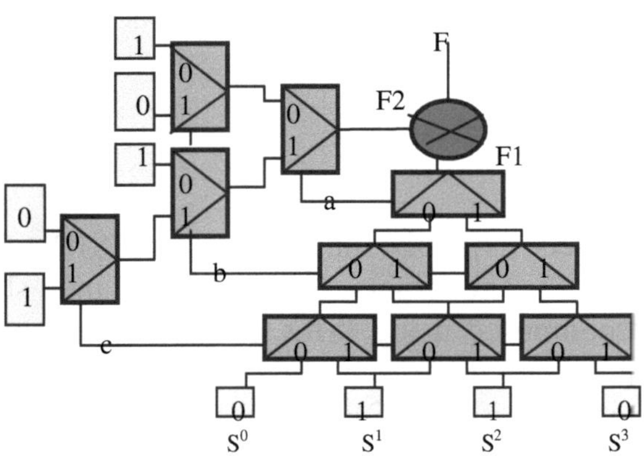

Fig. 4.30. Fully regular ⊗-ISID Shannon lattice structure using 2-to-1 multiplexers. F1 is a lattice structure that realizes symmetric function, and F2 is a decision tree (DT) that implements the binary Shannon expansion to realize the error function.

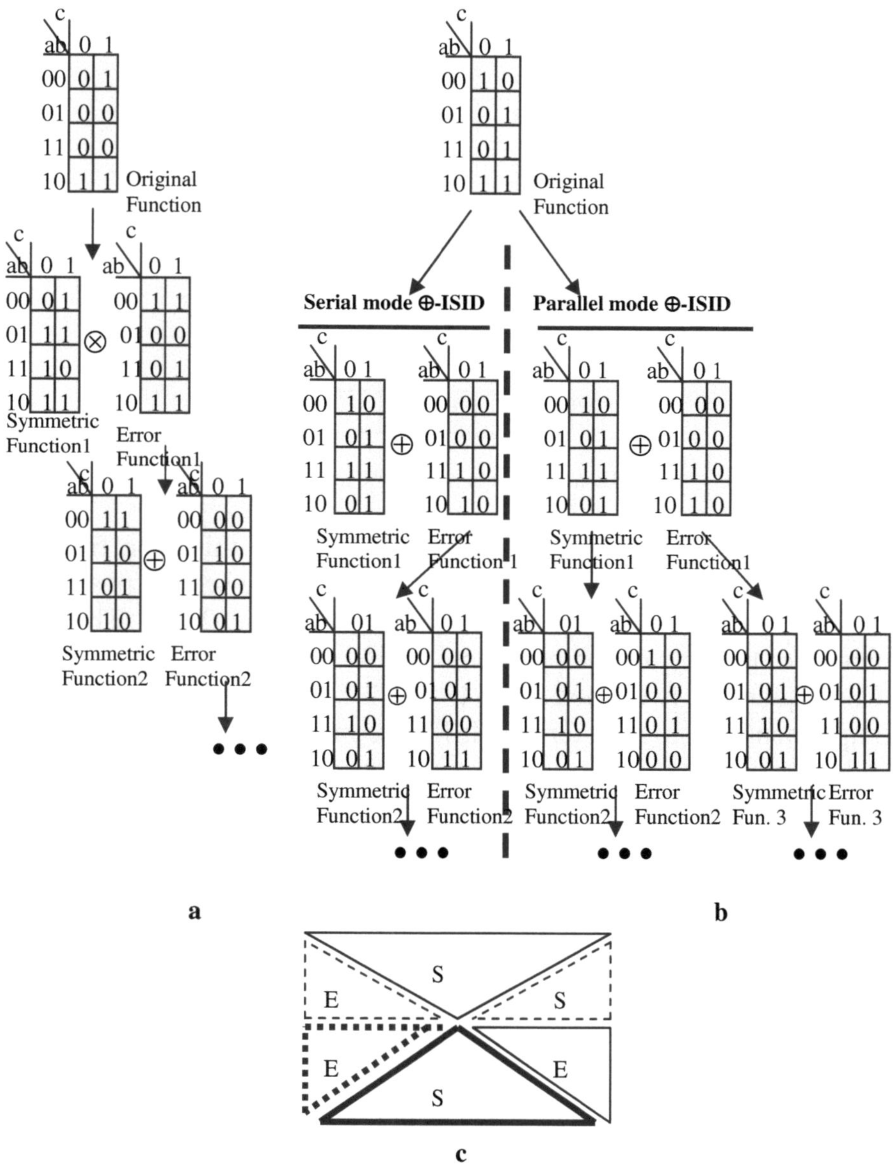

Fig. 4.31. ISID for the decomposition of non-symmetric Boolean functions: **a** iterative procedure using both ⊕ and ⊗ operations, **b** serial-mode ⊕-ISID versus parallel-mode ⊕–ISID, and **c** rectangular grid layout as a result of iterative implementation of ISID, where the original Boolean function is decomposed into error function E1 and symmetric function S1, E1 is then decomposed into E2 and S2, and E2 is decomposed into E3 and S3.

ab \ c	0	1	2
00	1	2	0
01	2	0	2
02	0	2	0
10	2	0	2
11	0	1	2
12	2	2	0
20	0	2	1
21	2	2	0
22	1	0	1

Fig. 4.32. Map for non-symmetric ternary function.

Consequently there are two possibilities for this. The first one follows Eq. (4.22): $1 +_{GF(3)} 1 +_{GF(3)} 1 = 0$, and the second possibility follows the algebraic rule from Fig. 2.1c: $1 +_{GF(3)} 2 = 0$. Thus two possible ternary decompositions follow as shown in Fig. 4.33. Consequently two possible (3,3) 3-D lattice structures are implemented using the ternary ISID decomposition as shown in Fig. 4.34. One notes that for the non-symmetric function in Fig. 4.32 one needs to repeat one of the variables in order to realize it using (3,3) 3-D lattice structure. This will impose the addition of many new internal cells, depending on the repeated variables chosen. On the other hand, the structures in Fig. 4.34 do not need to repeat variables because the ternary error part is realized in GFSOP form. This reduces substantial number of nodes needed for highly non-symmetric ternary functions.

The semi-regular realization of two-valued and multiple-valued lattice structures requires a "good" functional minimizer to minimize the non-lattice part of the total logic structure. One can use available minimizers for this purpose, or the GFSOP minimizer that utilizes the IF polarity from Chapt. 3. Such new structures will be used to produce reversible lattice structures for ternary functions in Chapt. 6. The new three dimensional lattice structures will be of a natural fit for the emerging 3-D nano technologies and optical devices as shown in [24].

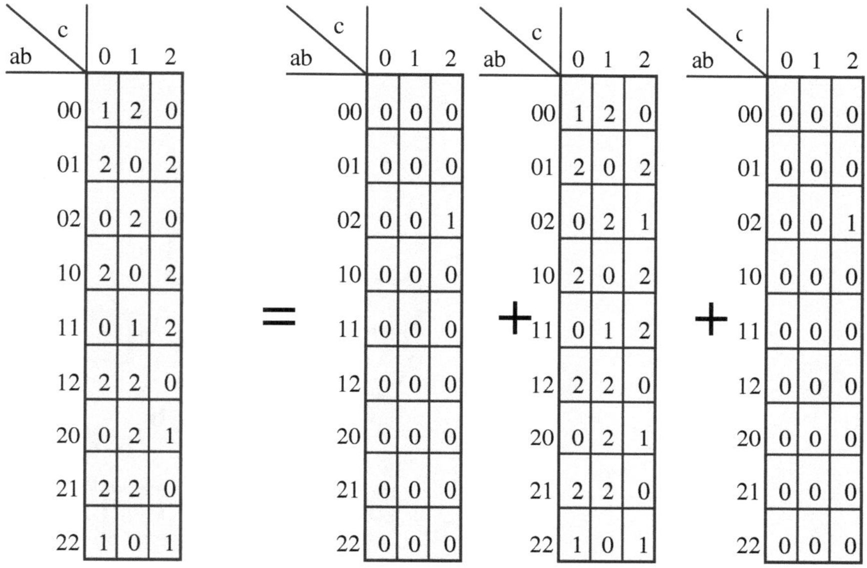

a

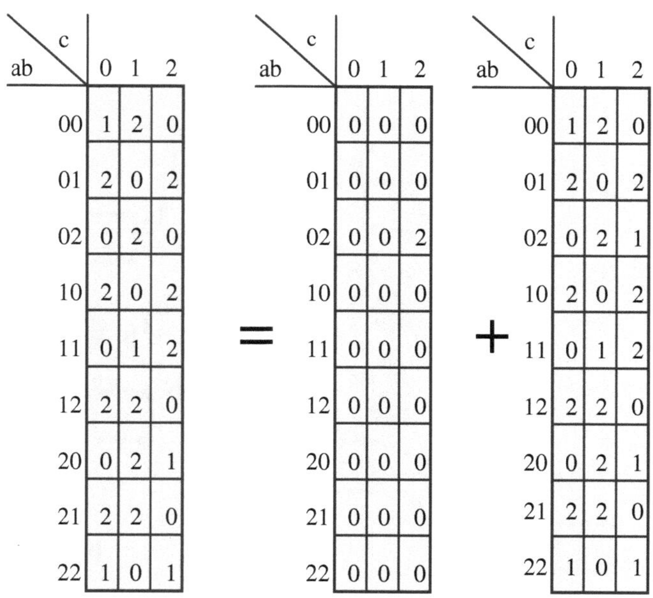

b

Fig. 4.33. Two possible ISID decompositions: **a** according to the algebraic rule: $1 +_{GF(3)} 1 +_{GF(3)} 1 = 0$, and **b** the second possibility follows the algebraic rule from Fig. 2.1c: $1 +_{GF(3)} 2 = 0$. (All additions and multiplications here are performed using Figs. 2.1c and 2.1d.)

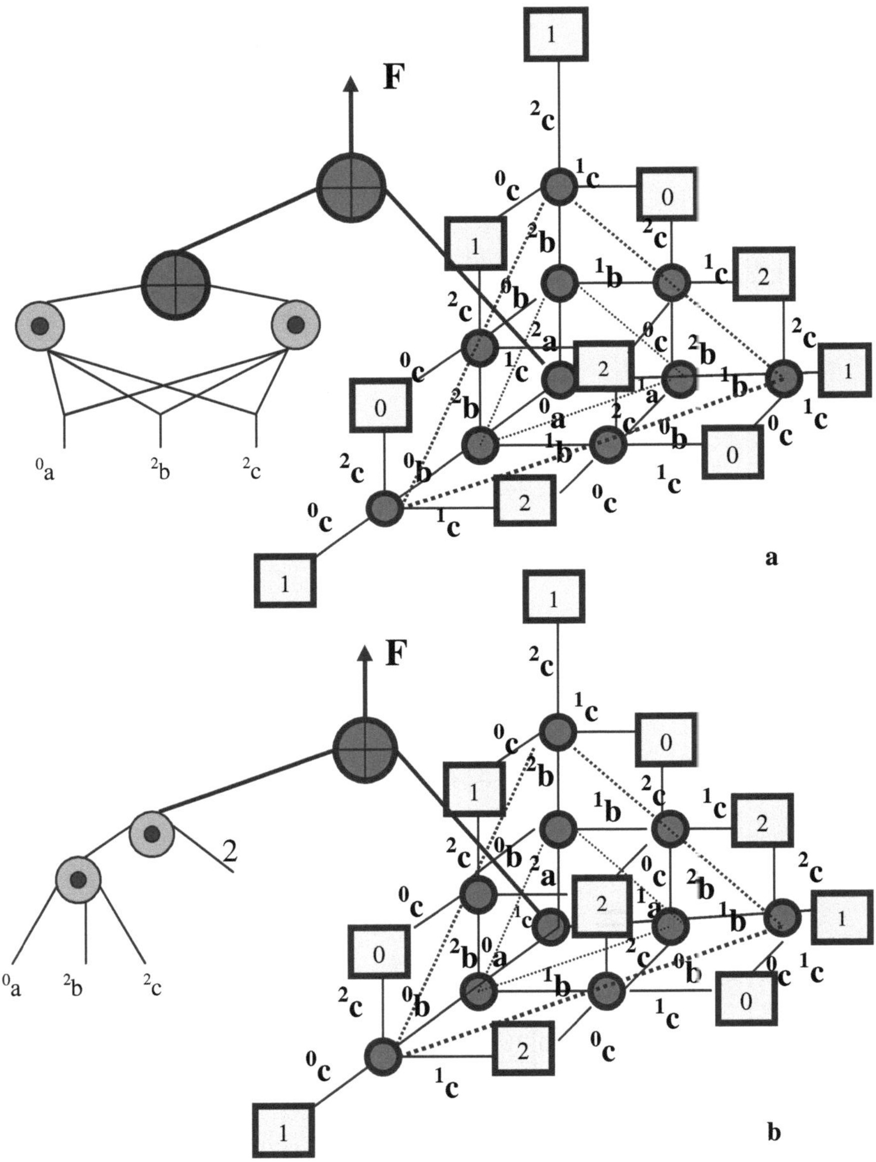

Fig. 4.34. (3,3) three-dimensional lattice structures using the ternary ISID: **a** realization that corresponds to Fig. 4.33a, and **b** realization that corresponds to Fig. 4.33b. All additions and multiplications are performed using $GF_3(+)$ and $GF_3(*)$, respectively.

Also, the spatial operations of the (2,2) 2-D and (3,3) 3-D sub-lattices will be mapped to be implemented in temporal operations of sub-lattices in quantum circuits and quantum computing as will be illustrated in Chapts. 10 and 11, respectively.

As observed from this Sect., the method of ISID is useful for maintaining interconnects of equal length for a large size lattice structure that does not fit specific 2-D or 3-D layout boundaries. Consequently, and analogously to the classical case, the ISID algorithm can play an important role in the minimization of the size of 2-D and 3-D reversible circuits, such as 2-D and 3-D reversible lattice structures, and therefore the minimization of the total size of the corresponding quantum circuits, and thus the minimization of the consequent number of basic quantum operations (that are performed by the corresponding quantum circuits) which will be illustrated in Chapts. 10 and 11, respectively.

The general result of using 3-D ISID iteratively is the decomposition of a total large (3,3) 3-D lattice structure into superimposed smaller (3,3) 3-D lattice structures as shown in Fig. 4.35a. The iterative use of ISID in a serial-mode or parallel-mode will "chop" the 3-D cube of lattice into 3-D pyramids as illustrated in Fig. 4.35b.

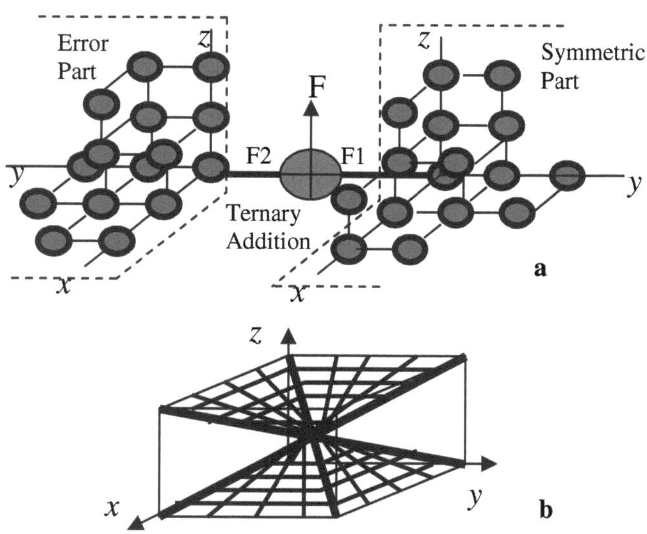

Fig. 4.35. **a** ISID for three-dimensional (3,3) lattice structures for the decomposition of ternary non-symmetric function F, and **b** pyramid grid layout as a result of iterative implementation of ISID for the decomposition of a ternary function.

4.8 Summary

In this Chapt. we presented 3-D lattice implementation of the new multi-valued invariant Shannon and Davio spectral transforms. We introduced the generalization of the concept of planar 4-neighbor lattice structures into 3-D (solid) 6-neighbor lattices. The 3-D joining rules of the new generalized sets of invariant multi-valued Shannon and Davio canonical expansions were derived, and the corresponding lattice structures were constructed. Lattice structures possess a very important property of high regularity, which is useful in many applications including fault-related issues; fault diagnosis (testing), fault localization, and fault self-repair. Other advantages of the new 3-D lattice structures include: (1) no need for layout routing and placement in 3-D space, (2) one-to-one mapping to regular isomorphic 3-D hardware structure (such as 3-D FPGAs), (3) regularity leads to the comparable ease of manufacturability, (4) 3-D lattices do not have intersecting edges which make them very suitable for quantum logic that will be presented in Chapts. 10 and 11, and (5) the 3-D new lattices are especially well suited for deep sub-micron technologies and future nano-technologies where the intrinsic physical delay of the irregular and lengthy interconnections limits the device performance (i.e., high power consumption and high delay in the interconnects especially at high frequencies (speeds) of operation).

A new decomposition called Iterative Symmetry Indices Decomposition (ISID) for Boolean and multiple-valued logic is introduced. This decomposition superimposes iteratively the symmetric part and the error part of a non-symmetric Boolean or multiple-valued functions. It has been shown [229] that most of circuit area is occupied by local and global interconnects, and the delay of interconnects is responsible for about 40-50% or more of the total delay associated with a circuit. Thus maintaining, as possible, equal length local inter-connects in a large size lattice structure, for specific layout constraints, will minimize the total length of wire used and consequently minimize the delay and power consumed. This idea of maintaining interconnects of equal length for a large size lattice structure that does not fit specific two-dimensional or three-dimensional layout boundaries can be achieved using ISID. ISID algorithm can play an important role in the minimization of the size of reversible lattice structures and thus the minimization of the total size of the corresponding quantum circuits that will be illustrated in Chapt. 10.

5 Reversible Logic: Fundamentals and New Results

Due to the anticipated failure of Moore's law around the year 2020, quantum computing will hopefully play an increasingly crucial role in building more compact and less power consuming computers [93,107,167,248,253]. Due to this fact, and because all quantum computer gates (i.e., building blocks; primitives) must be reversible [37,38,39,73,74,75,95,97,139,150,167,203,245,246], reversibility in computing will have increasing importance in the future design of regular, compact, and universal structures and machines (systems). (n,k) reversible circuits are circuits that have (n) inputs and (k) outputs and are one-to-one mappings between vectors of inputs and outputs, thus the vector of input states (values) can be always uniquely reconstructed from the vector of output states (values). (k,k) reversible circuits are circuits that have the same number of inputs (k) and outputs (k) and are one-to-one mappings between vectors of inputs and outputs, thus the vector of input states (values) can be always uniquely reconstructed from the vector of output states (values). Conservative circuits [98,210,211,212] are circuits that have the same number of values in inputs and outputs (e.g., the same number of ones in inputs and outputs for binary, the same number of ones and twos in inputs and outputs for ternary, etc). Conservativeness exists naturally in physical laws where no energy is created or destroyed.

As was proven in [37,139] it is a necessary but not sufficient condition for not dissipating power in any physical circuit that all system circuits must be built using fully reversible logical components. An important argument for power-free computation in a computer that "pushes information around" using reversible logic is given in [139], using the model of a particle in a bistable potential well, as follows:

> ... *Let us arbitrarily label the particle in the left-hand well as the ZERO state. When the particle is in the right-hand well, the device is in the*

ONE state. Now consider the operation RESTORE TO ONE, which leaves the particle in the ONE state, regardless of its initial location. if we are told that the particle is in the ONE state, then it is easy to leave it in the ONE state, without spending energy. If on the other hand we are told that the particle is in the ZERO state, we can apply a force to it, which will push it over the barrier, and then, when it has passed the maximum, we can apply a retarding force, so that when the particle arrives at ONE, it will have no excess kinetic energy, and we will not have expended any energy in the whole process, since we extracted energy from the particle in its downhill motion. Thus at first sight it seems possible to RESTORE TO ONE without any expenditure of energy ...

As a consequence, the statement *"information is physical"*, and consequently the famous Eq. *"information loss = energy loss"* are appropriate [139]. For this reason, different technologies have been investigated that implement reversible logic in hardware. Fully reversible digital systems will greatly reduce the power consumption (theoretically eliminate) through three conditions: (1) logical reversibility: the vector of input states (values) can always be uniquely reconstructed from the vector of output states (values), (2) physical reversibility: the physical switch operates backwards as well as forwards, and (3) the implementation using "ideal-like" switches that have no parasitic resistances.

To achieve reversible computing, different technologies have been studied to implement reversible logic in hardware including CMOS [68,70,71,72,131,143,152,153,206,250], optical [17,24,62,63,64,65,156,190,197,222], magnetic [132], mechanical [154], and quantum [31,53,55,111,112,156,259], respectively. Complete complex reversible circuits were fabricated [35].

Bit-permutations are a special case of reversible functions, that is, functions which permute the set of possible input values. Consequently, and in addition to logic synthesis, reversible computing was applied in areas where computational tasks are important enough to justify new microprocessor instructions and instruction sets where bit-permutation instructions greatly improve the performance of several standard algorithms as matrix transposition. These applications include digital signal processing, communications, computer graphics, and cryptography [225], where it is required that all of the information encoded in the input must be preserved in the output. Figure 5.1 illustrates the inclusion relationship between various classes of reversible circuits, where the shaded areas indicate the sub-sets of reversible logic synthesis that have been worked with throughout this Book.

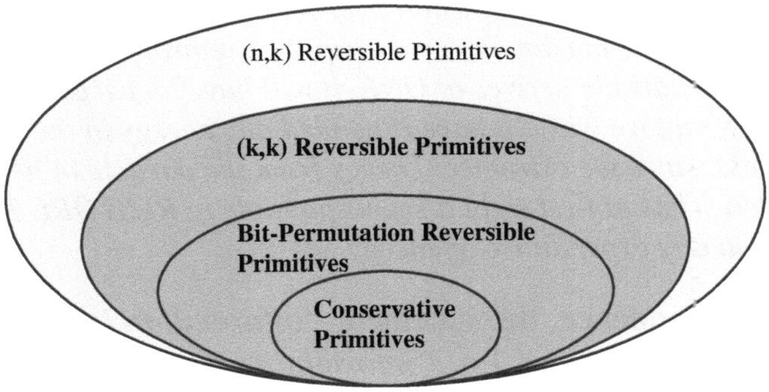

Fig. 5.1. Set-theoretic relationship between various classes of reversible logic.

Billiard Ball Model (BBM) is one of the fundamental models for reversible computing [98]. Cellular Automata (CA), which are computationally universal as they can compute any function, were modeled as a BBM [98]. Consider the elements in the BBM in Fig. 5.2 where billiard balls move on a lattice with unit velocity, and scatter off of each other and from walls. Two balls colliding generate the AND function, and if one of the streams of balls is continuous it generates the NOT function of the other input. These two elements are sufficient to build up all of Boolean logic. Specific memory structures can be implemented by delays, and wiring by various walls to guide the balls. The balls can be represented by four bits per site (one for each direction), with one extra bit per site needed to represent the walls. This kind of computing developed in [98] has many interesting features; no information is ever destroyed, which means that it is reversible (it can be run backwards to produce inputs from outputs), and which in turn means that it can be performed (in theory) with arbitrarily little dissipation [139].

Reversibility is also essential for designing quantum Cellular Automata (QCA), since quantum evolution is reversible. A quantum CA is much like a classical CA, but it permits the sites to be in a superposition of their possible states [145]. This is a promising conceptual architecture model for building quantum computers [101].

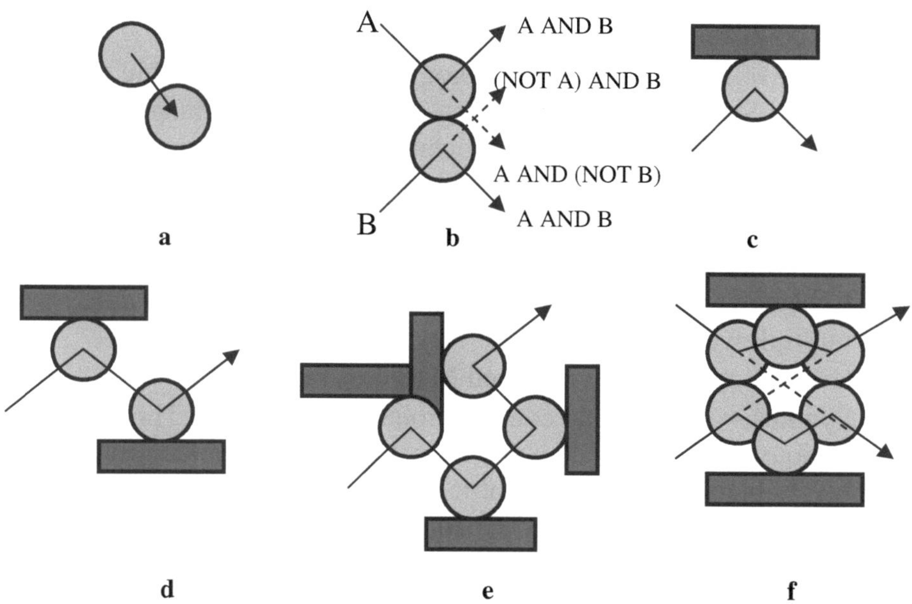

Fig. 5.2. Billiard Ball Logic (BBL): **a** transport, **b** scattering (logic), **c** reflection, **d** shift, **e** delay (memory), and **f** crossover.

In this Chapt., fundamental reversible primitives are presented, new theorems for reversible logic are introduced, and the corresponding reversible gates are created. Figure 5.3 demonstrates the link between the continuation of the theoretical development from this Chapt. and the following Chapts.

Fig. 5.3. A graph illustrates the theoretical development of theorems for reversible logic in this Chapt.

The new results and theorems that are obtained from this Chapt. are built on top of the results that were introduced in Chapt. 2, and will be utilized in Chapt. 6 in the construction of reversible lattice structures, and in Chapt. 10 where the quantum circuits for the corresponding reversible lattice structures will be introduced. The main contributions of this Chapt. are:

- The invention of new methodology to generate reversible multiple-valued Shannon decompositions (that includes the binary case as a special case): Latin Square Property of the Generalized Basis Function Matrix.
- The generation of reversible multiple-valued Davio decompositions.
- Exhaustive classification and count of all possible reversible multiple-valued Shannon and Davio gates into classes.
- Generalizations of two-valued Margolus primitive.
- Synthesis of reversible combinational logic circuits. This includes various reversible code converters, reversible barrel shifter, reversible sorter, and reversible MIN/MAX tree.

The remainder of this Chapt. is organized as follows. Basic reversible gates and circuits are presented in Sect. 5.1. The elimination of the garbage outputs in two-valued reversible circuits is presented in Sect. 5.2. Examples of combinational reversible circuits are given in Sect. 5.3. A new general methodology for the creation and classification of new Galois-based reversible spectral transforms, expansions, and examples of such transforms and expansions are presented in Sect. 5.4. The process of eliminating garbage outputs in multiple-valued reversible circuits is introduced in Sect. 5.5. A Summary of the Chapt. is presented in Sect. 5.6.

5.1 Fundamental Reversible Logic Primitives and Circuits

Reversible circuits which are hierarchically composed of reversible primitives have two types of outputs in general: (1) functionality outputs, and (2) outputs that are needed only to achieve reversibility which are called "garbage" [98]. Many reversible gates have been proposed as building blocks for reversible (and consequently quantum) computing. Figure 5.4 shows some of the binary (k,k) reversible gates that are commonly used in the synthesis of reversible logic circuits [6,14,95,126,127,128,129]. It is noted from Fig. 5.4 that while Wire (Buffer), Not, and Swap gates are naturally reversible, others are not, and thus "garbage" has to be added.

Multiple-valued counterparts of similar reversible primitives and some of their applications were introduced in [6,11,14,190,191,192,193]. Figure 5.5 illustrates the multiple-valued gate from [191]. More multiple-valued gates and the systematic methodology for their creation and classification will be introduced in Sect. 5.4.

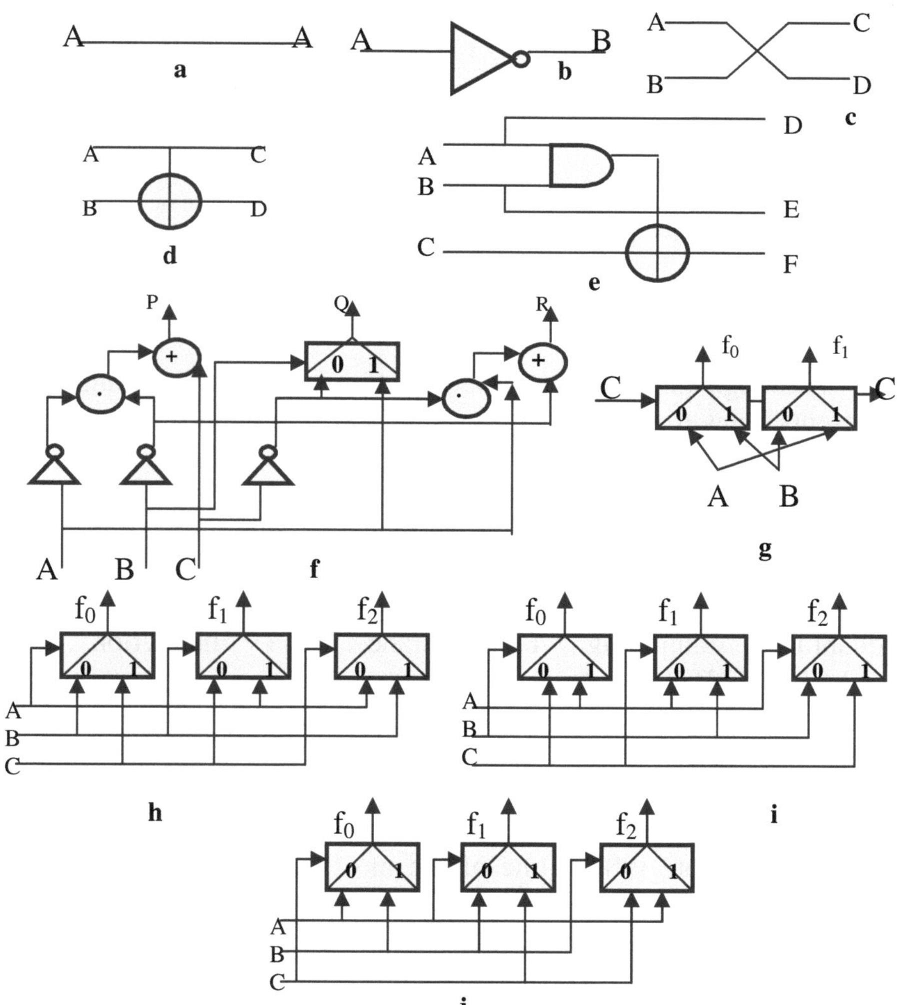

Fig. 5.4. Binary reversible gates: **a** (1,1) Wire, **b** (1,1) Inverter, **c** (2,2) Swap, **d** (2,2) Feynman gate (also known as quantum XOR, Controlled-NOT), **e** (3,3) Toffoli gate (also known as Controlled-Controlled-NOT), **f** Maximum Cofactor (MC) gate, **g** (3,3) Fredkin gate, **h** Margolus$_0$ gate, **i** Margolus$_1$ gate, and **j** Margolus$_2$ gate.

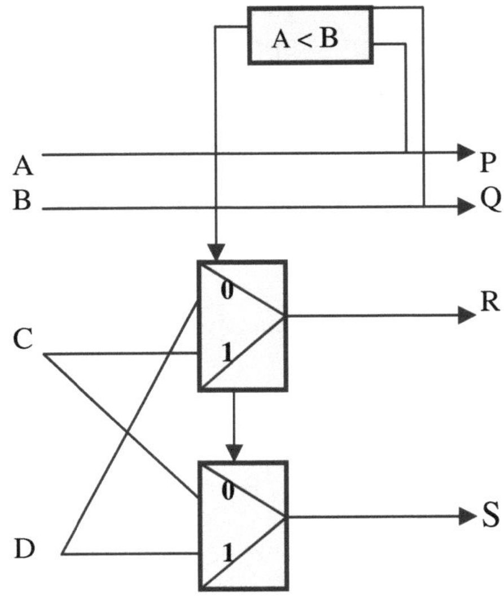

Fig. 5.5. Multi-valued logic primitive: Picton gate.

Although most of available literature on reversible computing presents gates that are (k,k) reversible, other literature has reported the conceptual need for (n,k) reversible primitives in general. The need for (n,k) reversible primitives stems from the fact that the logical model must fit the physical reality of computing, and not to be disjoint from the physical laws of computing as it was in the previous abstract mathematical logics before reversible (and thus quantum) computing. For instance, the Interaction gate [62,63,64,222] has been reported to be of a good fit to reversible computing in optics. Figure 5.6 illustrates some of the (n,k) reversible gates. (It is important to note that here fan-out and feedback are not allowed in reversible computing applications using either (k,k) reversible gates or (n,k) reversible gates.)

Fredkin gate [98] is one of the most basic building blocks in reversible and quantum computing. Many propositions have been proposed to realize the Fredkin gate in various technologies: Optical, Electrical, Mechanical (nano-technology), and Quantum. The Fredkin gate belongs to a group of gates that each represents a fundamental family of logic gates in reversible computing. These families of reversible gates are Fredkin-like, Toffoli-like, and Feynman-like gates. It will be shown in Sect. 5.4 how to formally generalize the Fredkin gate to any multiple-valued logic radix.

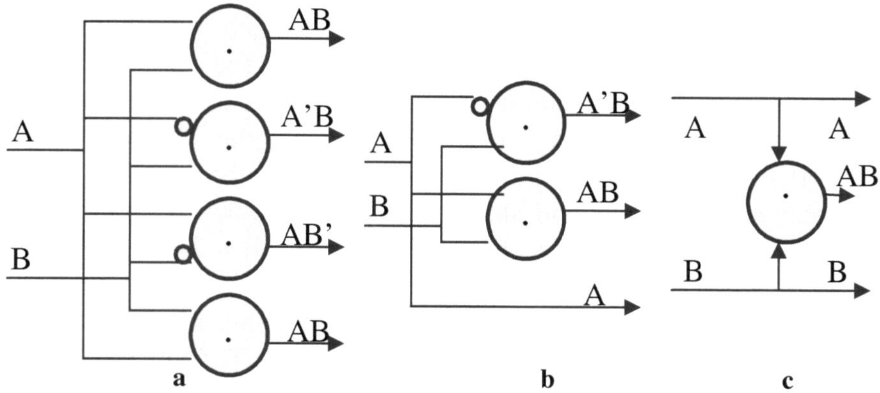

Fig. 5.6. Some (n,k) reversible gates: **a** (2,4) Interaction gate, **b** (2,3) Switch (Priese) gate, and **c** (2,3) AND gate.

This is especially important for ternary logic since ternary logic is the logic for regular lattice realization in three-dimensional space. It can be observed in Fig. 5.4 that while Feynman gate is not universal since it is composed only of a linear (EXOR) part, the Toffoli gate is universal as it is composed of a linear (EXOR) and nonlinear (AND) parts.

The multiple-valued gate in Fig. 5.5 is just one example of families of multiple-valued reversible gates. For example, Feynman gate from Fig. 5.4 can be extended to any radix over a Galois field using the same topological circuits, and the only difference will be the type of AND and OR operations that have to be performed over the corresponding radix of a Galois field. Also, note that in Fig. 5.4, the Fan-Out gate (Copying gate) is built using a single Feynman gate with constant "0" at the XOR control-input, and the Swap gate which is not realizable in quantum circuits is built using three serially inter-connected Feynman gates (this will be illustrated in Chapt. 11). Also, one can note that while a Maximum Cofactor (MC) gate (which is a member of a bigger family of related gates) produces a maximum number of cofactors (18 cofactors) for three binary inputs, other gates do not. The production of cofactors for various (k,k) reversible primitives is shown in Fig. 5.7 [8]. It has been shown in [98] that for a (k,k) reversible gate to be universal the gate should have at least three inputs (i.e., (3,3) gate). Various binary (3,3) reversible gates that are universal in two inputs have been also shown in [126].

Balanced primitives are primitives for which each output value appears a number of times which is equal to the number of times that each of the other output values appears. For example, in ternary (4,4) reversible primitive which is balanced, the number of times that 0's appear in the outputs is equal to the number of times that 1's appear in the outputs and also equal to the number of times that 2's appear in the outputs, and this is equal to (81/3) = 27 times for each output. One can note that while the (k,k) reversible gates: Wire (Buffer), Inverter, Swap, Fredkin, and Margolus are balanced and conservative, other reversible gates: Feynman, Toffoli, and MC are balanced but not conservative. In general, reversible primitives can be classified into families according to the corresponding functional properties of such gates [126], like being (or not being) conservative, balanced, cyclic, 0-1 preserving, invertible, etc.

5.2 The Elimination of Garbage in Two-Valued Reversible Circuits

In reversible logic, it is important to construct the inverse of the gate to eliminate the garbage outputs [95,98,262,263].

Feynman (6 cofactors)	**Toffoli (7 cofactors)**	**Fredkin (12 cofactors)**	**Margolus (15 cofactors)**	**MC (18 cofactors)**
F0(0)=0	F2(0,b,c)=c	F0(0,b,c)=bc	F1(0,b,c)=b	P(0,B,C)=(BC)'
F0(1)=1	F2(1,b,c)=b⊕c	F0(1,b,c)=c'+b	F1(1,b,c)=c	P(1,B,C)=B'C'
F1(0,b)=b	F2(a,1,c)= a⊕c	F0(a,0,c)=c'a	F1(a,0,c)=ac	P(A,0,C)=(AC)'
F1(1,b)=b'	F2(a,b,0)=ab	F0(a,1,c)=c+a	F1(a,1,c)=a'+c	P(A,1,C)=A'C'
F1(a,0)=a	F2(a,b,1)=(ab)'	F0(a,b,0)=a	F1(a,b,0)=a'b	P(A,B,0)=(AB)'
F1(a,1)=a'	F1(0)=0	F0(a,b,1)=b	F1(a,b,1)=a+b	P(A,B,1)=A'B'
	F1(1)=1	F1(0,b,c)=c'b	F2(0,b,c)=b'c	Q(0,B,C)=CB'
		F1(1,b,c)=c+b	F2(1,b,c)=c+b	Q(1,B,C)=B'+C
		F1(a,0,c)=ca	F2(a,1,c)=a	Q(A,0,C)=A+C
		F1(a,1,c)=c'+a	F2(a,b,0)=ab	Q(A,1,C)=AC
			F2(a,b,1)=b'+a	Q(A,B,0)=AB'
			F3(0,b,c)=bc	Q(A,B,1)=B'+A
			F3(1,b,c)=c'+b	R(0,B,C)=BC'
			F3(a,0,c)=c'a	R(1,B,C)=C'+B
			F3(a,1,c)=c+a	R(A,0,C)=AC'
				R(A,1,C)=C'+A
				R(A,B,0)=A+B
				R(A,B,1)=AB

Fig. 5.7. Demonstration of the number of cofactors for various (k,k) reversible gates.

This is achieved by taking the outputs of the first reversible circuit and produce from them "inversely" the inputs. This is important especially in quantum computing where garbage is not allowed [98]. Also, it is important in certain techniques in reversible CMOS computing [262,263]. Figure 5.8 shows reversible circuit (in white color) and its reversible "mirror" (inverse) (in shaded color).

A gate (F) is said to be inverse of itself when $\mathbf{FF^{-1}} = \mathbf{I}$. For instance, it has been shown [98] that a Fredkin gate is the inverse of itself and a Toffoli gate is also the inverse to itself. The following examples show circuits and their inverses.

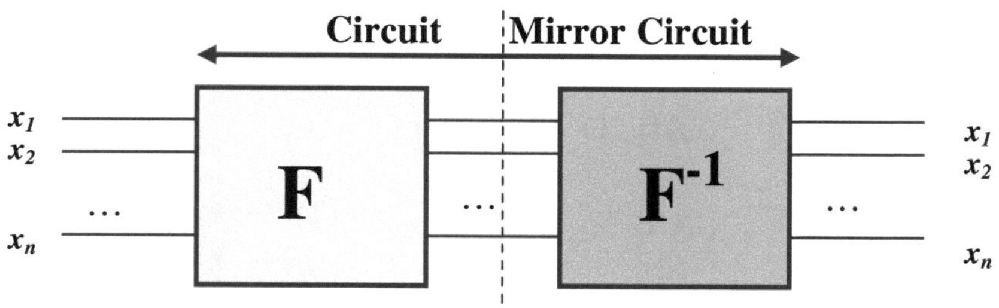

Fig. 5.8. The reversible circuit and its reversible mirror to eliminate garbage.

Example 5.1. Figures 5.9 and 5.10 generate the inverse gate for the Interaction gate and the Switch (Priese) gate, respectively. To measure the state of the hidden functions within the total network of the reversible circuit and its inverse, one has to use the "spy" circuit [98]. (This has been shown to be of special importance in QC [98].)

Note that in the examples for two-valued and multiple-valued reversible logic synthesis through this Book, two commonly applied constraints for reversible logic synthesis are imposed: (1) feedback is not allowed, and (2) fan-out is not allowed (i.e., fan-out = 1). One explanation for not using fan-out in reversible logic is as follows: in the "forward" conventional logic synthesis combining wires is not allowed, so in reversible logic synthesis branching wires will not be allowed since branching of a signal, if looked at in reverse, will appear to be as combining signals. Consequently, Feynman gate is used as a copier (i.e., fan-out generator) by setting the value of the control input to value "0". Also, one can observe that, when the garbage outputs are eliminated by cascading the forward reversible circuit with the inverse reversible circuit and therefore the inputs are generated as outputs from the whole garbage-free netowork, although it is possible in some technologies to connect the inputs that are generated at the output of the garbage-free network to the inputs of the whole network using wires (e.g., CMOS technology), it is not possible to do so in quantum circuits since the concept of “wire” does not exist physically.

Input		Output			
A	B	1	2	3	4
0	0	0	0	0	0
0	1	0	1	0	0
1	0	0	0	1	0
1	1	1	0	0	1

a

Inverse →

Input				Output	
1	2	3	4	A	B
0	0	0	0	0	0
0	1	0	0	0	1
0	0	1	0	1	0
1	0	0	1	1	1

c

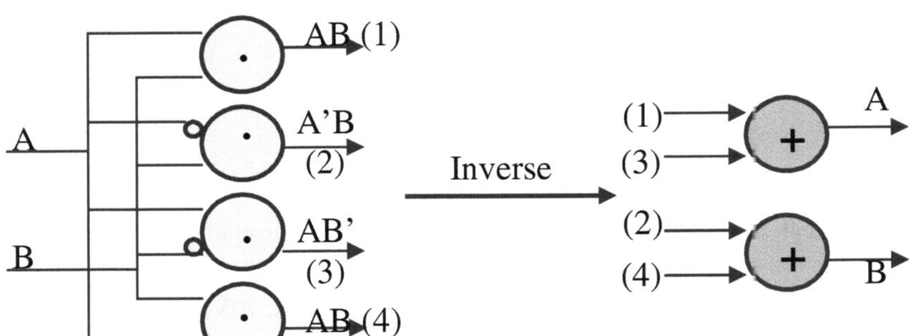

b **d**

Fig. 5.9. Interaction gate: **a** truth table, **b** logic circuit, **c** the inverse truth table, and **d** the inverse logic circuit.

Input		Output		
A	B	1	2	3
0	0	0	0	0
0	1	1	0	0
1	0	0	0	1
1	1	0	1	1

a

Inverse →

Input			Output	
1	2	3	A	B
0	0	0	0	0
1	0	0	0	1
0	0	1	1	0
0	1	1	1	1

c

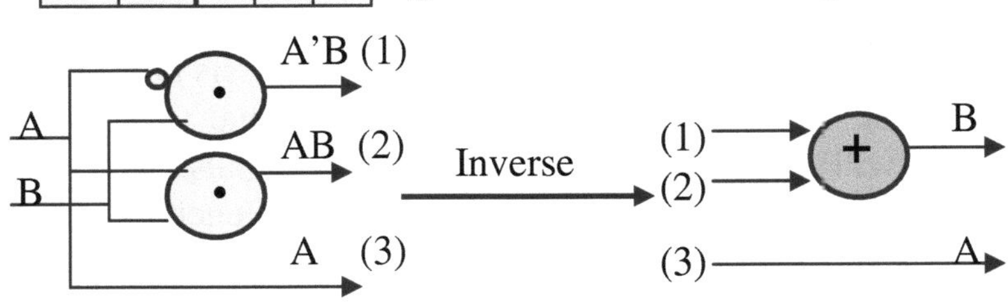

b **d**

Fig. 5.10. Switch (Priese) gate: **a** truth table, **b** logic circuit, **c** inverse truth table, and **d** inverse logic circuit.

Example 5.2. This example shows the process of obtaining two-valued reversible inverse gates for fundamental reversible (forward) gates. These gates are illustrated in Figs. 5.11 through 5.13. One can note that Fredkin gate in Fig. 5.11 is the inverse to itself. The same note can be observed for the case of two-valued Feynman and Toffoli gates in Figs. 5.12 and 5.13, respectively.

a	b	c	f_0	f_1	c
a	b	0	a	b	0
a	b	1	b	a	1

a

→

f_0	f_1	c	a	b	c
a	b	0	a	b	0
b	a	1	a	b	1

b

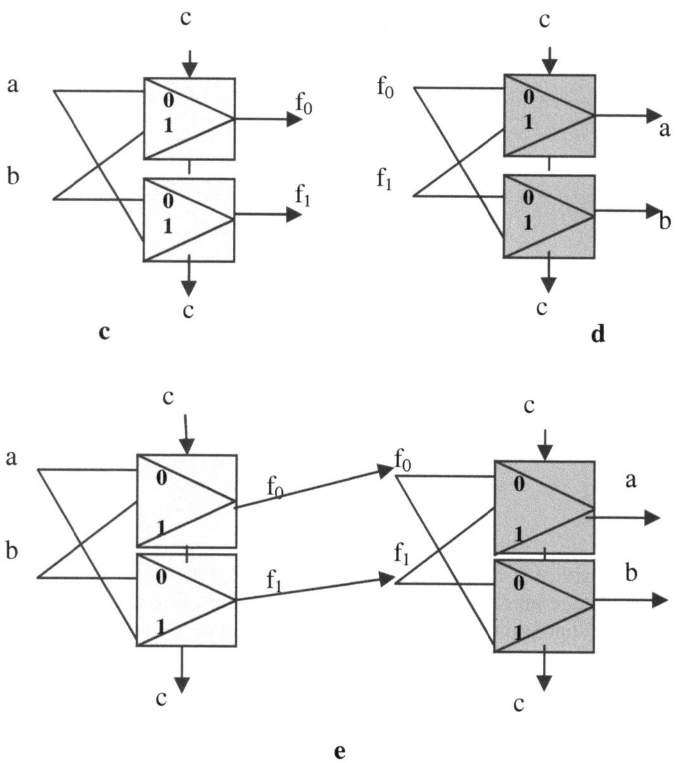

Fig. 5.11. a Truth table for reversible forward Fredkin gate, **b** truth table for reversible inverse Fredkin gate, **c** reversible forward Fredkin circuit, **d** reversible inverse Fredkin circuit, and **e** the elimination of garbage for Fredkin gate by combining the reversible forward Fredkin circuit with the reversible inverse Fredkin circuit.

a	b	c	d
0	0	0	0
0	1	0	1
1	0	1	1
1	1	1	0

a

→

c	d	a	b
0	0	0	0
0	1	0	1
1	1	1	0
1	0	1	1

b

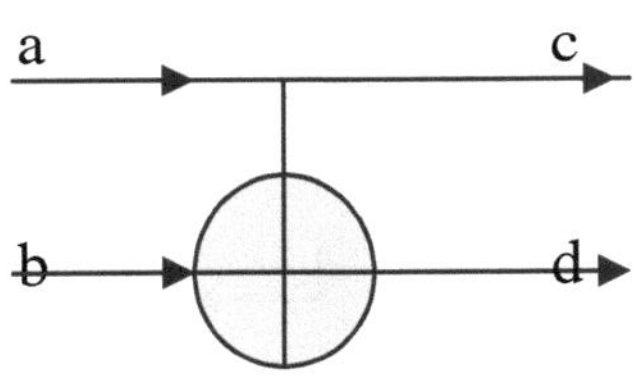

c

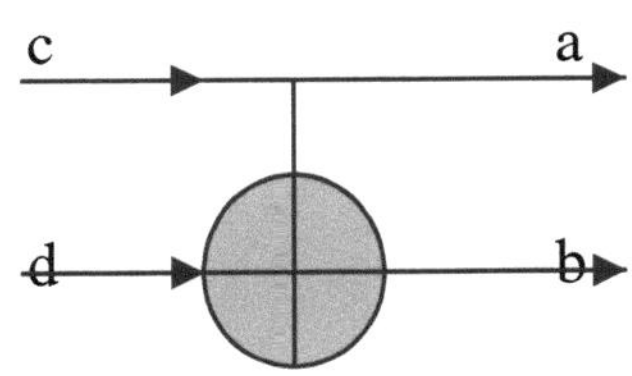

d

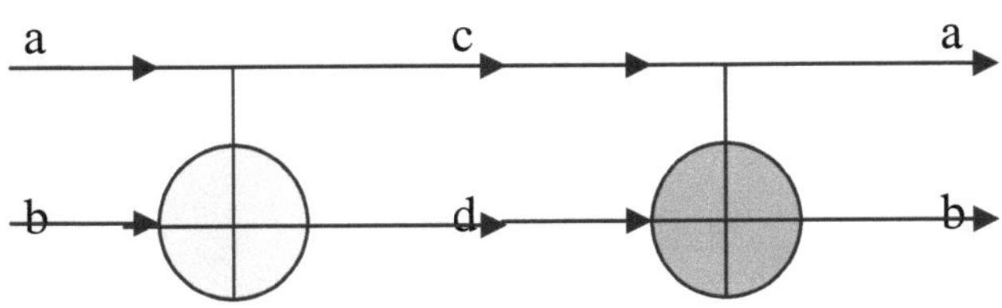

e

Fig. 5.12. a Truth table for reversible forward Feynman gate, **b** truth table for reversible inverse Feynman gate, **c** reversible forward Feynman circuit, **d** reversible inverse Feynman circuit, and **e** the elimination of garbage for Feynman gate by combining the reversible forward Feynman circuit with the reversible inverse Feynman circuit.

a	b	c	d	e	f
0	0	0	0	0	0
0	0	1	0	0	1
0	1	0	0	1	0
0	1	1	0	1	1
1	0	0	1	0	0
1	0	1	1	0	1
1	1	0	1	1	1
1	1	1	1	1	0

a

d	e	f	a	b	c
0	0	0	0	0	0
0	0	1	0	0	1
0	1	0	0	1	0
0	1	1	0	1	1
1	0	0	1	0	0
1	0	1	1	0	1
1	1	1	1	1	0
1	1	0	1	1	1

b

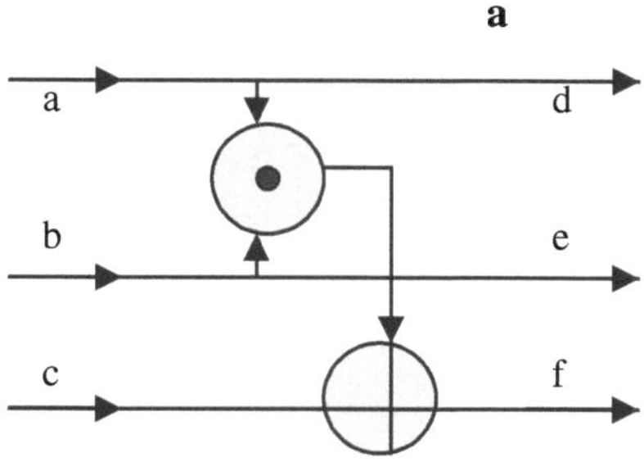

c

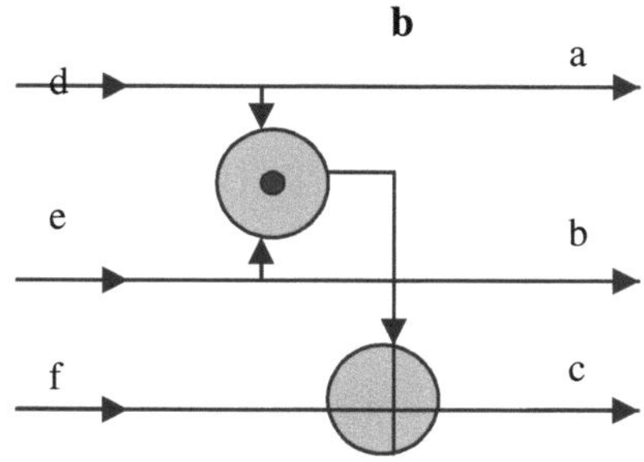

d

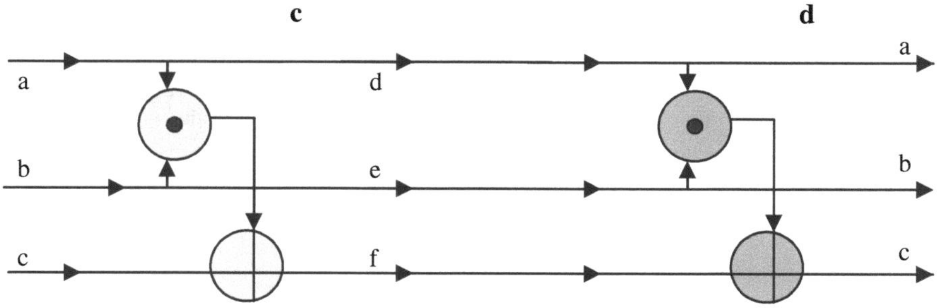

e

Fig. 5.13. **a** Truth table for reversible forward Toffoli gate, **b** truth table for reversible inverse Toffoli gate, **c** reversible forward Toffoli circuit, **d** reversible inverse Toffoli circuit, and **e** the elimination of garbage for Toffoli gate by combining the reversible forward Toffoli circuit with the reversible inverse Toffoli circuit.

The following example illustrates the concept of the creation of the total network that consists of the reversible circuit and its mirror image, and the use of "spy" circuit to measure the hidden functionalities within the total network.

Example 5.3. Figure 5.14 shows the network composed of the forward reversible part, inverse reversible part, and the "spy" circuit.

Function: **F**

ABCD	XYZV
0000	0011
0001	1011
0010	0010
0011	1010
0100	0000
0101	0111
0110	0001
0111	0110
1000	1111
1001	1000
1010	1110
1011	1001
1100	1101
1101	0101
1110	1100
1111	0100

a

Function^{-1}: $\mathbf{F^{-1}}$

XYZV	ABCD
0000	0100
0001	0110
0010	0010
0011	0000
0100	1111
0101	1101
0110	0111
0111	0101
1000	1001
1001	1011
1010	0011
1011	0001
1100	1110
1101	1100
1110	1010
1111	1000

b

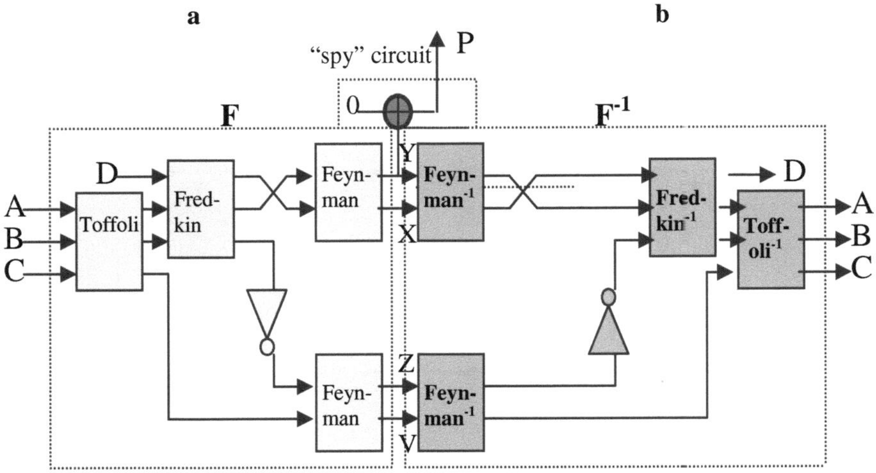

c

Fig. 5.14. An illustration of the creation of the reversible circuit and its mirror image: **a** truth table for the forward reversible circuit, **b** truth table for the inverse reversible circuit, and **c** the total network including the "spy" circuit.

5.3 Combinational Reversible Circuits

Reversible circuits can be synthesized using careful design methodologies where one utilizes the outputs from a previous stage as inputs to the next stage. Various reversible circuits have been synthesized using this methodology [8,200]. This Sect. introduces some of these circuits. Figure 5.15 illustrates the creation of all of the 16 possible binary logic functions of two variables (cf. Fig. G.3 in Appendix G) using certain reversible logic primitives.

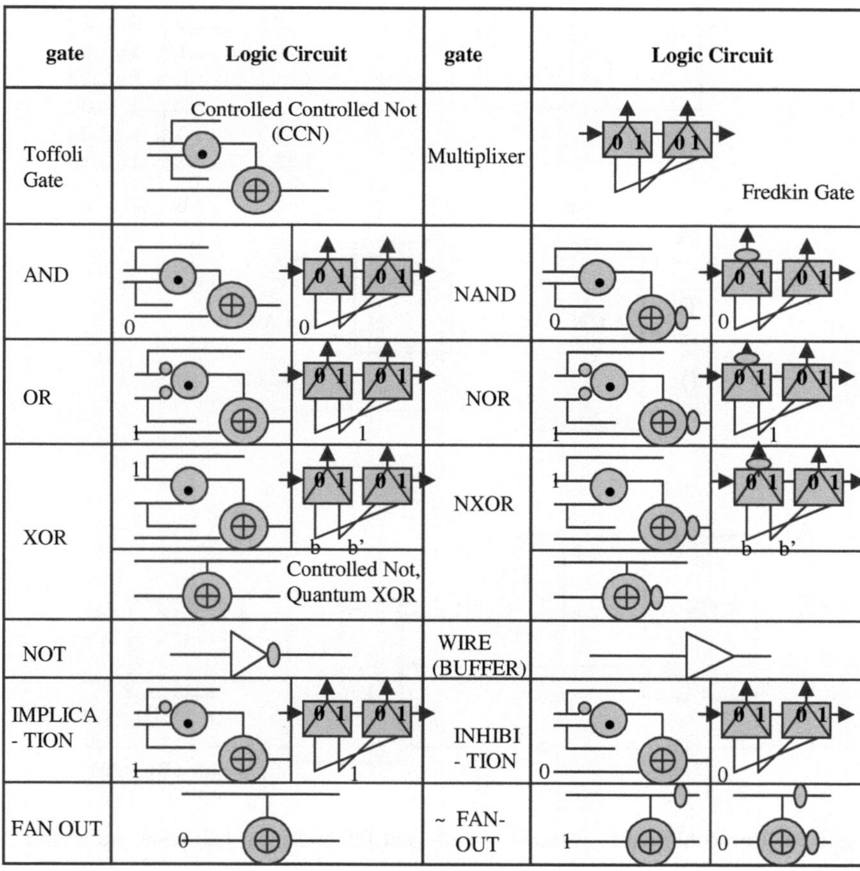

Fig. 5.15. Synthesis of various Boolean functions using some reversible logic primitives.

Note that such constructions are not unique, and thus the optimization criteria should be (1) minimum size and (2) have mimimum garbage used in the synthesis. Figure 5.16 illustrates the synthesis of half-adder and full–adder using Feynman (Controlled-NOT: CN) and Toffoli (Controlled-Controlled-NOT: CCN) gates.

Inputs		Outputs	
A	B	C_o	S
0	0	0	0
0	1	0	1
1	0	0	1
1	1	1	0

a

Inputs			Outputs	
A	B	C_i	C_o	S
0	0	0	0	0
0	0	1	0	1
0	1	0	0	1
0	1	1	1	0
1	0	0	0	1
1	0	1	1	0
1	1	0	1	0
1	1	1	1	1

b

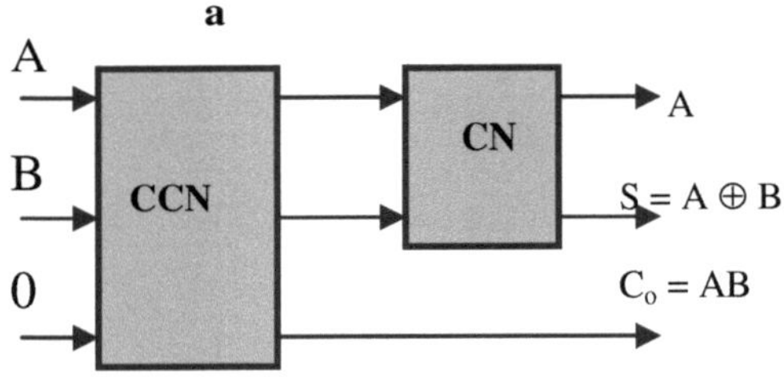

c

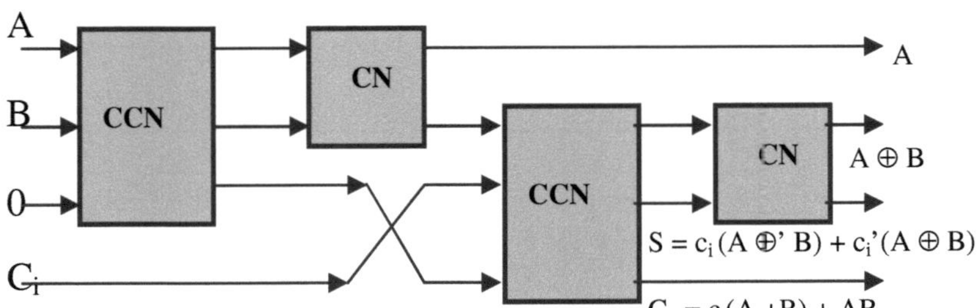

d

Fig. 5.16. Reversible logic synthesis of half- and full-adders: **a** half-adder truth table, **b** full-adder truth table, **c** half-adder reversible circuit, and **d** full adder reversible circuit.

Figure 5.17 demonstrates one possible reversible realization of various coding schemes (Natural, Gray, and Aiken) [8] using Feynman and Toffoli gates.

Natural	Gray	Aiken
0000	0000	0000
0001	0001	0001
0010	0011	0010
0011	0010	0011
0100	0110	0100
0101	0111	0101
0110	0101	0110
0111	0100	0111
1000	1100	1110
1001	1101	1111

a

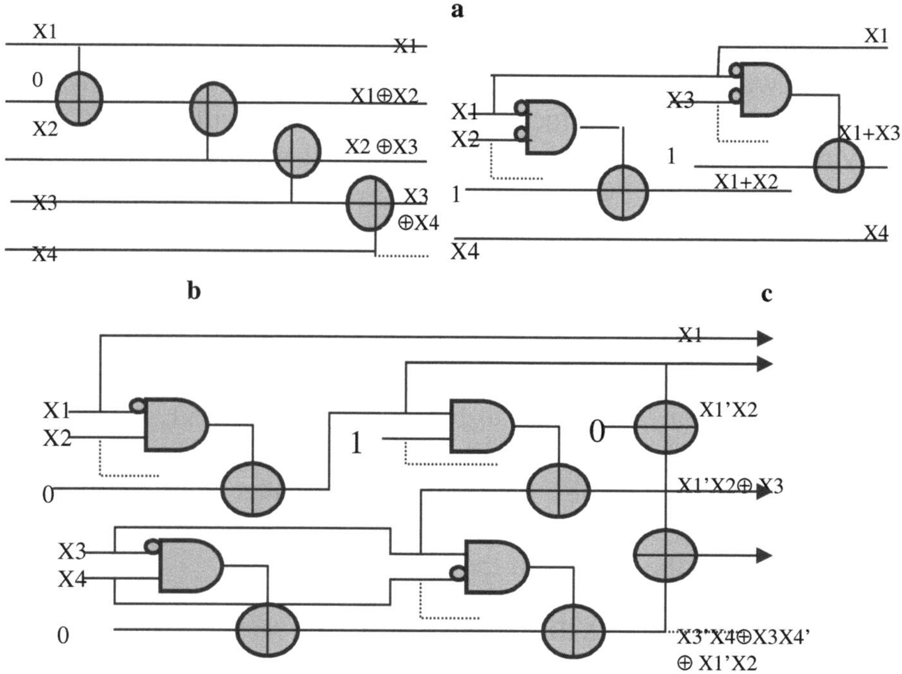

d

Fig. 5.17. a truth tables for Natural, Gray, and Aiken codes, **b** reversible cascade of Natural code, **c** reversible circuit of Gray code, and **d** reversible circuit of Aiken code.

As shifters are important in combinational and sequential logic synthesis, it is important to produce a reversible logic shifter. Figure 5.18 illustrates a novel reversible Barrel shifter design from [8]. Figure 5.18 represents one possible design of concurrent shift-left and shift-right reversible Barrel shifter, which shows a fundamental concept in the design of reversible logic circuits: *the idea of the use of reversibility to perform multiple operations using the same design while retaining reversibility* [8].

Note that by controlling the value of the variable in the first level, the Barrel shifter operates in the shift-left mode by setting the value of variable X in the first level to value "0" and collecting the shifted-left outputs from the locations that are marked by (X) at the outputs of Fredkin gates, or the shift-right mode by setting the value of variable X in the first level to value "1" and collecting the shifted-right outputs from the locations that are marked by (+) at the outputs of Fredkin gates, respectively. The first level of the reversible Barrel shifter will shift the inputs by one location, the second level will shift the inputs by two locations, the third level will shift the inputs by three locations, and the fourth level will shift the inputs by four locations (i.e., full cycle or rotation).

Figure 5.19 illustrates the use of MIN/MAX gate, which is synthesized from Picton gate, to realize a multiple-valued Sorter [8]. By following the paths, from the inputs to the outputs, one will obtain the sorted values of the inputs at the outputs.

Figure 5.20 illustrates a MIN/MAX Tree [8] using MIN/MAX from Fig. 5.19a. The inputs in Fig. 5.20b are (A0, A1, A2, A3, B0, B1, B2, B3). The outputs in Fig. 5.20b are denoted as (1), (2), (3), (4), (5), (6), (7), and (8). By following the paths from inputs to the outputs in Fig. 5.20b, one obtains the following MIN and MAX expressions at the outputs of the MIN/MAX tree:

Output (1): MAX [MIN (A_3,B_3), MIN (A_2,B_2)],
Output (2): MIN [MAX (A_3,B_3), MAX (A_2,B_2)],
Output (3): MAX [MIN (A_1,B_1), MIN (A_0,B_0)],
Output (4): MIN [MAX (A_1,B_1), MAX (A_0,B_0)],

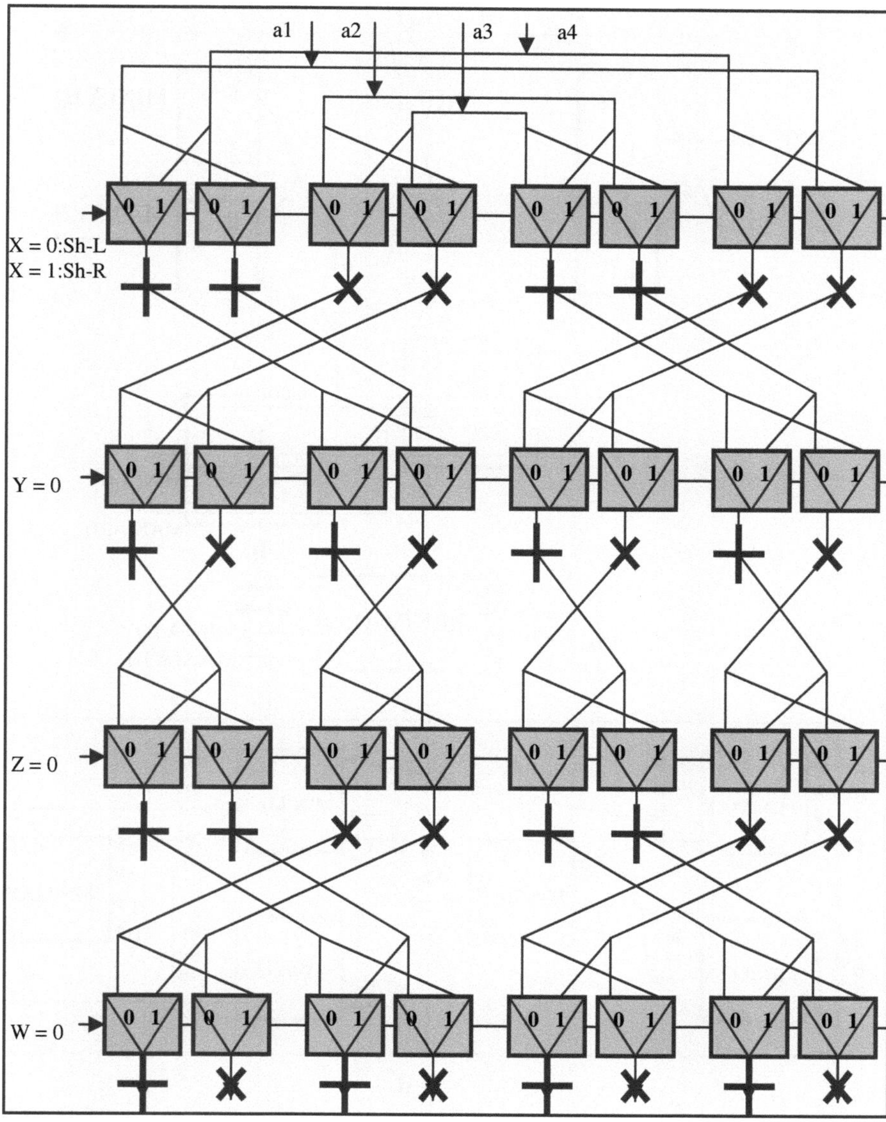

Fig. 5.18. Concurrent shift-left (X) and shift-right (+) Barrel shifter.

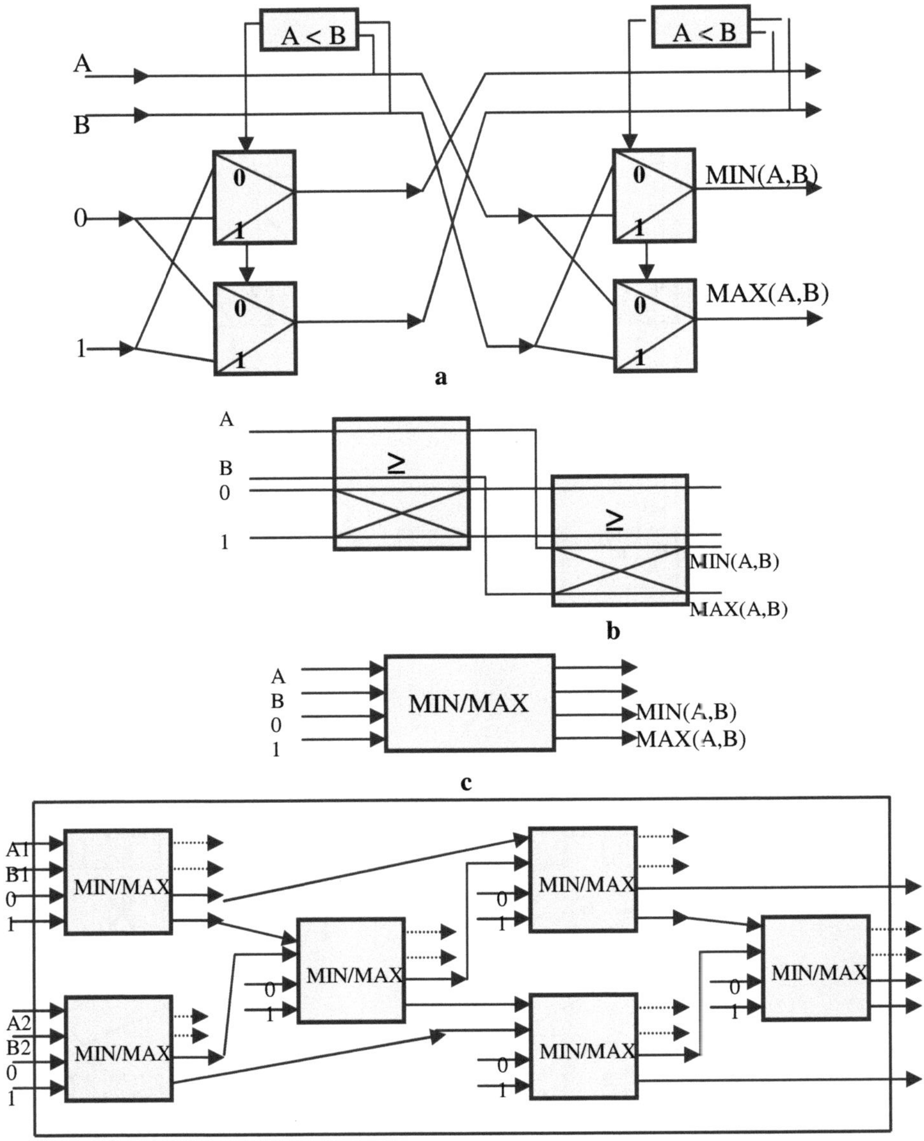

Fig. 5.19. Reversible pipelined 4-input binary combinational sorter: **a** two interconnected Picton gates, **b** symbol for the gate from (a), **c** symbol for (b), and **d** binary and multi-valued reversible Sorter.

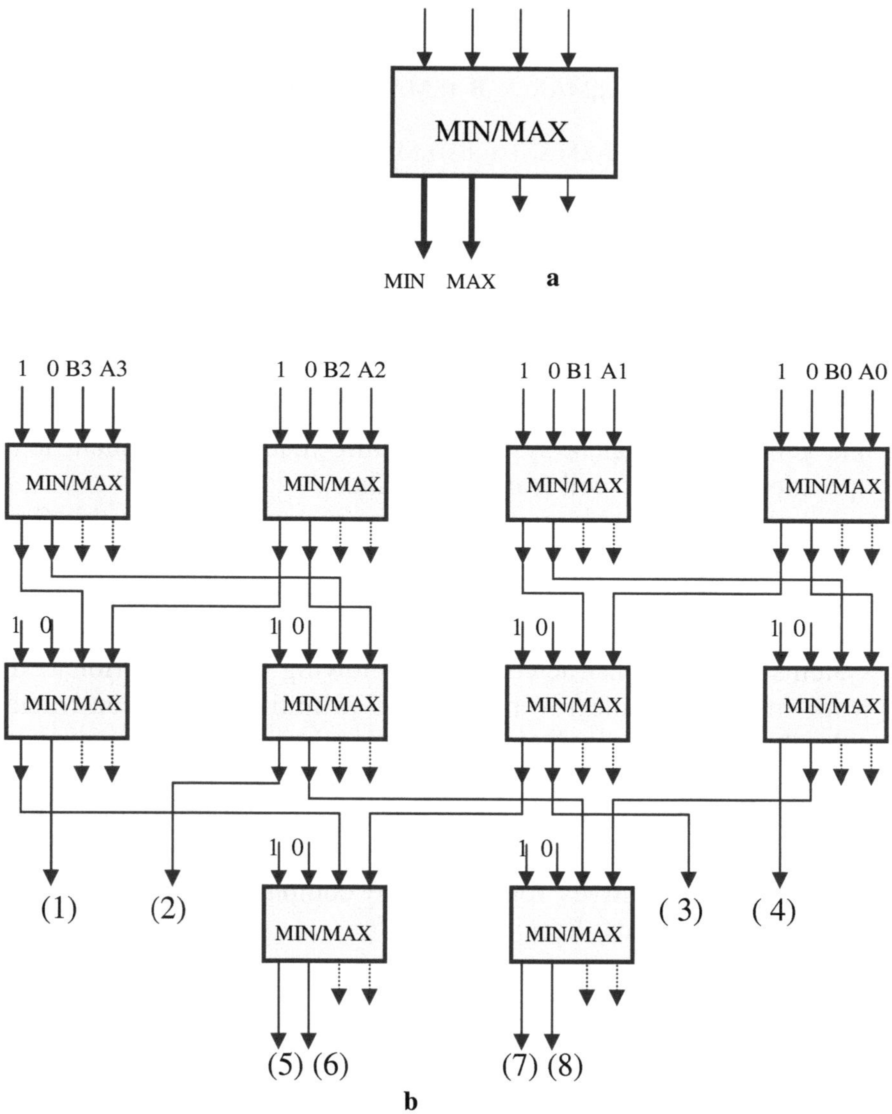

Fig. 5.20. Reversible MIN/MAX tree using MIN/MAX gate: **a** Symbol, and **b** reversible MIN/MAX tree.

Output(5):
MIN[MIN{MIN(A_3,B_3),MIN(A_2,B_2)},MIN{MIN(A_1,B_1),MIN(A_0,B_0)}],
Output (6):
MAX[MIN{MIN(A_3,B_3),MIN(A_2,B_2)},MIN{MIN(A_1,B_1),MIN(A_0,B_0)}],
Output (7):
MIN[MAX{MAX(A_3,B_3),MAX(A_2,B_2)},MAX{MAX(A_1,B_1),MAX (A_0,B_0)}],
Output (8):
MAX[MAX{MAX(A_3,B_3),MAX(A_2,B_2)},MAX{MAX(A_1,B_1),MAX(A_0,B_0)}].

Figure 5.21 [8] implements one-way production of the logic polynomial: $F = a_n \cdot x + a_{n-1}$. Figure 5.22b [8] implements the two-way production of the logic polynomial: $F = a_{in} \cdot (W_2+W_1+W_0)$. The circuits in Figs. 5.21 and 5.22b apply reversible pipelined and reversible systolic operations. A logic cell that can be systolically connected to neighbor cells to implement reversible functions is the octagon systolic cell from Fig. 5.23a [8]. Figure 5.23b [8] resembles one general topological systolic structure that can implement logic operations using the cell from Fig. 5.23a.

Although new reversible circuits were synthesized in this Sect. by using the outputs from a previous stage as inputs to the next stage, no mathematical theory has been yet established to construct the reversible building blocks (from which more complex reversible systems will be constructed in the following Chapts.). Prior to the work in [6,7,14], the reversible logic primitives were constructed either by ad-hoc methods or using exhaustive computer programs to generate all possible reversible gates for certain radix and certain number of variables as in [126]. Consequently, the novel construction of a systematic mathematical formalism for the creation of reversible primitives for reversible computing from [6,7,14] is introduced in the next Sect.

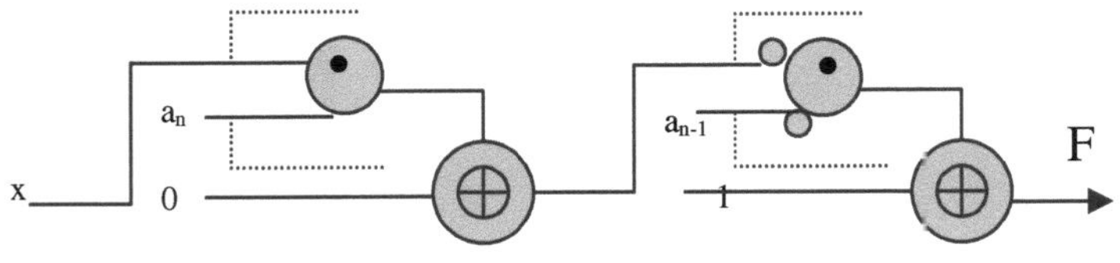

Fig. 5.21. One-directional reversible polynomial generator.

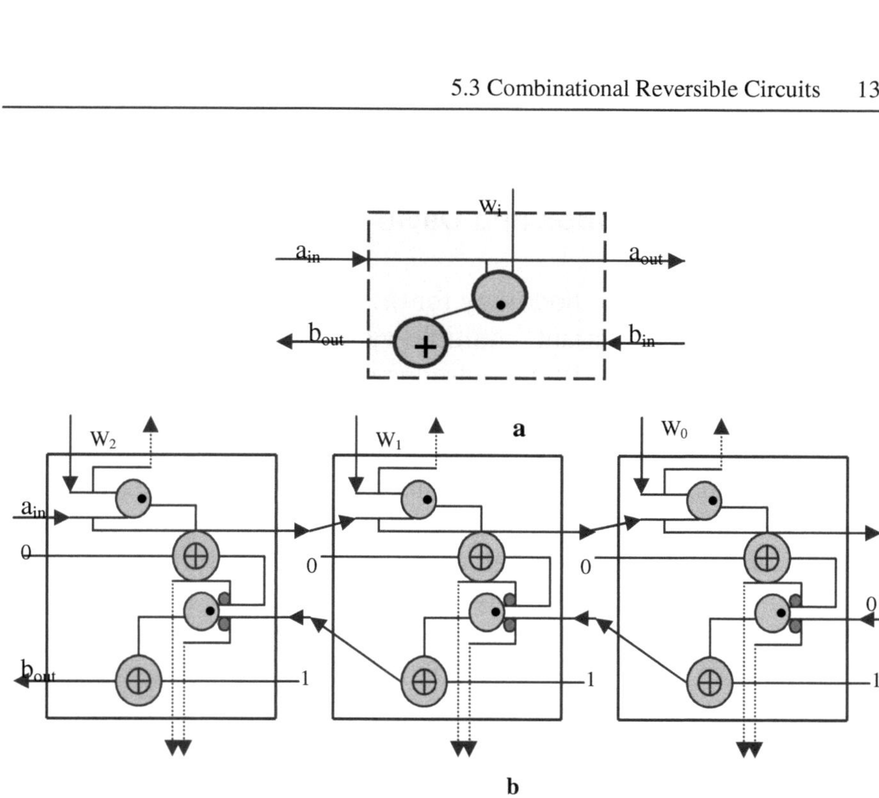

Fig. 5.22. Two-directional polynomial generator: **a** cell, and **b** reversible logic circuit.

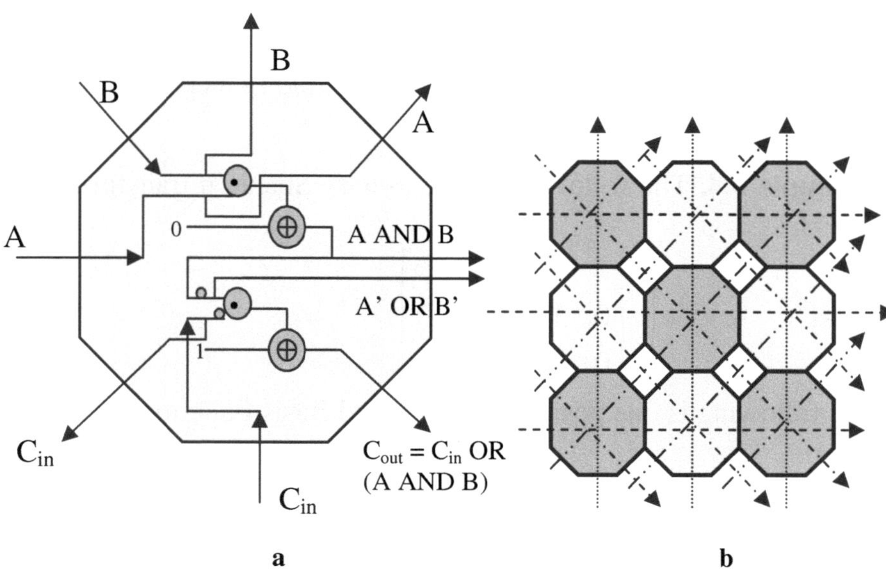

Fig. 5.23. Logic systolic implementation: **a** octagon systolic cell, and **b** topological systolic structure that uses the logic systolic cell from (a).

5.4 Novel General Methodology for the Creation and Classification of New Families of Reversible Invariant Multi-Valued Shannon and Davio Spectral Transforms

In this Sect. a new methodology for the creation and classification of new reversible invariant multiple-valued GF-based families of spectral transforms will be introduced [6,7,14]. In this Sect. we present and prove new theorems to systematically generate and classify the new families of multiple-valued invariant reversible GF-based Shannon and Davio expansions from Sect. 2.2.2. These theorems are the only theorems introduced so far for reversible expansions, and they stand alone so far as the only formalism and methodology introduced in the available literature for the creation and classification of multiple-valued reversible Shannon and Davio expansions.

Definition 5.1. The matrix that is constructed from the permutations of many basis functions of the same type of the corresponding spectral transform is called Generalized Basis Functions Matrix (GBFM).

Definition 5.2. From the total state-space of all possible Generalized Basis Functions Matrices the matrices that produce reversible spectral transforms are called Reversible Generalized Basis Functions Matrices (RGBFM).

Example 5.4. The following is the ternary Shannon transform over GF(3).

$$f = \begin{bmatrix} {}^{0}c & {}^{1}c & {}^{2}c \end{bmatrix} \begin{bmatrix} 1 & 0 & 0 \\ 0 & 1 & 0 \\ 0 & 0 & 1 \end{bmatrix} \begin{bmatrix} f_0 \\ f_1 \\ f_2 \end{bmatrix}.$$

The following is one possible Generalized Basis Functions Matrix:

$$\begin{bmatrix} {}^{0}c & {}^{1}c & {}^{2}c \\ {}^{0}c & {}^{2}c & {}^{1}c \\ {}^{2}c & {}^{1}c & {}^{0}c \end{bmatrix}.$$

Yet as will be demonstrated in the following theorem, the upper Generalized Basis Functions Matrix is not reversible; it does not produce a reversible spectral transform. One possible Reversible Shannon Generalized Basis Functions Matrix that leads to a reversible spectral transform is the following matrix:

$$\begin{bmatrix} {}^{0}c & {}^{1}c & {}^{2}c \\ {}^{1}c & {}^{2}c & {}^{0}c \\ {}^{2}c & {}^{0}c & {}^{1}c \end{bmatrix}.$$

In Example 5.4, note that the Shannon set of basis functions $\{{}^{0}c, {}^{1}c, {}^{2}c\}$ will always appear in any of the rows of the corresponding Shannon RGBFM that would accordingly produce a reversible Shannon spectral transform (in this case a Shannon set of basis functions $\{{}^{0}c, {}^{1}c, {}^{2}c\}$ appears in the first row).

Theorem 5.1. A necessary and sufficient condition to generate the reversible invariant multi-valued Shannon expansions is that the order of the permuted basis functions in the Generalized Basis Functions Matrix should satisfy the Latin Square Property (Cyclic Group Property): in any given row or column the elements in that row or column are different than the elements in the corresponding positions of the other rows or columns.

Proof. Since the Shannon spectral transform matrix is orthogonal, then the multi-input multi-output multiple-valued Shannon expansion that is shown in Eq. (5.1):

$$\vec{f} = \begin{bmatrix} {}^{2}c & {}^{0}c & {}^{1}c \\ {}^{1}c & {}^{2}c & {}^{0}c \\ {}^{0}c & {}^{1}c & {}^{2}c \end{bmatrix} \begin{bmatrix} 1 & 0 & 0 \\ 0 & 1 & 0 \\ 0 & 0 & 1 \end{bmatrix} \begin{bmatrix} f_0 \\ f_1 \\ f_2 \end{bmatrix} = \begin{bmatrix} f_{r0} \\ f_{r1} \\ f_{r2} \end{bmatrix}, \tag{5.1}$$

is reversible (as an example we are using in this proof the Shannon set of basis functions $\{{}^{0}c, {}^{1}c, {}^{2}c\}$ that appears in the third row, but as discussed earlier, a Shannon set of basis functions $\{{}^{0}c, {}^{1}c, {}^{2}c\}$ can appear in any of the rows of the corresponding Reversible Generalized Basis Functions Matrix). Eq. (5.1) is reversible since it satisfies the following reversibility restrictions:

(1) For number of inputs $\{f_0, f_1, f_2, c\}$ is equal to the number of outputs $\{f_{r0}, f_{r1}, f_{r2}, c\}$.

(2) We can uniquely reconstruct any set of inputs from the set of outputs and vice versa. This stems from the fact that for any reduced

Post literal ${}^{k}c$ the set of outputs are uniquely selected for the specified value of ${}^{k}c$ in the corresponding ternary truth table of the ternary inputs/ternary outputs (since ${}^{k}c = 1$ iff c = k). This allows for the unique selection of I/O such that for any combination of inputs there is only one corresponding combination of outputs. **Q.E.D.**

The resulting outputs in Eq. (5.1) are fully balanced; for 4-outputs the number of times that 0's appear is equal to the number of times that 1's appear and also equal to the number of times that 2's appear, and this is equal to (81/3) = 27. Also the reversible Shannon spectral transform in Eq. (5.1) is conservative; it has the same number of values in inputs and outputs (i.e., for ternary logic the number of ones and twos in every input vector is equal to the number of ones and twos in the corresponding output vector). Therefore the circuits (gates) that are constructed from Eq. (5.1) are both reversible and conservative. Utilizing the same methodology used in Theorem (5.1), all possible permutations of a multi-valued Shannon spectral transform can be converted to be a reversible permuted Shannon spectral transforms.

Theorem 5.2. In general for any invariant Shannon spectral transform matrix

$$\begin{bmatrix} \alpha & 0 & 0 \\ 0 & \beta & 0 \\ 0 & 0 & \gamma \end{bmatrix},$$

the following is a reversible expansion:

$$\vec{f} = \begin{bmatrix} \hat{\alpha}\,{}^{2}c & \hat{\beta}\,{}^{0}c & \hat{\gamma}\,{}^{1}c \\ \hat{\alpha}\,{}^{1}c & \hat{\beta}\,{}^{2}c & \hat{\gamma}\,{}^{0}c \\ \hat{\alpha}\,{}^{0}c & \hat{\beta}\,{}^{1}c & \hat{\gamma}\,{}^{2}c \end{bmatrix} \begin{bmatrix} \alpha & 0 & 0 \\ 0 & \beta & 0 \\ 0 & 0 & \gamma \end{bmatrix} \begin{bmatrix} f_0 \\ f_1 \\ f_2 \end{bmatrix} = \begin{bmatrix} f_{r0} \\ f_{r1} \\ f_{r2} \end{bmatrix}, \qquad (5.2)$$

where $\alpha\hat{\alpha} = 1, \beta\hat{\beta} = 1, \gamma\hat{\gamma} = 1$.

Proof. the proof is similar to the proof of Theorem 5.1. **Q.E.D.**

The following is an example of some of the total possible Shannon reversible expansions over GF(2) and GF(3), respectively.

Example 5.5. Let's produce the reversible Shannon gates for binary logic as well as ternary logic.

5.5a. In binary logic there are only two Reversible Shannon gates as follows:

$$\vec{f} = \begin{bmatrix} \bar{c} & c \\ c & \bar{c} \end{bmatrix} \begin{bmatrix} 1 & 0 \\ 0 & 1 \end{bmatrix} \begin{bmatrix} f_0 \\ f_1 \end{bmatrix} = \begin{bmatrix} f_{r0} \\ f_{r1} \end{bmatrix}, \tag{5.3}$$

$$\vec{f} = \begin{bmatrix} c & \bar{c} \\ \bar{c} & c \end{bmatrix} \begin{bmatrix} 1 & 0 \\ 0 & 1 \end{bmatrix} \begin{bmatrix} f_0 \\ f_1 \end{bmatrix} = \begin{bmatrix} f_{r0} \\ f_{r1} \end{bmatrix}. \tag{5.4}$$

The following are logic circuit realizations of Eqs. (5.3) and (5.4).

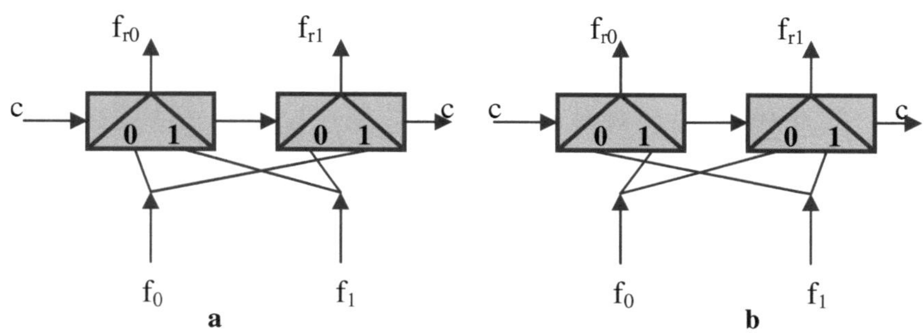

Fig. 5.24. Logic circuit realizations of reversible Shannon primitives: **a** two-valued reversible Shannon (Fredkin gate), and **b** the flipped reversible Shannon (flipped Fredkin gate). Both gates are composed of two 2-to-1 multiplexers.

Note that the function of the gates in Fig. 5.24 is the permutation of inputs (cofactors) to produce outputs (that are merely a permutation of inputs). Figure 5.25 illustrates such property.

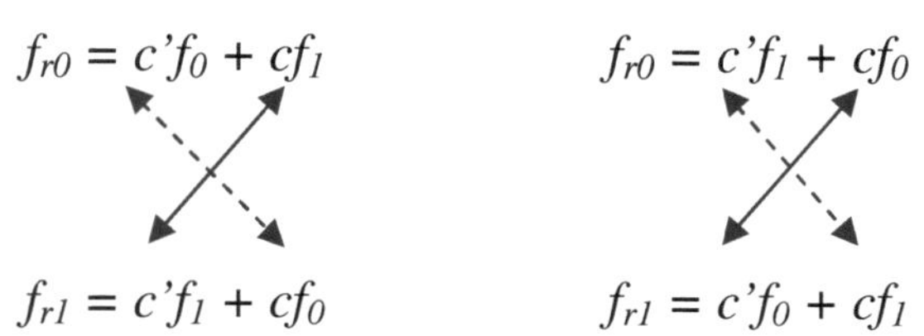

Fig. 5.25. Permutation of cofactors.

5.5b. Utilizing Theorem 5.1, Let's produce all the possible permutations of the Reversible Generalized Basis Functions Matrix for the invariant Shannon transform matrix $\begin{bmatrix} 2 & 0 & 0 \\ 0 & 1 & 0 \\ 0 & 0 & 2 \end{bmatrix}$ to produce the corresponding reversible invariant ternary Shannon expansions over GF(3).

$$\vec{f} = \begin{bmatrix} 2^0c & {}^1c & 2^2c \\ 2^1c & {}^2c & 2^0c \\ 2^2c & {}^0c & 2^1c \end{bmatrix} \begin{bmatrix} 2 & 0 & 0 \\ 0 & 1 & 0 \\ 0 & 0 & 2 \end{bmatrix} \begin{bmatrix} f_0 \\ f_1 \\ f_2 \end{bmatrix} = \begin{bmatrix} f_{r0} \\ f_{r1} \\ f_{r2} \end{bmatrix}, \tag{5.5}$$

$$\vec{f} = \begin{bmatrix} 2^0c & {}^1c & 2^2c \\ 2^2c & {}^0c & 2^1c \\ 2^1c & {}^2c & 2^0c \end{bmatrix} \begin{bmatrix} 2 & 0 & 0 \\ 0 & 1 & 0 \\ 0 & 0 & 2 \end{bmatrix} \begin{bmatrix} f_0 \\ f_1 \\ f_2 \end{bmatrix} = \begin{bmatrix} f_{r0} \\ f_{r1} \\ f_{r2} \end{bmatrix}, \tag{5.6}$$

$$\vec{f} = \begin{bmatrix} 2^1c & {}^2c & 2^0c \\ 2^0c & {}^1c & 2^2c \\ 2^2c & {}^0c & 2^1c \end{bmatrix} \begin{bmatrix} 2 & 0 & 0 \\ 0 & 1 & 0 \\ 0 & 0 & 2 \end{bmatrix} \begin{bmatrix} f_0 \\ f_1 \\ f_2 \end{bmatrix} = \begin{bmatrix} f_{r0} \\ f_{r1} \\ f_{r2} \end{bmatrix}, \tag{5.7}$$

$$\vec{f} = \begin{bmatrix} 2^2c & {}^0c & 2^1c \\ 2^0c & {}^1c & 2^2c \\ 2^1c & {}^2c & 2^0c \end{bmatrix} \begin{bmatrix} 2 & 0 & 0 \\ 0 & 1 & 0 \\ 0 & 0 & 2 \end{bmatrix} \begin{bmatrix} f_0 \\ f_1 \\ f_2 \end{bmatrix} = \begin{bmatrix} f_{r0} \\ f_{r1} \\ f_{r2} \end{bmatrix}, \tag{5.8}$$

$$\vec{f} = \begin{bmatrix} 2^1c & {}^2c & 2^0c \\ 2^2c & {}^0c & 2^1c \\ 2^0c & {}^1c & 2^2c \end{bmatrix} \begin{bmatrix} 2 & 0 & 0 \\ 0 & 1 & 0 \\ 0 & 0 & 2 \end{bmatrix} \begin{bmatrix} f_0 \\ f_1 \\ f_2 \end{bmatrix} = \begin{bmatrix} f_{r0} \\ f_{r1} \\ f_{r2} \end{bmatrix}, \tag{5.9}$$

$$\vec{f} = \begin{bmatrix} 2^2c & {}^0c & 2^1c \\ 2^1c & {}^2c & 2^0c \\ 2^0c & {}^1c & 2^2c \end{bmatrix} \begin{bmatrix} 2 & 0 & 0 \\ 0 & 1 & 0 \\ 0 & 0 & 2 \end{bmatrix} \begin{bmatrix} f_0 \\ f_1 \\ f_2 \end{bmatrix} = \begin{bmatrix} f_{r0} \\ f_{r1} \\ f_{r2} \end{bmatrix}. \tag{5.10}$$

The following reversible logic circuits represent logic circuit realizations for Eqs. (5.6) and (5.9), respectively, where all inputs and outputs can have any of the ternary values (i.e., 0, 1, or 2).

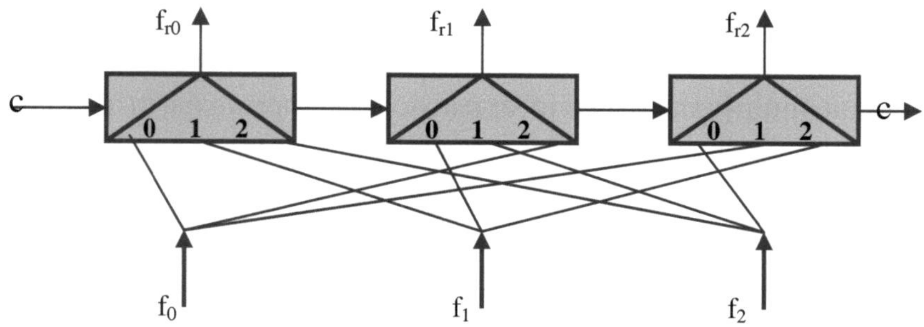

Fig. 5.26. Logic circuit realization of the reversible expansion in Eq. (5.6).

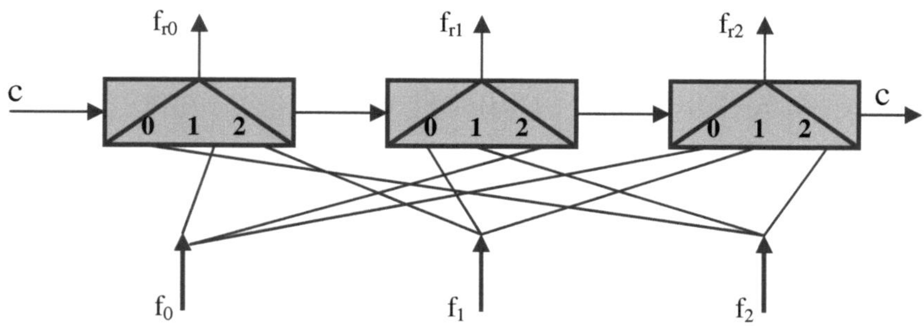

Fig. 5.27. Logic circuit realization of the reversible expansion in Eq. (5.9).

Where:

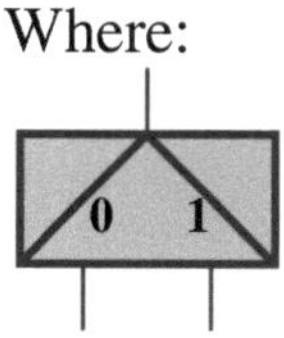

Is a 2-to-1 multiplexer in binary logic

Is a 3-to-1 multiplexer in ternary logic

Note that Eqs. (5.5) through (5.10) lead through the application of the Latin Square Property of the Generalized Basis Functions Matrix to the permutation of cofactors to achieve reversibility. Also, generalizations of the reversible gates (as in Figs. 5.26 and 5.27, for instance) are possible by using ternary inverters (shifters) (which are intrinsically reversible) interchangeably at the inputs and/or outputs of the reversible gates.

Theorem 5.3. For each reversible invariant multi-valued Shannon expansion over GF(n) there exist n^2 fixed reversible invariant multi-valued Davio expansions of all types. For each type of reversible invariant multi-valued Davio expansion D_n there exist n reversible invariant multi-valued Davio expansions of that type (i.e., D_n).

Proof. The proof of Theorem 5.3 provides also the systematic methodology of generating all possible reversible invariant multi-valued Davio expansions. Let us prove this theorem for the ternary case over GF(3), for one possible reversible invariant multi-valued Shannon expansion. Using the following reversible invariant multi-valued Shannon expansion:

$$\vec{f} = \begin{bmatrix} {}^0c & {}^1c & {}^2c \\ {}^1c & {}^2c & {}^0c \\ {}^2c & {}^0c & {}^1c \end{bmatrix} \begin{bmatrix} 1 & 0 & 0 \\ 0 & 1 & 0 \\ 0 & 0 & 1 \end{bmatrix} \begin{bmatrix} f_0 \\ f_1 \\ f_2 \end{bmatrix}. \tag{5.11}$$

For Eq. (5.11), there exists three Davio types for each row of the Generalized Reversible Shannon Basis Functions Matrix. Utilizing the derivation of ternary Davio expansion, the following are the D_0-type expansions for the first row, second row, and third row of the above Generalized Reversible Shannon Basis Functions Matrix $\begin{bmatrix} {}^0c & {}^1c & {}^2c \\ {}^1c & {}^2c & {}^0c \\ {}^2c & {}^0c & {}^1c \end{bmatrix}$, respectively:

$$f_{D0,row0} = [1 \;\; c \;\; c^2] \begin{bmatrix} 1 & 0 & 0 \\ 0 & 2 & 1 \\ 2 & 2 & 2 \end{bmatrix} \begin{bmatrix} f_0 \\ f_1 \\ f_2 \end{bmatrix}, \tag{5.12}$$

$$f_{D0,row1} = [1 \;\; c \;\; c^2] \begin{bmatrix} 0 & 0 & 1 \\ 2 & 1 & 0 \\ 2 & 2 & 2 \end{bmatrix} \begin{bmatrix} f_0 \\ f_1 \\ f_2 \end{bmatrix}, \tag{5.13}$$

$$f_{D0,row2} = [1 \;\; c \;\; c^2]\begin{bmatrix} 0 & 1 & 0 \\ 1 & 0 & 2 \\ 2 & 2 & 2 \end{bmatrix}\begin{bmatrix} f_0 \\ f_1 \\ f_2 \end{bmatrix}. \tag{5.14}$$

To produce one form of the reversible Davio$_0$-type functional expansion, let us choose the transform matrix $\begin{bmatrix} 0 & 0 & 1 \\ 2 & 1 & 0 \\ 2 & 2 & 2 \end{bmatrix}$ in Eq. (5.13) to produce the corresponding reversible invariant multi-valued Davio$_0$ expansion. Note that we have two other choices of $\begin{bmatrix} 1 & 0 & 0 \\ 0 & 2 & 1 \\ 2 & 2 & 2 \end{bmatrix}$ and $\begin{bmatrix} 0 & 1 & 0 \\ 1 & 0 & 2 \\ 2 & 2 & 2 \end{bmatrix}$ transform matrices in Eqs. (5.12) and (5.14), respectively. The utilization of $\begin{bmatrix} 0 & 0 & 1 \\ 2 & 1 & 0 \\ 2 & 2 & 2 \end{bmatrix}$ transform matrix as our representation matrix will impose the following conditions to produce the overall correct Davio$_0$ expansion:

$$\text{Condition 1:} \begin{bmatrix} a & b & c \\ d & e & f \\ g & h & i \end{bmatrix} \begin{bmatrix} 0 & 0 & 1 \\ 2 & 1 & 0 \\ 2 & 2 & 2 \end{bmatrix} = \begin{bmatrix} 1 & 0 & 0 \\ 0 & 2 & 1 \\ 2 & 2 & 2 \end{bmatrix}.$$

$$\text{Condition 2:} \begin{bmatrix} a & b & c \\ d & e & f \\ g & h & i \end{bmatrix} \begin{bmatrix} 0 & 0 & 1 \\ 2 & 1 & 0 \\ 2 & 2 & 2 \end{bmatrix} = \begin{bmatrix} 0 & 1 & 0 \\ 1 & 0 & 2 \\ 2 & 2 & 2 \end{bmatrix}.$$

Solving condition 1 over GF(3) produces the solution:

$$\begin{bmatrix} a & b & c \\ d & e & f \\ g & h & i \end{bmatrix} = \begin{bmatrix} 1 & 1 & 1 \\ 0 & 1 & 2 \\ 0 & 0 & 1 \end{bmatrix}.$$

Solving condition 2 over GF(3) produces the solution:

$$\begin{bmatrix} a & b & c \\ d & e & f \\ g & h & i \end{bmatrix} = \begin{bmatrix} 1 & 2 & 1 \\ 0 & 1 & 1 \\ 0 & 0 & 1 \end{bmatrix}.$$

The above solutions (for Eq. (5.13)) for the two conditions produce the following ternary Davio expansion:

$$\vec{f}_{D0} = \begin{bmatrix} 1 & 1+c & 1+2c+c^2 \\ 1 & c & c^2 \\ 1 & 2+c & 1+c+c^2 \end{bmatrix} \begin{bmatrix} 0 & 0 & 1 \\ 2 & 1 & 0 \\ 2 & 2 & 2 \end{bmatrix} \begin{bmatrix} f_0 \\ f_1 \\ f_2 \end{bmatrix}, \qquad (5.15)$$

which can be written in the form:

$$f_{r0,D0} = 1 \cdot f_0 + c \cdot (2f_1+f_2) + (c)^2(2f_0+2f_1+2f_2), \qquad (5.16)$$

$$f_{r1,D0} = 1 \cdot f_2 + c \cdot (2f_0+f_1) + (c)^2(2f_0+2f_1+2f_2), \qquad (5.17)$$

$$f_{r2,D0} = 1 \cdot f_1 + c \cdot (2f_2+f_0) + (c)^2(2f_0+2f_1+2f_2). \qquad (5.18)$$

The above Davio$_0$ expansion is reversible, since Eq. (5.15) satisfies the following reversibility restrictions:
(1) For the number of inputs $\{f_0, f_1, f_2, c\}$ is equal to the number of outputs $\{f_{r0}, f_{r1}, f_{r2}, c\}$.
(2) We can uniquely reconstruct any set of inputs from the set of outputs and vice versa (for any unique selection of inputs/outputs such that for any combination of inputs there is only one unique corresponding combination of outputs). **Q.E.D.**

Each of the result outputs in Eq. (5.15) is fully *balanced*; the number of times that 0's appear is equal to the number of times that 1's appear and also equal to the number of times that 2's appear, and this is equal to (81/3) = 27. Also the reversible Davio$_0$ spectral transform in Eq. (5.15) is *conservative*; it has the same number of values in inputs and outputs (i.e., the number of ones and twos in inputs is equal to the number of ones and twos in the outputs). Therefore the circuits that are constructed using Eq. (5.15) are both reversible and conservative. Reversible Davio gates (that result from the corresponding reversible Davio decompositions) can be constructed either using individual GF(*) and GF(+) gates or using multiplexers.

The previous theorem provides a very general methodology of creating reversible Davio spectral transforms of *any type* from the corresponding reversible Shannon spectral transforms. Theorem 5.3 shows also that for each reversible Shannon expansion one constructs the corresponding reversible Davio expansions. Utilizing the same methodology used in Theorem 5.3, all possible permutations of multi-valued Davio spectral transforms can be converted to be reversible permuted Davio spectral transforms.

Note that Eqs. (5.1), (5.2), and (5.15) can be written in the general form:

$$\begin{aligned} \vec{f_r} &= \left[M_{rgbf}\right] \left[M_t\right] \vec{f}, \\ &= \left[M_c\right] \vec{f}, \end{aligned} \tag{5.19}$$

where M_{rgbf} is the reversible generalized basis function symbolic matrix, M_t is the transform matrix, and M_c is the combined matrix (i.e., $[M_c] = [M_{rgbf}][M_t]$). One can also obtain the output vector $\vec{f_r}$ from knowing the input vector $\vec{f}$ by using the inverse of Eq. (5.19) as follows:

$$\begin{aligned} \vec{f} &= \left[M_t\right]^{-1} \left[M_{rgbf}\right]^{-1} \vec{f_r}, \\ &= \left[M_c\right]^{-1} \vec{f_r}. \end{aligned} \tag{5.20}$$

Analogously to the binary case, the Matrices M_t and M_{rgbf} can be generated recursively for an arbitrary number of variables using a Kronecker-like product. Analogously to the result in Theorem 5.2, reversible expansions can be created for any multi-valued invariant Davio spectral transform. Also, the methodology of obtaining reversible decompositions that has been introduced in this Sect. can be adapted for other functional representations like K-maps, maps, etc. Figure 5.28 presents a tree-based analysis of the relationship of the reversible spectral transforms that are presented in this Book and other spectral transforms.

An extensive treatment for the count of all possible families of binary and multiple-valued reversible Shannon and Davio decompositions is presented in Appendix F. The general theories of producing reversible Shannon and Davio expansions will be used hierarchically to obtain the corresponding reversible regular lattice structures in Chapt. 6. Also such reversible primitives will be used to obtain the quantum Shannon and Davio logic primitives that will be presented in Chapt. 11 in order to perform quantum computing using such new quantum primitives.

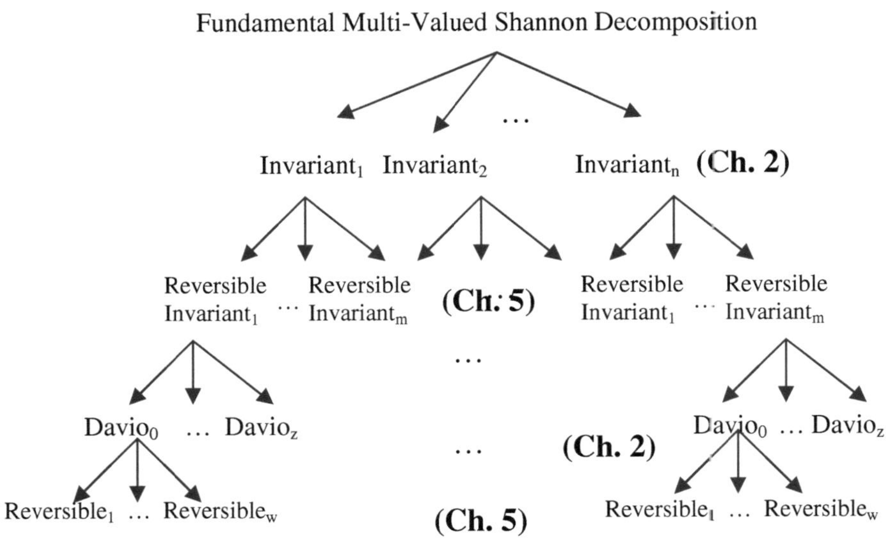

Fig. 5.28. A tree-based relationship between various non-reversible and reversible decompositions. Generalizations of each level of reversible Shannon and Davio expansions are possible through using shifters (inverters) at the input and/or output of the corresponding reversible gates (circuits). Also, all permutations of the expansions in each level can be done to yield the corresponding permuted decompositions.

5.5 The Elimination of Garbage in Multiple-Valued Reversible Circuits

The process of the elimination of output garbage in multiple-valued reversible circuits follow exactly the same methodology used for two-valued circuits. Example 5.6 illustrates the elimination of garbage for ternary reversible Shannon and Feynman gates, respectively.

Example 5.6. This example illustrates the process of obtaining three-valued reversible inverse gates to some fundamental reversible (forward) gates. These gates are illustrated in Figs. 5.29 through 5.30, respectively.

5.6 Summary

In this Chapt. we introduced a new and general methodology to generate, classify, and count new reversible multiple-valued expansions and their corresponding primitives. The new results in this Chapt. utilize the multiple-valued families of spectral transforms that were introduced in Sect. 2.2.2 in order to produce the corresponding reversible decompositions that encompass reversible invariant multiple-valued Shannon and reversible multiple-valued Davio expansions. Such new reversible decompositions will play an important role for the synthesis of logic functions into reversible 3-D regular structures as will be shown in Chapt. 6. Since the basic requirement for logic synthesis for several new technologies are: (1) reversibility, (2) no wire intersections, and (3) three-dimensionality (to utilize the atomic 3-D structures), the new families of multiple-input multiple-output multiple-valued reversible decompositions that are presented in this Chapt. can be used to create new category of reversible regular structures and will be used to synthesize quantum logic circuits in Chapt. 10 and their corresponding quantum computations in Chapt. 11. Various new reversible circuits were also created in this Chapt. by the careful design using basic reversible gates. These include reversible barrel shifter, reversible code converters, reversible sorter, and reversible MIN/MAX tree. Such new reversible circuits can be used to start building a library for reversible logic synthesis. The process of eliminating garbage outputs that can be produced in two-valued and multiple-valued reversible circuits is also shown. This is done through the synthesis of the reversible inverse (mirror image) circuit and then cascading this circuit to the reversible forward circuit. The intermediate functions are measured using the "spy" circuit which is merely a copy circuit made up of a Feynman primitive that uses "0" value in the input to the XOR gate. The process of garbage elimination is important as quantum circuits do not allow for garbage in the outputs. This point will be further illustrated in Chapts. 10 and 11, where quantum logic circuits and quantum computations are implemented for the new reversible logic structures that will be created in Chapts. 6, 7, and 8.

f_0	f_1	f_2	c	f_{r0}	f_{r1}	f_{r2}	c
f_0	f_1	f_2	0	f_0	f_1	f_2	0
f_0	f_1	f_2	1	f_1	f_2	f_0	1
f_0	f_1	f_2	2	f_2	f_0	f_1	2

a

f_{r0}	f_{r1}	f_{r2}	c	f_0	f_1	f_2	c
f_0	f_1	f_2	0	f_0	f_1	f_2	0
f_1	f_2	f_0	1	f_0	f_1	f_2	1
f_2	f_0	f_1	2	f_0	f_1	f_2	2

b

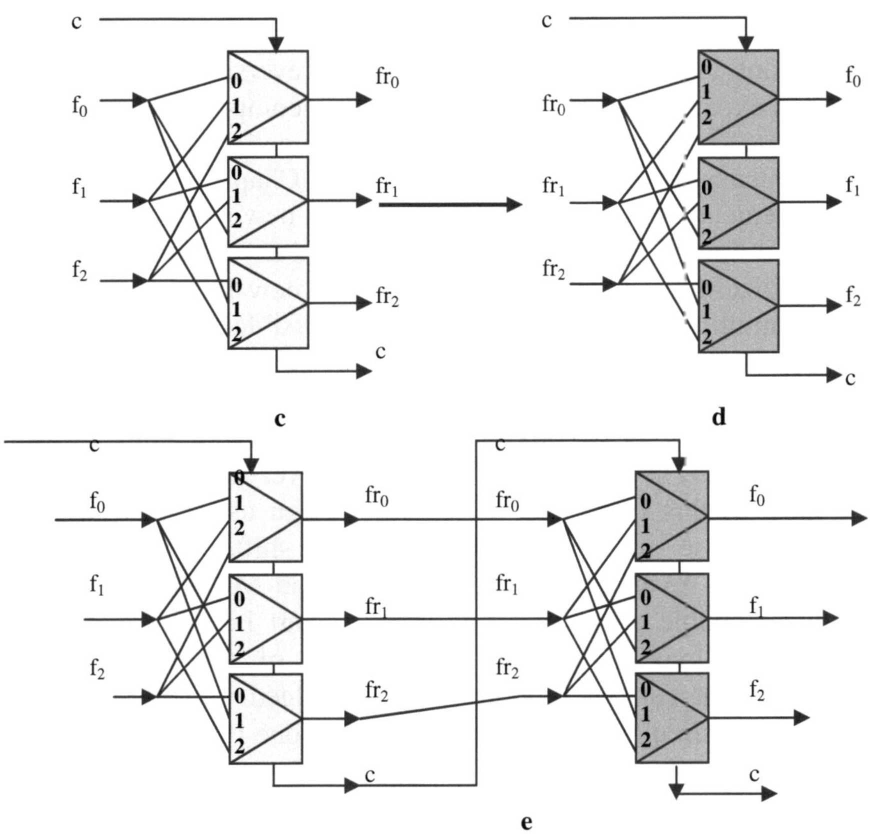

Fig. 5.29. a Truth table for ternary reversible forward Shannon gate, **b** truth table for ternary reversible inverse Shannon gate, **c** ternary reversible forward Shannon circuit, **d** ternary reversible inverse Shannon circuit, and **e** the elimination of garbage for ternary Shannon gate by combining the ternary reversible forward Shannon circuit with the ternary reversible inverse Shannon circuit.

a	b	c	d
0	0	0	0
0	1	0	1
0	2	0	2
1	0	1	1
1	1	1	2
1	2	1	0
2	0	2	2
2	1	2	0
2	2	2	1

a

c	d	a	b
0	0	0	0
0	1	0	1
0	2	0	2
1	1	1	0
1	2	1	1
1	0	1	2
2	2	2	0
2	0	2	1
2	1	2	2

b

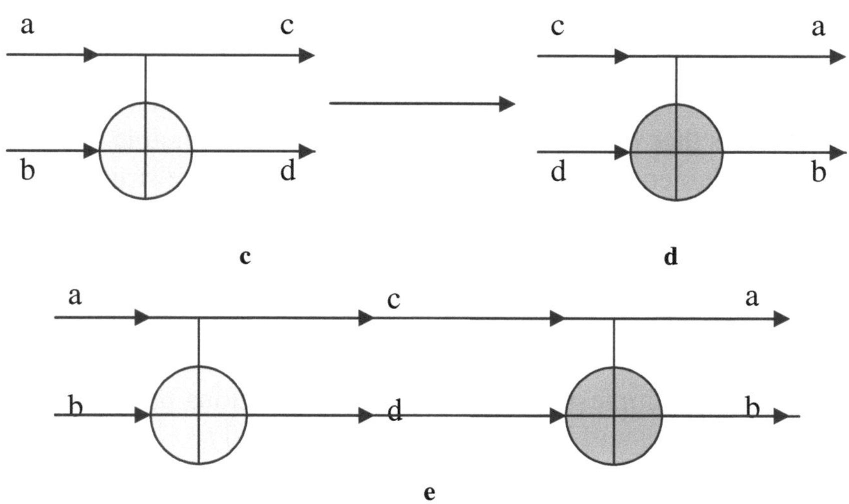

Fig. 5.30. Ternary reversible Feynman gate: **a** truth table for forward gate, **b** truth table for inverse gate, **c** forward circuit, **d** inverse circuit, and **e** the elimination of garbage.

Since a "forward" function is defined if one-to-one mapping exists between the set of values in Domain **D** and set of values in Range **R**, and a "forward" relation is defined if one-to-many mapping exists between **D** and **R**, one can define "reversible" function if a unique one-to-one mapping exists reversely between **R** and **D**, and "reversible" relation if a unique many-to-one mapping exists reversely between **R** and **D**. This conclusion is general (i.e., valid in Discrete or Continuos **D** and **R**) and can be helpful in order to use reversibility-based *symmetries* in other mathematical formalisms, besides spectral techniques from Sect. 5.4 in this Chapt., such as Group Theory for example.

6 Reversible Lattice Structures

This Chapt. will introduce a methodology of synthesizing binary and multiple-valued logic functions using a regular reversible structure. This will be done by utilizing the new reversible binary and multiple-valued Shannon primitives that were introduced in Theorem 5.1, and the use of the process of permutation of cofactors that resulted from the new reversible Shannon primitives. The new idea in this Chapt. is:

- The hierarchical application of the process of permutation of cofactors that will produce the corresponding reversible lattices.
- The implementation of the inverse reversible lattice structure (mirror image) to produce structure that is suited for quantum computing since in quantum logic garbage is not allowed.

The new two-valued and multiple-valued reversible lattice structures will be used to obtain their counterparts of quantum logic circuits in Chapt. 10, and the methodologies of implementing two-valued and multiple-valued quantum computations using reversible lattice structures will be implemented in Chapt. 11.

Section 6.1 of this Chapt. presents a new general algorithm for the production of reversible lattice structures for any radix of Galois logic, and the introduction of the idea of mirror image of reversible lattice structures to eliminate the output garbage. Summary of the results that are introduced in this Chapt. will be presented in Sect. 6.2.

6.1 A General Algorithm for the Creation of Two-Valued and Multiple-Valued Reversible Lattice Structures

It has been shown in Fig. 5.25 that the application of the processs of permutation of cofactors will lead to the reversible Shannon primitives. The algorithm for the synthesis of reversible lattice structures depends on the hierarchical application of this process of permutation of cofactors (as a result of the application of the Latin square property onto the Generalized Basis Function Matrix (GBFM)) that has been presented in Sect. 5.4. A general procedure for the construction of reversible Shannon lattice structure over n^{th} radix logic is as follows [6,14,182]:

Synthesis Stage:

(1) Utilizing a reversible Shannon primitive (From Sect. 5.4), assign the multi-valued map of the function (that is needed to be realized in the reversible lattice structure) for one output of the reversible Shannon primitive in the first level, and assign don't care maps for the rest of the primitive outputs at the first level. Also assign don't care maps to the "garbage" outputs of the primitives in each level of the reversible lattice structure. These "garbage" outputs are needed only for the purpose of satisfying reversibility. The process of assigning don't cares to multi-valued maps stems from the fact that one does not know a priori what will be the values of the leaves of the corresponding reversible lattice structure.

(2) Following the output-to-input paths of the reversible Shannon primitive in the first level of the reversible lattice structure, going from outputs-to-inputs, and using the reverse of the method of permutation of cofactors from Sect. 5.4 (e.g., constructing inputs from outputs in Figs. 5.24, 5.26, and 5.27, for instance), construct new maps at the input of the reversible Shannon primitive by permuting the output cofactors (in the output maps) that correspond to the expansion variable in the first level. This process of permuting the output cofactors will result in new maps at the inputs of the reversible Shannon primitive at the first level. Thus, the contents of the input maps will result from the permutation of the values of the cofactors in maps at the output of the same reversible Shannon primitives at the first level.

(3) Going from top-to-bottom and left-to-right of the reversible lattice structure, repeat step 2 for each expansion variable in each level (i.e., for each reversible Shannon primitive in every level) until one reaches multi-valued maps at the bottom of the reversible lattice structure with each map having only a constant value from the set $\{0, 1, 2\}$.

Analysis Stage: This is an opposite process to the process of synthesis.

(4) Following the input-to-output paths of the reversible Shannon primitives at the last level of the reversible lattice structure, going from inputs-to-outputs, and using the forward method of permutation of cofactors from Sect. 5.4, construct new maps at the output of the reversible Shannon primitives by permuting the input cofactors (in the input maps) that correspond to the expansion variable in the last level. This process of permuting the input cofactors will result in multi-valued maps at the outputs of the reversible Shannon primitives at the last level. Thus, the contents of the output maps (at the last level) will result from the permutation of the values of the cofactors in maps at the inputs of the same reversible Shannon primitives.
(5) Going from bottom-to-top and right-to-left of the reversible lattice structure, repeat step 4 for each expansion variable in each level (i.e., for each reversible Shannon primitive in every level) until one reaches completely specified maps, in all wires throughout the reversible lattice structure from bottom-to-top and right-to-left, with no don't cares.

The following examples illustrate the concept of reversible lattice structures.

Example 6.1. This example illustrates the creation of the reversible binary lattice structure for the Boolean function (F) in Figs. 6.1 and 6.2. Note that in Figs. 6.1 and 6.2 the desired output function is denoted as F and the "garbage" outputs (that are necessary only for reversibility) are denoted as G1-G5.

Example 6.2. Figures 6.3 and 6.4 illustrate the creation of the reversible ternary lattice structure for the ternary function (F). Note in that Figs. 6.3 and 6.4 the desired output function is denoted as F and the "garbage" outputs (that are necessary only for reversibility) are denoted as G1-G8.

Note that the regular lattice structures in Figs. 6.2 and 6.4 are fully reversible as the vector of input values (9 inputs in Fig. 6.2, and 11 inputs in Fig. 6.4) can be always uniquely reconstructed from the vector of output values (9 outputs in Fig. 6.2, and 11 outputs in Fig. 6.4), respectively. One can note that the main advantage of such reversible structures is that they possess regularity. The disadvantage is that such lattice structures produce big garbage in the outputs.

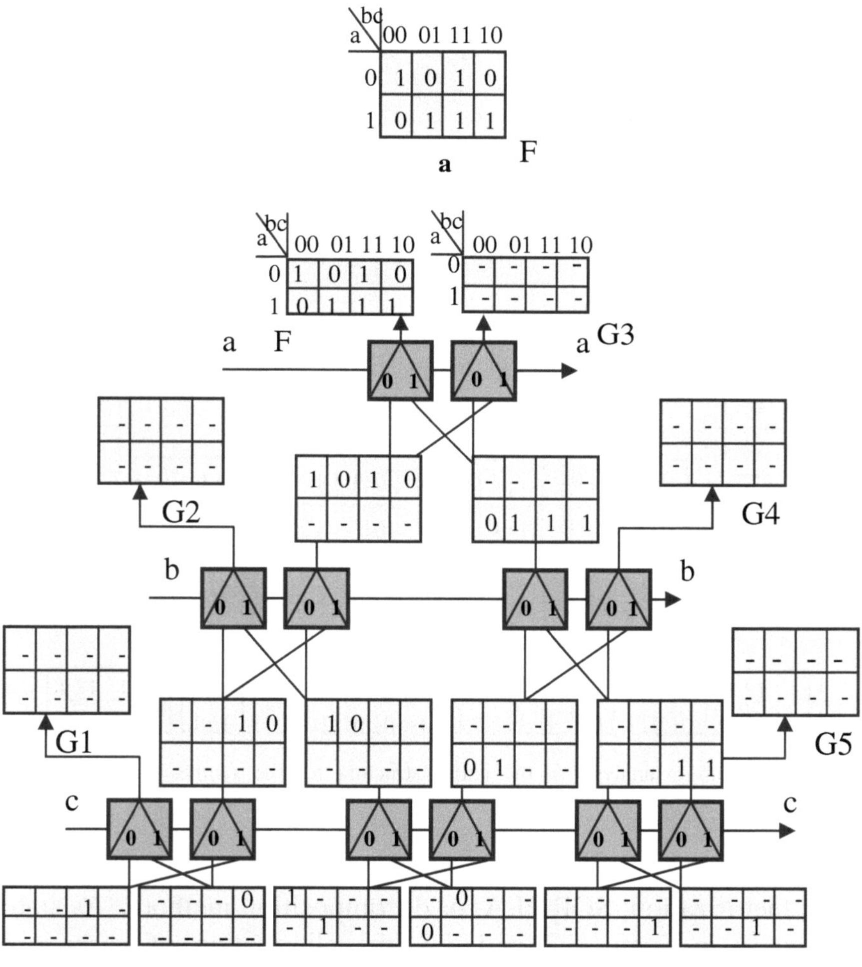

Fig. 6.1. Synthesis of a reversible lattice structure: **a** Boolean function to be realized in 2-D reversible lattice structure, and **b** top-to-down and left-to-right algorithm to produce the values of leaves of the 9 input/9 output two-dimensional reversible lattice structure.

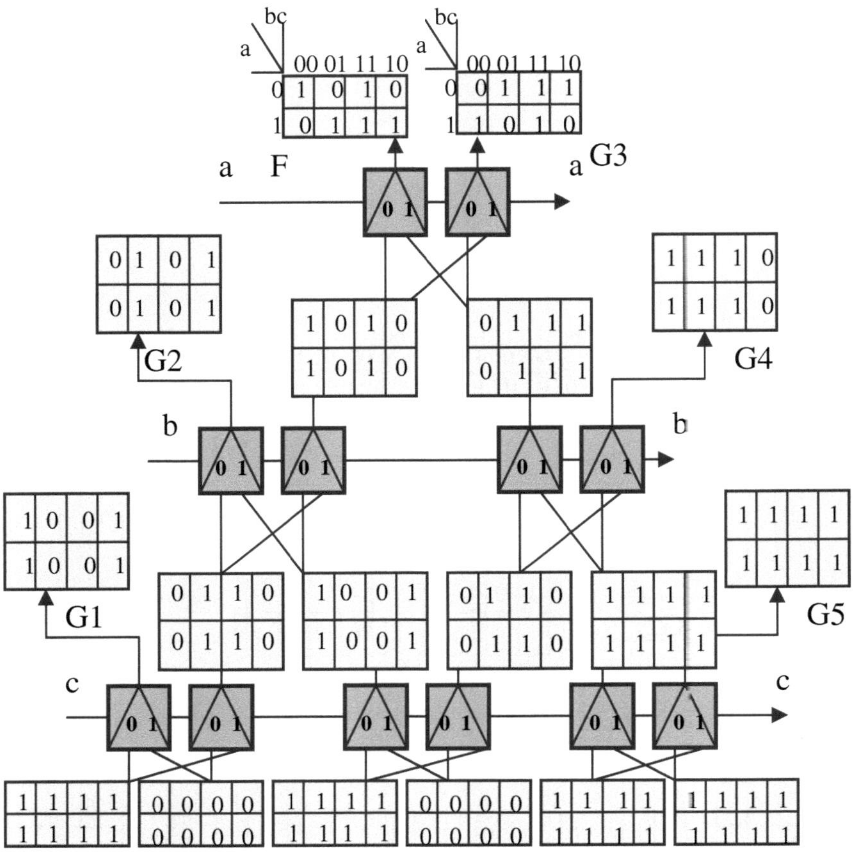

Fig. 6.2. Bottom-up analysis of the lattice structure for the synthesis of the Boolean function (F) in Example 6.1.

As verified on small functions, this garbage is still smaller than using factorization, SOP-PLA, or decomposition methods. As stated previously in Sects. 5.2 and 5.5, the elimination of garbage in reversible structures is done by using the reversible mirror image of the forward reversible circuit. Figure 6.5 illustrates the process of garbage elimination by cascading (i.e., serially interconnecting) the reversible forward lattice structure to the reversible inverse lattice structure. The functionality in the outputs can be measured using the "spy" circuit which is a Feynman primitive with "0" value to the input of its Galois addition gate.

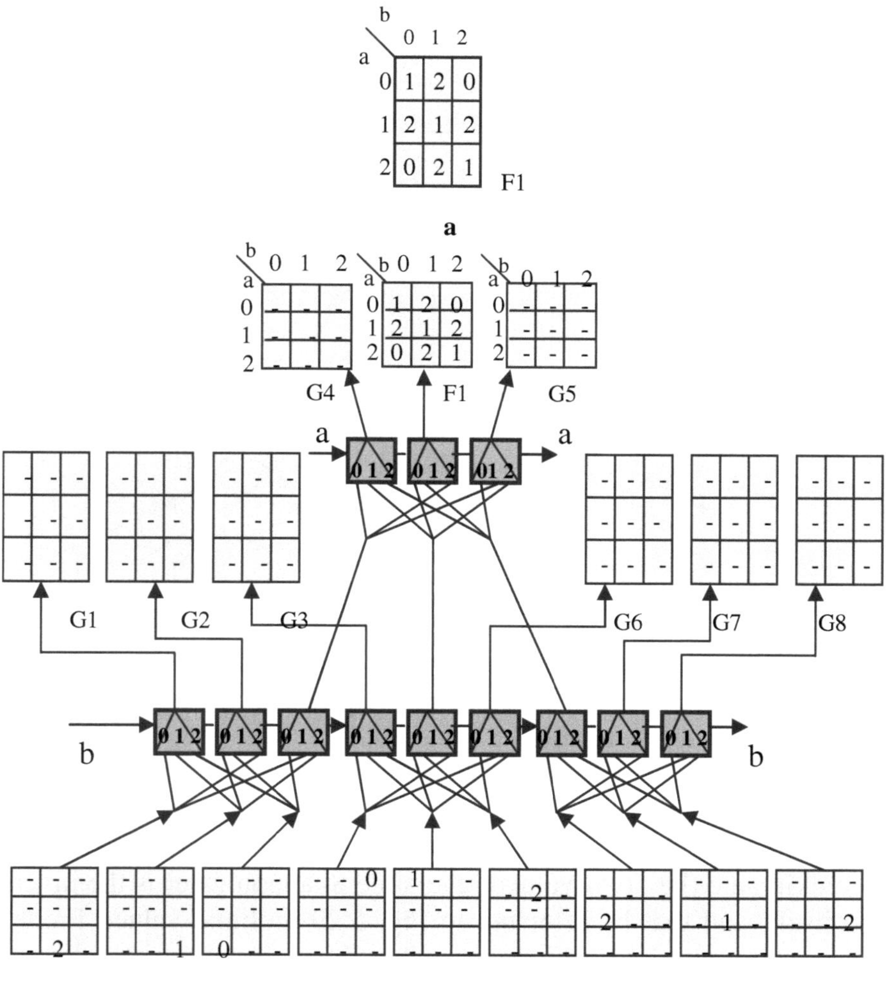

Fig. 6.3. Synthesis of a ternary reversible lattice structure: **a** ternary function to be realized in a reversible lattice structure, and **b** top-to-down and left-to-right algorithm to produce the values of leaves of the 11 input/11 output reversible lattice structure.

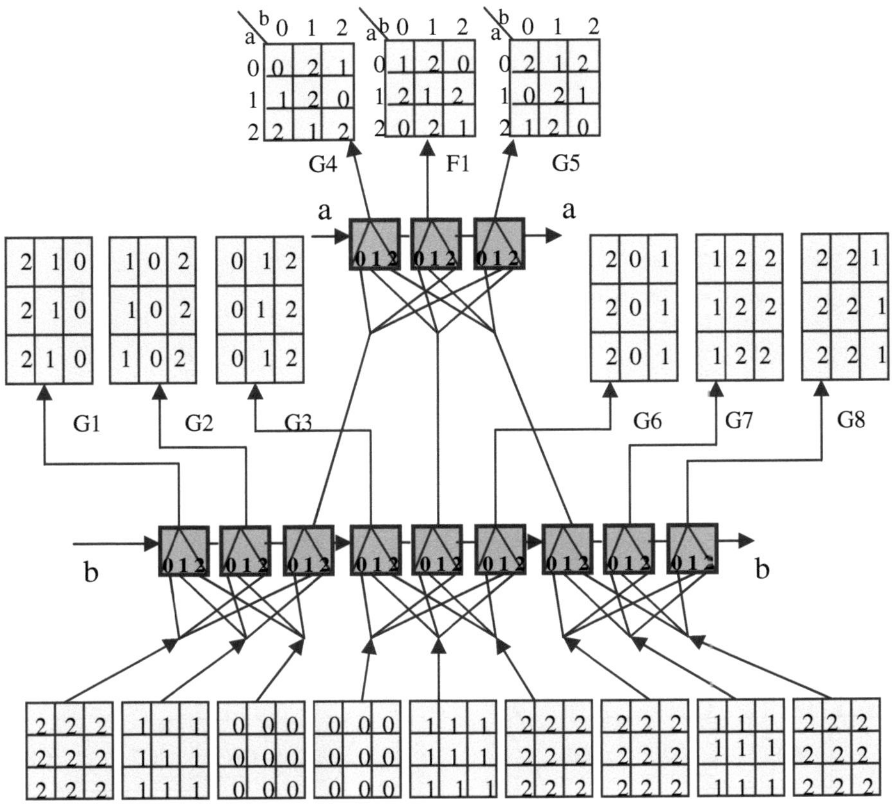

Fig. 6.4. Bottom-up analysis of the resulting reversible lattice structure for the synthesis of the ternary function (F) in Example 6.2.

In Fig. 6.5, the gate K^{-1} is the inverse to the gate K, and the gate denoted by 0 is the spy circuit. Note that by referring to Fig. 5.11 Fredkin gate (i.e., two-valued reversible Shannon gate) is the inverse to itself, and that by referring to Fig. 5.29 multiple-valued Fredkin gate (i.e., multiple-valued reversible Shannon gate) is also the inverse to itself.

6.2 Summary

A general new algorithm has been presented in this Chapt. to produce reversible lattice structures. This algorithm depends on the hierarchical application of the process of permutation of cofactors that has been presented in Sect. 5.4. Since garbage is not allowed in quantum computing, the concept of reversible inverse lattice structure has been also presented. The intermediate functionalities from the total circuit that is composed of cascading the forward reversible lattice structure and the inverse reversible lattice structure are measured using the "spy" circuit, which consists of a Feynman gate with value "0" to its Galois addition gate in order to make the copy primitive. The total network of the forward and inverse reversible lattice structures will be used to implement the corresponding quantum circuits from Chapt. 10 and quantum computations in Chapt. 11.

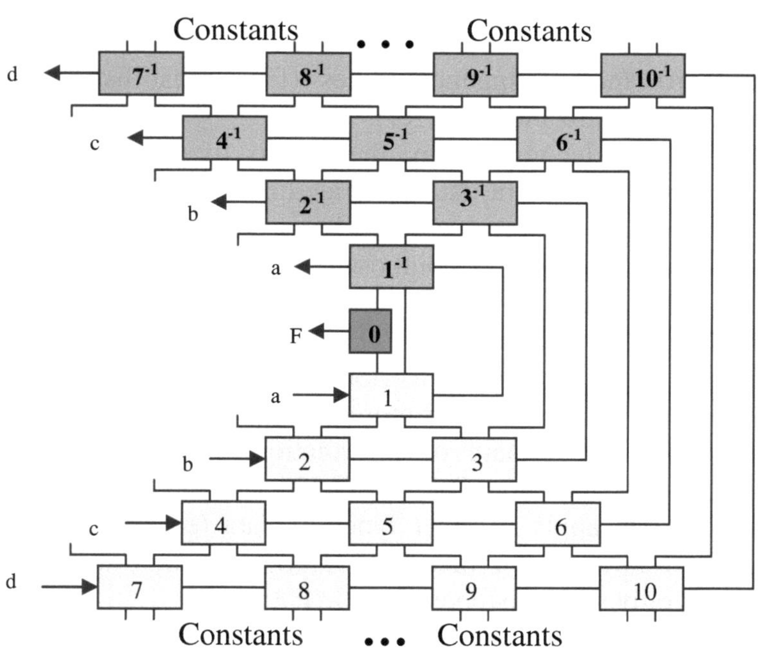

Fig. 6.5. Garbageless reversible lattice structure.

7 Novel Reconstructability Analysis Circuits and their Reversible Realizations

This Chapt. will introduce another new type of reversible structures called Reversible Modified Reconstructability Analysis (RMRA). Reconstructability Analysis (RA) is an important decomposition technique that is used widely in system science area to decompose qualitative data [133,134,135,138,273,275]. This kind of decomposition is commonly used in the decomposition of data obtained in the social and system science fields, and overlaps with other decomposition techniques used in the social sciences as well like the Log-Linear (LL) decomposition method [138]. RA decomposition aims at the simplest decomposition of qualitative data using Lattice-Of-Structure (LOS) (cf. Fig. 7.1) as representation (generation) and contingency tables (for probabilistic data) for evaluation (minimization of error).

In lossless decomposition, the aim is to obtain the simplest decomposed model from data (saturated model) without the loss of any information (i.e., error = 0). In lossy decomposition, the aim is to obtain the simplest decomposed model from data (saturated model) with an acceptable amount of error. RA data is typically either a set-theoretic relation [271,272] (or mapping) or it is a probability (or frequency) distribution. The former case is the domain of set-theoretic RA or more precisely crisp possibilistic RA. The latter is the domain of information-theoretic RA, or more precisely probabilistic RA [133,134,135,138,274]. The RA framework can apply to other types of data (e.g., fuzzy data) via generalized information theory [135,223]. In this work, we are concerned only with crisp possibilistic RA.

New RA decomposition, called the Modified RA (MRA) decomposition [10,20,21,22,27,28,29,30] is introduced in this Chapt., and the lossless MRA decomposition is used to decompose logic functions. A comparison of the complexities obtained from the resulting decompositions from MRA decomposition with the complexities obtained through lossless Ahsenhurst-Curtis (AC) decomposition and Bi-decomposition for the same Boolean logic functions will be also provided in Appendix H. Although the comparisons, which will be presented in this Chapt., and in Appendix H serve only as a first step, since we consider only the 256 3-input Boolean functions, the results will provide a first useful "glimpse" of the comparative complexities obtained from the new MRA decomposition.

For three variables, the LOS for RA decomposition consists of a total of 5 decomposed structures and 9 decomposed models (as model is a structure applied on data) (See Fig. 7.1). The RA decomposition is a general graph-based decomposition which applies recursively level by level the following rules: (1) remove one relation from the previous model, and (2) restore embedded relations if they are not already present. Composition process in Reconstructibility Analysis requires opposite rules to the rules of the RA decomposition process. The lattice of relations, for three-variables, is illustrated in Fig. 7.1, where each box represents a relation between the variables that are represented as input lines. This representation that is used for RA decomposition is complementary to the general graph-based representation, which uses the general graph-based notation in which a relation is represented as a line connecting variables which are boxes (nodes).

Each variable in Fig. 7.1 represents an input variable to a function or a general object. Objects {A,B,C} can be for instance: A can be a father, B be a mother, and C be a child, and the boxes represents the general interaction (relation) between the three members of the family (i.e., father, mother, and child). Richer models of interaction (i.e., higher dimensional functions) can be constructed by considering other elements like D to be the fourth argument (as a family cat for instance). For 4 objects (elements) similar lattice of relations of 4 variables (objects, elements) is constructed.

For logic synthesis, the lattice of structure in Fig. 7.1 represents circuit decomposition, where elements A, B, and C in Fig. 7.1 are the input variables to the circuit, and the resulting structure (in the lattice of structures in Fig. 7.1) is the decomposed circuit. This Chapt. introduces the following new results:

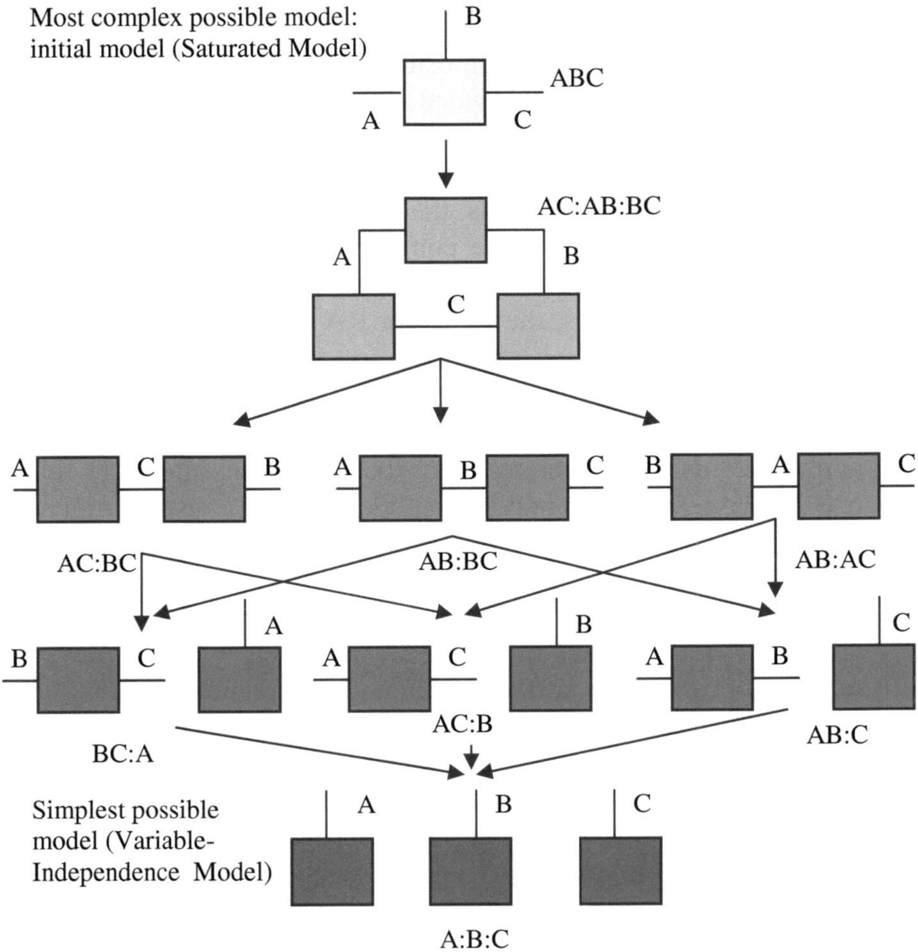

Fig. 7.1. Three-variable lattice of relations for RA (graph-based) decomposition: lines represent variables (elements or objects), and boxes represent the interactions between the associated variables.

- Novel type of two-valued decomposition based on Reconstructability Analysis called Modified Reconstructability Analysis (MRA). Two variants of MRA for Boolean logic are provided: 1-MRA and 0-MRA. The superiority of MRA over the Conventional Reconstructability Analysis (CRA) in terms of the decomposability of Boolean functions and the reduction of complexity is also provided.
- The generalization of two-valued MRA into multiple-valued MRA.

- New type of reversible structures which is based on reversible MRA (R-MRA).
- Evaluations for MRA versus Ashenhurst-Curtis (AC) decomposition and Bi-decomposition in terms of the decomposability of Boolean functions and the reduction of complexity is provided in Appendix H.

The remainder of this Chapt. is organized as follows. Section 7.1 introduces the new two-valued Modified Reconstructability Analysis (MRA). Multiple-valued MRA is introduced in Sect. 7.2. Section 7.3 introduces the Reversible realization of MRA (R-MRA). A Summary of this Chapt. is presented in Sect. 7.4. (New results of complexity comparisons of MRA versus AC-like decompositions will be also introduced in Appendix H.)

7.1 New Type of Reconstructability Analysis: Two-Valued Modified Reconstructability Analysis (MRA)

This Sect. introduces an innovation in set-theoretic RA, which we call "modified" RA (or MRA) as opposed to the conventional set-theoretic RA (or CRA). This innovation will be illustrated by Examples 7.1 and 7.2. The main idea of MRA stems from the following fact: While the conventional RA (CRA) decomposes on the set of all functional values of the corresponding function, the modified RA (MRA) decomposes on the set of minimum functional values from which the function can be totally reconstructed. In general, the procedure for the lossless MRA decomposition follows the following steps:

(1) Using the lattice-of-relations, decompose for one value only of the Boolean function into the simplest error-free decomposed structure:

(1a) Remove one relation between variables from the previous level.

(1b) Add the embedded relation(s) between variables, in the current level, if they are not already present in the new model.

(2) As a result of step 1, one obtains MRA decomposition for value "1" of the Boolean function (denoted as 1-MRA decomposition), and MRA decomposition for value "0" of the Boolean function (denoted as 0-MRA decomposition). Select the simplest model from

the resulting 1-MRA decomposition and the 0-MRA decomposition, respectively.
(3) In the resulting simplest decomposed data model from step 2, generate the corresponding sub-functional values for each interaction (relation) between the variables that exist in the decomposed model.
(4) Generate the total functional values using the intersection between all possible sub-functional values for 1-MRA, and the union between all possible sub-functional values for 0-MRA.
Example 7.1. Figure 7.2 illustrates decomposed structures using both CRA and MRA decompositions, respectively for the logic function: $F = x_1x_2 + x_1x_3$.

CRA decomposition is illustrated in the upper half of the Fig., while MRA decomposition is illustrated in the lower half of the Fig. As one can observe, MRA decomposition yields a much simpler logic circuit than the corresponding CRA decomposition, while retaining complete information about the decomposed function. For CRA as shown in the top middle part of the Fig., the calculated function for the model $x_1x_2f_1{:}x_1x_3f_2{:}x_2x_3f_3$ (i.e., $\alpha{:}\beta{:}\gamma$) is defined as follows:

$$x_1x_2x_3F_{x1x2f1:x1x3f2:x2x3f3} \equiv (x_1x_2f_1 \otimes x_3) \cap (x_1x_3f_2 \otimes x_2) \cap (x_2x_3f_3 \otimes x_1),$$

where $\otimes$ here means the Cartesian product of sets. (For lossless CRA decomposition, this equals the original function $x_1x_2x_3F$ that is shown at the top left of the Fig., and for *lossy* CRA $x_1x_2x_3F_{x1x2f1:x1x3f2:x2x3f3}$ would not be equivalent to $x_1x_2x_3F$). The CRA model can be interpreted by the circuit shown at the top right of the Fig., where different projections of F are labeled f_1, f_2, and f_3. In Fig. 7.2, while CRA decomposes for all values of Boolean functions, MRA decomposes for an arbitrarily chosen value of the Boolean functions (e.g., for value “1”).

From Fig. 7.2, one notes that 1-MRA has two advantages over CRA for the decomposition of two-valued functions: (1) the resulting decomposed structures from 1-MRA are less complex than the corresponding decomposed structures from CRA, and (2) the resulting decomposed structures from 1-MRA are directly realizable in Boolean-based circuits, while the resulting decomposed structures from CRA are not directly realizable in Boolean circuits.

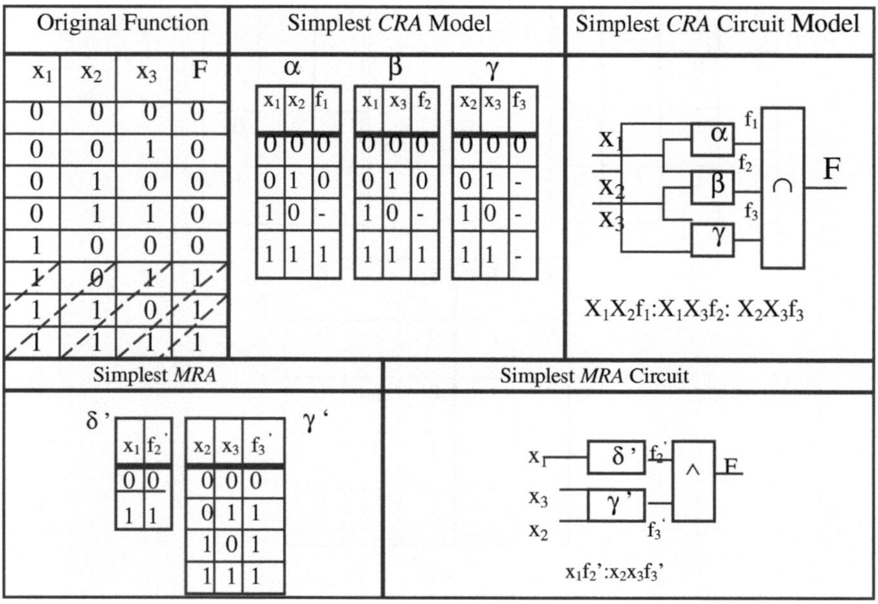

Fig. 7.2. CRA versus 1-MRA decompositions for the Boolean function F = x1x2 + x1x3.

The completely specified Boolean function can be retrieved if one knows the MRA decomposition for the Boolean function being equal either to "1" or to "0". As can be observed from Fig. 7.2, 1-MRA decomposition yields much simpler logic circuit than the corresponding CRA decomposition, while retaining complete information about the decomposed logic function. MRA simplifies the decomposition problem by focusing, in the original function F, on the three shaded tuples ("cubes") for which F = 1. The procedure for 1-MRA in Fig. 7.2 is as follows:

(1) Using the Lattice-Of-Structures (LOS), decompose the Boolean function of value "1" into the simplest lossless CRA decomposition.

(2) For a particular model, selected from the LOS, get the projections.

(3) Assign value "1" (for 1-MRA) to tuples in the resulted projections. Add all tuples that are missing in the projections which will have the functional value "0".

(4) Perform the AND operation for 1-MRA in the output block to obtain the total functionality.

Steps (2)-(4) are illustrated as follows:

x_1	x_2	x_3	F
1	0	1	1
1	1	0	1
1	1	1	1

→

x_1	x_2	x_3
1	0	1
	1	0
	1	1

→

x_1	f_2'	x_2	x_3	f_3'
0	0	0	0	0
1	1	0	1	1
		1	0	1
		1	1	1

The output function in step (4) is the (logical) AND of the two sub-functions, i.e., $F = f_2'(x1) \wedge f_3'(x2,x3)$. Set-theoretically. this is illustrated as F = (x1⊗(1∪x1') ⊗0) ∩ (x2x3⊗(1∪(x2x3)') ⊗0). The idea of 0-MRA versus 1-MRA is illustrated in the following Example 7.2.

Example 7.2. For the logic function $F = x_1x_2 + x_1x_3$. Figure 7.3 illustrates the simplest model using both 1-MRA and 0-MRA. In this example, the completely specified Boolean function can be retrieved if one knows the MRA decomposition for the Boolean function being equal either to "1" (that is 1-MRA) or to "0" (that is 0-MRA). 0-MRA simplifies the decomposition problem by focusing, in the original function F, on the five shaded tuples ("cubes") for which F = 0. The procedure used to obtain the 0-MRA in Fig. 7.3 is as follows:

(1) Using the Lattice-Of-Structures (LOS), decompose the Boolean function of value "0" into the simplest lossless CRA decomposition.

(2) For a particular model, selected from the LOS, get the projections.

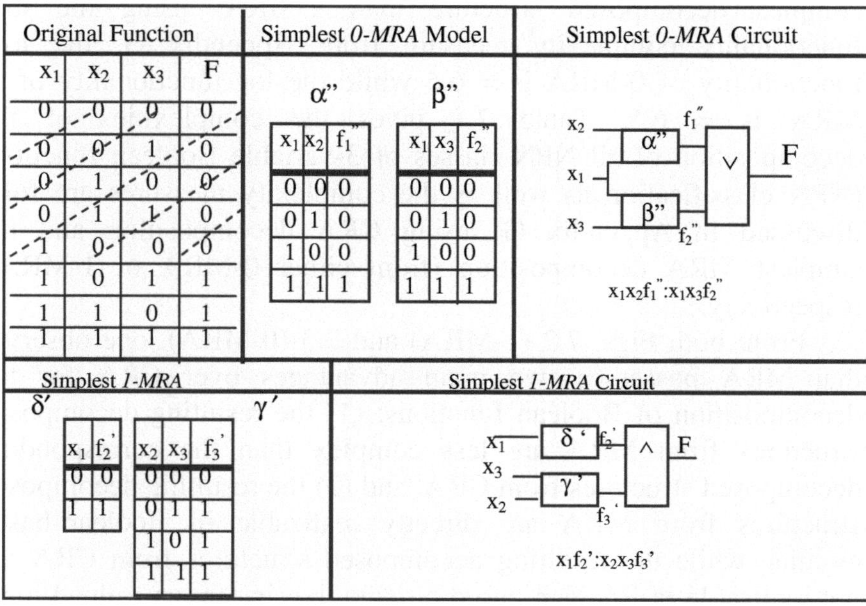

Fig. 7.3. 0-MRA versus 1-MRA decompositions for the Boolean function $F = x_1x_2 + x_1x_3$.

(3) Assign value “0” (for 0-MRA) to tuples in the resulted projections. Add all tuples that are missing in the projections which will have the functional value “1”.

(4) Perform the OR operation for 0-MRA in the output block to obtain the total functionality.

Steps (2)-(4) are illustrated in the following flow diagram where the output function in step (4) is the (logical) OR of the two sub-functions. This is illustrated as $F = f1''(x1,x2) \vee f_2''(x1,x3)$.

x1	x2	x3	F
0	0	0	0
0	0	1	0
0	1	0	0
0	1	1	0
1	0	0	0

→

x1	x2	x1	x3
0	0	0	0
0	1	0	1
1	0	1	0

→

x1	x2	f1”	x1	x3	f2”
0	0	0	0	0	0
0	1	0	0	1	0
1	0	0	1	0	0
1	1	1	1	1	1

As can be observed from Fig. 7.3, 0-MRA produces more complex decomposed structure than 1-MRA: using the log-functionality complexity measure from Appendix G, the log-functionality of 0-MRA is = 6.6 while the log-functionality of 1-MRA is = 6.5. Table 7.1 gives the complexities of the decomposition of all NPN-classes of 3-variable Boolean functions (NPN classification as well as the complexity measures are fully discussed in Appendix G) using CRA decomposition and the simplest MRA decomposition (from either 0-MRA or 1-MRA), respectively.

From both Figs. 7.2 (1-MRA) and 7.3 (0-MRA), one observes that MRA possesses two main advantages over CRA for the decomposition of Boolean functions: (1) the resulting decomposed structures from MRA are less complex than the corresponding decomposed structures from CRA, and (2) the resulting decomposed structures from MRA are directly realizable in Boolean-based circuits, while the resulting decomposed structures from CRA are not realizable in Boolean-based circuits, but in ternary-valued logic circuits, and thus the resulting logic circuits from MRA are directly implementable using the current technologies.

Table 7.1 shows that, in terms of the log-functionality complexity measure, in five NPN classes (classes 1, 2, 6, 8, 9) MRA and CRA give equivalent complexity decompositions, but in the remaining five classes (classes 3, 4, 5, 7, 10) MRA is superior in complexity reduction.

One observes from Table 7.1 that, when decomposition exists, MRA produces four distinct structures, while CRA generates only two distinct structures. This observation for the capability of MRA to generate more different structural topologies can be very important when compared for instance with decompositions that are capable of producing only a specific structural topology as a result of functional decomposition (e.g., Ashenhurst-Curtis (AC) decomposition that is shown in Appendix H). The importance of this point is in the potential of additional design flexibility that one can have, when considering the synthesis of certain functions, in the form of having larger space of designs that one can choose from to meet certain design specifications.

Table 7.1. Conventional RA (CRA) versus Modified RA (MRA) for the decomposition of all NPN-classes of 3-variable Boolean functions (See Table G.1 in Appendix G). (Compare the right-most two columns.)

NPN-Representative Function	Simplest MRA model (0-MRA or 1-MRA)	Simplest CRA model	C (LUT)	C_{LF} (CRA)	C_{LF} (MRA)
Class 1 (8) $F = x_1x_2 + x_2x_3 + x_1x_3$	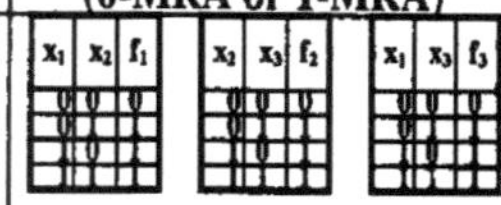	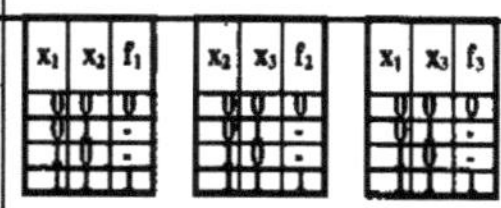	8	7.2	7.2
Class 2 (2) $F = x_1 \oplus x_2 \oplus x_3$	non-decomposable	non-decomposable	8	8	8
Class 3 (16) $F = x_1 + x_2 + x_3$	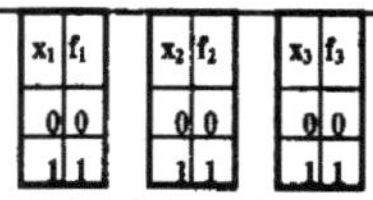	non-decomposable	8	8	4.3
Class 4 (48) $F = x_1(x_2 + x_3)$	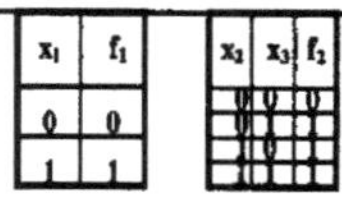	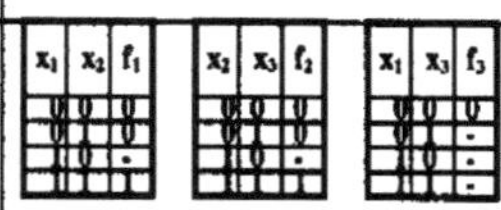	8	7.2	6.5
Class 5 (8) $F = x_1x_2x_3 + x_1'x_2'x_3'$	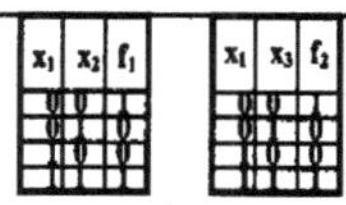	non-decomposable	8	8	6.6
Class 6 (24) $F = x_1'x_2x_3 + x_1x_2' + x_1x_3'$	non-decomposable	non-decomposable	8	8	8
Class 7 (24) $F = x_1(x_2x_3 + x_2'x_3')$	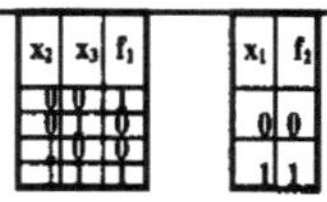	non-decomposable	8	8	6.5
Class 8 (24) $F = x_1x_2 + x_2x_3 + x_1'x_3$	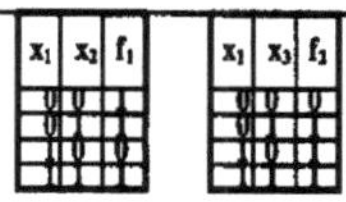	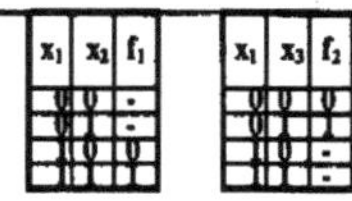	8	6.6	6.6
Class 9 (16) $F = x_1'x_2x_3 + x_1x_2'x_3 + x_1x_2x_3'$	non-decomposable	non-decomposable	8	8	8
Class 10 (48) $F = x_1x_2'x_3' + x_2x_3$	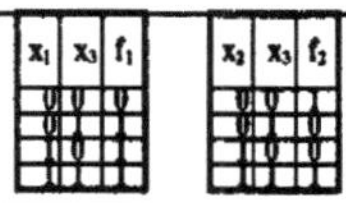	non-decomposable	8	8	6.6

By observing Table 7.1, one observes that the MRA decomposition possesses more advantages over the CRA decomposition when comparing the corresponding CRA versus MRA decompositions for all NPN-classess of 3-variable Boolean functions. For this purpose, we will use the MRA decomposition to compare RA decomposition versus the Ashenhurst-Curtis (AC) and Bi-decomposition (BD) (as will be illustrated in Appendix H).

Figure 7.4 provides a quantitative analysis of the decomposition of the NPN-classified functions using MRA and CRA, respectively. The analysis, in terms of complexity, of the results in Fig. 7.4 is as follows:

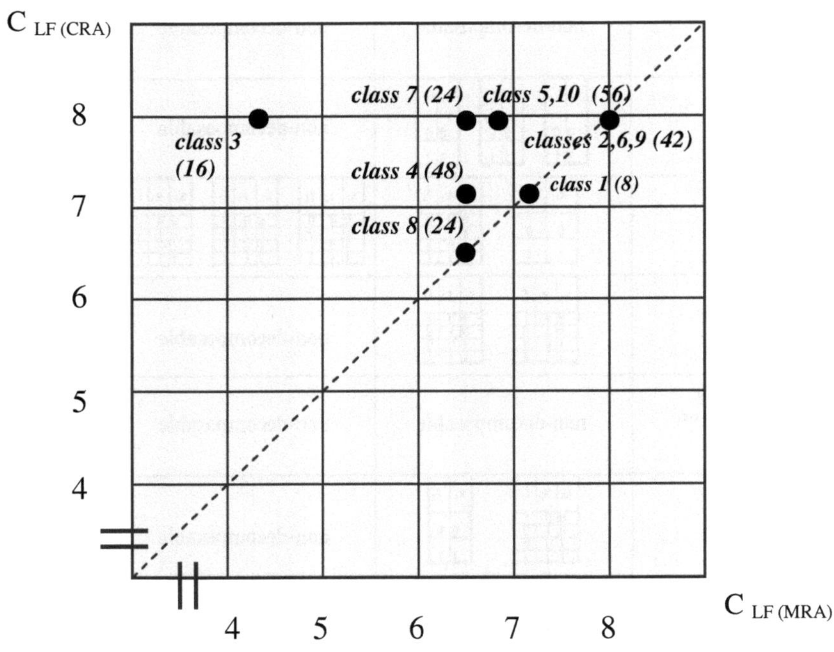

Fig. 7.4. Comparison of the Log-Functionality complexity measure between modified RA (MRA) and conventional RA (CRA) of 3-variable NPN-classified Boolean functions.

Total number of classes that CRA is better than MRA: 0.
Total number of functions that CRA is better than MRA: 0.
Total number of classes that MRA is better than CRA: 5 (3, 4, 5, 7, 10).
Total number of functions that MRA is better than CRA: 144.
Total number of classes for CRA is the same as MRA: 5 (1, 2, 6, 8, 9).
Total number of functions for CRA is the same as MRA: 74.

From the results in Fig. 7.4, one observes the clear superiority of MRA (whether it is 0-MRA or 1-MRA) over CRA in terms of the decomposition of Boolean functions. Logic circuits to realize all NPN-classified Boolean functions using CRA and MRA (the simplest decomposition from either 1-MRA or 0-MRA) are given in Table 7.2.

As can be observed that while the output block in CRA decomposition is the set-theoretic intersection operation ($\cap$), the output block in 1-MRA decomposition is the Boolean AND operation and the output block in 0-MRA decomposition is the Boolean OR operation. The importance of this is that both the AND and OR Boolean operations are directly realizable in current binary technologies such as CMOS, while the set-theoretic intersection operation ($\cap$) is only realizable in technologies that implement ternary logic.

While the log-functionality complexity measure that is used in Table 7.1 is a good cost measure for machine learning, it is not in general a good measure to measure the cost for the purpose of circuit design. An alternative acceptable cost measure for circuit design will be the count of the total number of two-input gates in the final circuit ($C_\#$).

Table 7.3 presents an initial evaluation for MRA using the $C_\#$ complexity measure. One can note that, for all of the classes of NPN-classified 3-variable Boolean functions in Table 7.3, the total $C_\#$ for all of the MRA Boolean circuits with inverters is 20, while the total $C_\#$ for all of the MRA Boolean circuits without inverters is 19, and this result is not surprising since usually Boolean circuits with inverters are more complex than the same Boolean circuits without counting inverters. While the results in Table 7.3 are technology-independent, the same results that are obtained in Table 7.3 can be viewed from technology-dependent point of view as well. This is because while the realization of certain two-input logic primitives (gates) from Fig. G.3 (in Appendix G) needs less number of physical primitives (devices) in certain types of technologies, the same gates can need more number of devices in other types of technologies.

Table 7.2. Conventional RA (CRA) circuits versus Modified RA (MRA) circuits for the decomposition of all NPN-classes of 3-variable Boolean functions.

NPN Representative Function	Simplest MRA Circuit	Simplest CRA Circuit
Class 1 (8) $F = x_1x_2 + x_2x_3 + x_1x_3$	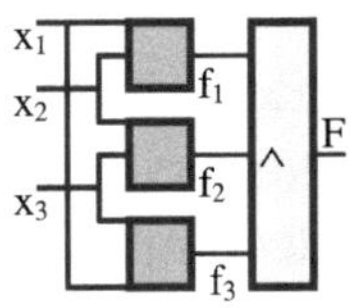	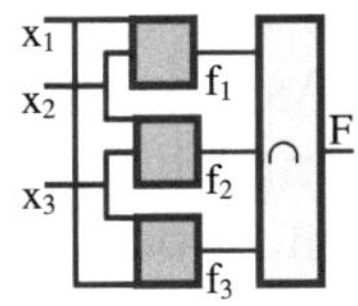
Class 2 (2) $F = x_1 \oplus x_2 \oplus x_3$	non-decomposable	non-decomposable
Class 3 (16) $F = x_1 + x_2 + x_3$	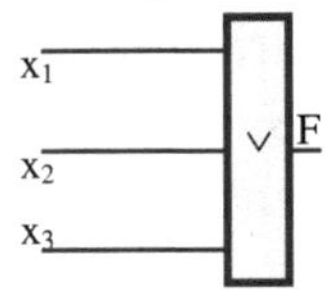	non-decomposable
Class 4 (48) $F = x_1(x_2 + x_3)$	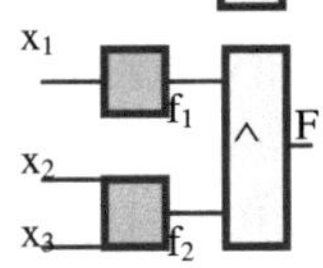	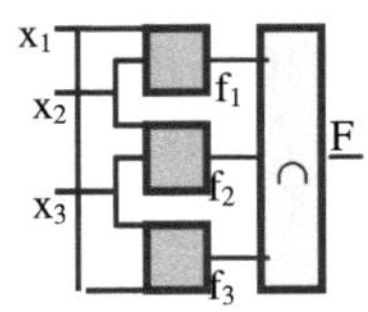
Class 5 (8) $F = x_1x_2x_3 + x_1'x_2'x_3'$	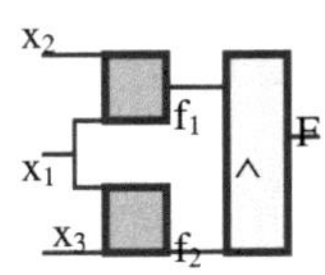	non-decomposable
Class 6 (24) $F = x_1'x_2x_3 + x_1x_2' + x_1x_3'$	non-decomposable	non-decomposable
Class 7 (24) $F = x_1(x_2x_3 + x_2'x_3')$	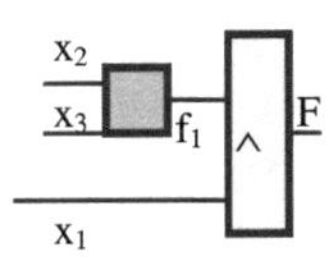	non-decomposable
Class 8 (24) $F = x_1x_2 + x_2x_3 + x_1'x_3$	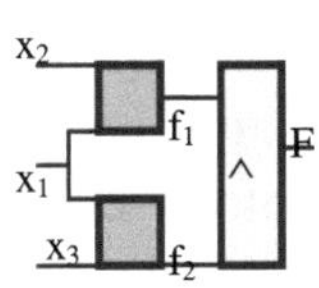	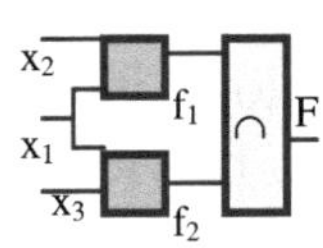
Class 9 (16) $F = x_1'x_2x_3 + x_1x_2'x_3 + x_1x_2x_3'$	non-decomposable	non-decomposable
Class 10 (48) $F = x_1x_2'x_3' + x_2x_3$	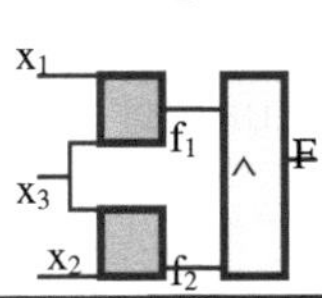	non-decomposable

Table 7.3. Evaluating two-valued MRA circuits using $C_{\#}$ cost measure.

Class	$C_{\#}$ with inverters (MRA)	$C_{\#}$ without inverters (MRA)
1	5	5
2	-	-
3	1	1
4	2	2
5	3	3
6	-	-
7	2	2
8	4	3
9	-	-
10	3	3

7.2 Multiple-Valued MRA

Real-life data are in general many-valued. Consequently, if MRA can decompose relations between many-valued variables it can have practical applications in machine learning (ML) and data mining (DM). Many-valued MRA is made up of two main steps which are common to two equivalent (intersection-based and union-based) algorithms as follows:

(1) partition the many-valued truth table into sub-tables, each contain only single functional value.

(2) Perform CRA on all sub-tables. Figure 7.5 illustrates the general pre-processing procedure for the two many-valued MRA algorithms, which will be explained in more detail below.

For an "n"-valued completely specified function one needs (n-1) values to define the function (i.e., to be able to retrieve the completely specified function without any loss of information). We thus do all n decompositions and use for our MRA model the (n-1) simplest of these. Figure 7.5 illustrates the general flow diagram of the multiple-valued MRA decomposition using the pre-processing steps that are common to both of the intersection-based and union-based algorithms.

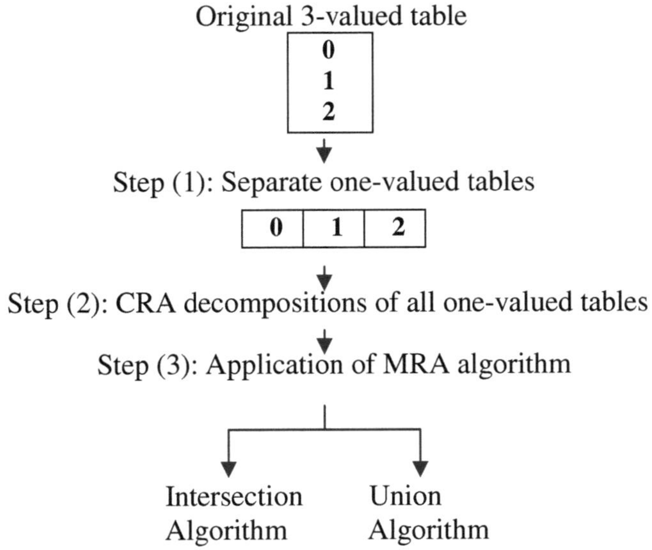

Fig. 7.5. Steps for many-valued MRA.

For example, using the lattice-of-structures, decompose the 3-valued function for each individual value. One then obtains the simplest lossless MRA decomposition for value “0” of the function (denoted as the 0-MRA decomposition), for value “1” (1-MRA decomposition), and for value “2” (2-MRA decomposition). By selecting the simplest two models from these 0-MRA, 1-MRA, and 2-MRA decompositions, one can generate the complete function.

In the intersection method, first CRA decompositions are expanded to include the full set of variable and function values, and these expanded decompositions are then intersected to yield the original table. Equivalently, one can use a union operation to generate the corresponding many-valued MRA as follows: **(1)** decompose the original table (function or relation) into sub-tables for each output value: e.g., $T = T_0 \cup T_1 \cup T_2$ for the corresponding output values O_0, O_1, and O_2 respectively, **(2)** do the 3-valued CRA decomposition on each sub-table. Let M_j be the decomposition of T_j, and **(3)** the reconstructed function or relation (T^*) is the union of all the sub-table decompositions, $T^* = \bigcup_{j=0}^{n-1} M_j \otimes O_j$, where ⊗ is the set-theoretic Cartesian product. The union procedure can also be done with (n-1) decompositions.

The following are two examples which illustrate many-valued MRA of three-valued logic functions. In the first example MRA can decompose the function for only two values, and one has no choice but to use both of these decompositions in the MRA model. In the second example, the function is decomposable for all three of its values, and the two simplest decompositions are chosen to define the model. In discussing the second example, we show that this approach is generalizable to set-theoretic relations, in addition to mappings.

Example 7.3. For the following ternary map:

X_1X_2 \ X_3	0	1	2
00	0	0	0
01	1	1	0
02	1	1	1
10	0	0	2
11	0	0	2
12	1	1	1
20	0	2	0
21	1	1	0
22	2	2	0

F

The following is the many-valued MRA intersection algorithm.

Step 1. decompose the ternary chart of the function into three separate tables each for a single function value. This will produce the following three sub-tables.

Value "0"	Value "1"	Value "2"
000	010	102
001	011	112
002	020	201
012	021	220
100	022	221
101	120	
110	121	
111	122	
200	210	
202	211	
212		
222		
D0	**D1**	**D2**

Step 2. Perform CRA for each sub-table.

Step 2a. The simplest error-free 0-MRA decomposition is the original "0"-subtable itself since it is not decomposable.

Step 2b. 1-MRA decomposition of D1 is as follows:

Table 1			**Table 2**	
X_1	X_2	:	X_2	X_3
0	1		1	0
0	2		1	1
1	2		2	0
2	1		2	1
			2	2
D11			**D12**	

Step 2c. The 2-MRA decomposition of D2 is as follows:

Table 3			**Table 4**	
X_1	X_3	:	X_2	X_3
1	2		0	2
2	1		1	2
2	0		0	1
			2	0
			2	1
D21			**D22**	

THE INTERSECTION ALGORITHM

Step 3.1. Select the two simplest error-free decomposed models. In this particular example, these are 1-MRA and 2-MRA decompositions. MRA thus gives the decomposition model of D11:D12:D21:D22 from which the original function can be reconstructed as follows.

Step 3.2. Note that, for Tables 1 and 2, the MRA decomposition is for the value "1" of the logic function. Therefore, the existence of the tuples in the decomposed model implies that the function has value "1" for those tuples, and the non-existence of the tuples in the decomposed model implies that the function does not have value "1" but "0" or "2" for the non-appearing tuples. This is shown in Tables 1' and 2', respectively. Similarly note that, for Tables 3 and 4, the MRA decomposition is for the value "2" of the logic function. Therefore, the existence of the tuples in the decomposed model implies that the function has value "2" for those tuples, and the non-existence of the tuples in the decomposed model implies that the function does not have value "2" but "0" or "1" for the non-appearing tuples. This is shown in Tables 3' and 4', respectively.

Table 1'			:	Table 2'		
X_1	X_2	F_1	:	X_2	X_3	F_2
0	0	0,2		0	0	0,2
0	1	1,0,2		0	1	0,2
0	2	1,0,2		0	2	0,2
1	0	0,2		1	0	1,0,2
1	1	0,2		1	1	1,0,2
1	2	1,0,2		1	2	0,2
2	0	0,2		2	0	1,0,2
2	1	1,0,2		2	1	1,0,2
2	2	0,2		2	2	1,0,2

Table 3'			:	Table 4'		
X_1	X_3	F_3	:	X_2	X_3	F_4
0	0	0,1		0	0	0,1
0	1	0,1		0	1	2,0,1
0	2	0,1		0	2	2,0,1
1	0	0,1		1	0	0,1
1	1	0,1		1	1	0,1
1	2	2,0,1		1	2	2,0,1
2	0	2,0,1		2	0	2,0,1
2	1	2,0,1		2	1	2,0,1
2	2	0,1		2	2	0,1

In Tables 1' and 2' (i.e., the decomposition for value "1" of the function), the existence of value "1" (of sub-relations F_1 and F_2) means that the value "1" appeared in the original non-decomposed function for the corresponding tuples that appear in each table, but *does not imply* that the values "0" or "2" (of sub-relations F_1 and F_2) did not exist in the original non-decomposed function for the same tuples. Therefore "0" and "2" are added to "1" as allowed values. In the remaining tuples, however, only "0" and "2" are allowed since the value "1" did not occur. Similarly, in Tables 3' and 4', the existence of the value "2" (of sub-relations F_3 and F_4) means that the value "2" appeared in the original non-decomposed function for the corresponding tuples that appear in each table, but *does not imply* that values "0" or "1" did not exist in the original non-decomposed function for the same tuples. Therefore "0" and "1" are added to "2" as allowed values. In the remaining tuples, however, only "0" and "1" are allowed since the value "2" did not occur. Set-theoretically, obtaining Tables 1', 2', 3', and 4' from Tables 1, 2, 3, and 4 is described as follows where ′ here means complement.

Table 1': (D11⊗(0,1,2))∪(D11′⊗(0,2)),
Table 2': (D12⊗(0,1,2))∪(D12′⊗(0,2)),
Table 3': (D21⊗(0,1,2))∪(D21′⊗(0,1)),
Table 4': (D22⊗(0,1,2))∪(D22′⊗(0,1)),

Step 3.3. Tables 1', 2', 3', and 4' are used to obtain the block diagram in Fig. 7.6, where the following set-theoretic Eqs. govern the outputs of the levels in the circuit shown in the Fig.: F = F5 ∩ F6, F5 = F1 ∩ F2, F6 = F3 ∩ F4, where F1 is given by Table 1', F2 by Table 2', F3 by Table 3', and F4 by Table 4', respectively.

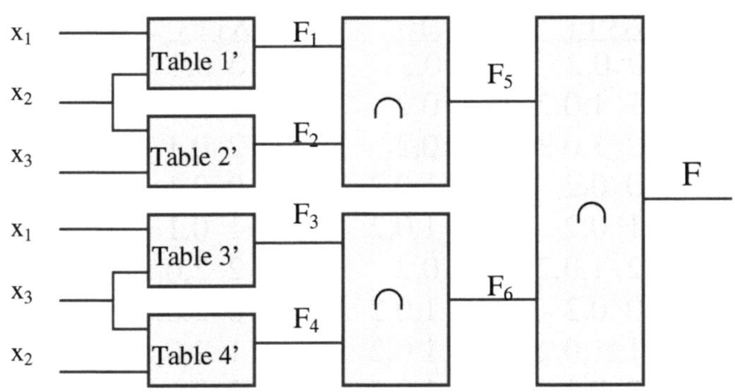

Fig. 7.6. The resulting decomposed structure by applying the multi-valued MRA decomposition.

The intermediate sub-functions, F5 and F6 are shown in the following maps, respectively.

x_1 \ x_2x_3	00	01	02	10	11	12	20	21	22
0	0,2	0,2	0,2	1	1	0,2	1	1	1
1	0,2	0,2	0,2	0,2	0,2	0,2	1	1	1
2	0,2	0,2	0,2	1	1	0,2	0,2	0,2	0,2

$F_5 = F_1 \cap F_2$

x_1 \ x_2x_3	00	01	02	10	11	12	20	21	22
0	0,1	0,1	0,1	0,1	0,1	0,1	0,1	0,1	0,1
1	0,1	0,1	2	0,1	0,1	2	0,1	0,1	0,1
2	0,1	2	0,1	0,1	0,1	0,1	2	2	0,1

$F_6 = F_3 \cap F_4$

Note that in Fig. 7.6 the intersection blocks in the second level and the intersection block at the third (output) level, are general and do not depend on the function being decomposed. Only the tables at the first level depend upon this function.

THE UNION ALGORITHM

Steps 1 and 2 are the same as in the intersection algorithm.
Step 3.1. Using the decomposition model D11:D12:D21:D22 obtain D1 and D2 by standard methods as follows:
$D1 = (D11 \otimes x3) \cap (D12 \otimes x1)$,
$D2 = (D21 \otimes x2) \cap (D22 \otimes x1)$,
$D0 = (D1 \cup D2)'$,
where D1 is the decomposition for function value "1", D2 for function value "2", and x1, x2, and x3 $\in \{0,1,2\}$.
Step 3.2. Perform the set-theoretic operations to obtain the total function from the decomposed sub-functions.
$x1x2x3F = (D1 \otimes 1) \cup (D2 \otimes 2) \cup ((D1 \cup D2)' \otimes (1 \cup 2)')$,
$= (D1 \otimes 1) \cup (D2 \otimes 2) \cup ((D1 \cup D2)' \otimes 0)$.
Alternatively, one can use all three decompositions:
$x1x2x3F = (D0 \otimes 0) \cup (D1 \otimes 1) \cup (D2 \otimes 2)$.

The function value of (x1,x2,x3) is determined by the following block diagram, where G performs the following operation:
F = 0 if (x1x2x3) $\in$ D0,
F = 1 if (x1x2x3) $\in$ D1,
F = 2 if (x1x2x3) $\in$ D2.

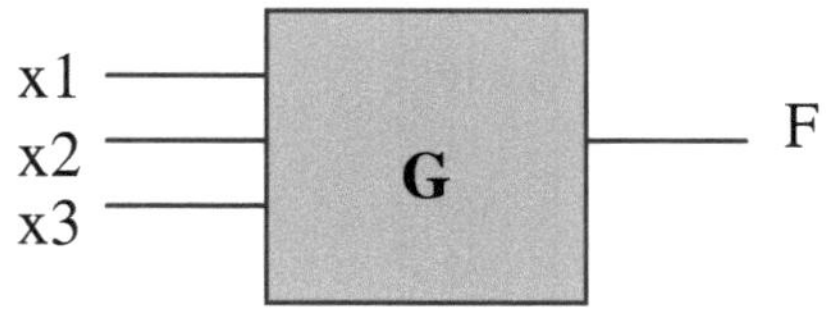

Note that the logic function in Example 7.3 is non-decomposable using CRA. Consequently, as can be seen from this example and analogously to the binary case, the new many-valued MRA is superior to CRA.

We now consider an example where CRA does decompose, and also where MRA decomposes for all three values.

Example 7.4. Let us generate the MRA decomposition for the ternary function specified by the following ternary map:

X_1X_2 \ X_3	0	1	2
0 0	0	0	0
0 1	1	1	1
0 2	1	1	1
1 0	0	0	2
1 1	0	0	2
1 2	1	1	1
2 0	0	2	0
2 1	1	1	1
2 2	2	2	0

F

Utilizing the intersection-based algorithm, one obtains the following results for MRA for the ternary function in Example 7.4.

Step 1. decompose the ternary chart of the function into three separate tables each for a single function value. This will produce the following three sub-tables.

Value "0"	Value "1"	Value "2"
000	010	102
001	011	112
002	012	201
100	020	220
101	021	221
110	022	
111	120	
200	121	
202	122	
222	210	
	211	
	212	
D0	**D1**	**D2**

Step 2. Perform CRA for each sub-table.

Step 2a. The 0-MRA decomposition of D0 is as follows:

Table 1 Table 2 Table 3

X_1X_2 :	X_2X_3 :	X_1X_3
0 0	0 0	0 0
1 0	0 1	0 1
1 1	0 2	0 2
2 0	1 0	1 0
2 2	1 1	1 1
	2 2	2 0
		2 2
D01	**D02**	**D03**

Step 2b. The 1-MRA decomposition of D1 is as follows:

Table 4 Table 5

X_1	X_2 :	X_3
0	1	0
0	2	1
1	2	2
2	1	
D11		**D12**

Step 2c: The 2-MRA decomposition of D2 is as follows:

Table 6 Table 7

X_1	X_3 :	X_2	X_3
1	2	0	2
2	1	1	2
2	0	0	1
		2	0
		2	1
D21		**D22**	

THE INTERSECTION ALGORITHM

Step 3.1. Select the two simplest decomposed models, namely 1-MRA and 2-MRA decompositions. These are at a lower level in the lattice of structures than 0-MRA.

Step 3.2. Analogously to Example 7.3, one obtains the following expanded tables:

Table 4'			**Table 5'**		**Table 6'**			**Table 7'**		
X_1	X_2	F_1 :	X_3	F_2	X_1	X_3	F_3 :	X_2	X_3	F_4
0	0	0,2	0	1,0,2	0	0	0,1	0	0	0,1
0	1	1,0,2	1	1,0,2	0	1	0,1	0	1	2,0,1
0	2	1,0,2	2	1,0,2	0	2	0,1	0	2	2,0,1
1	0	0,2			1	0	0,1	1	0	0,1
1	1	0,2			1	1	0,1	1	1	0,1
1	2	1,0,2			1	2	2,0,1	1	2	2,0,1
2	0	0,2			2	0	2,0,1	2	0	2,0,1
2	1	1,0,2			2	1	2,0,1	2	1	2,0,1
2	2	0,2			2	2	0,1	2	2	0,1

Set-theoretically, obtaining Tables 4'-7' is as follows:

Table 4': (D11⊗(0,1,2))∪(D11'⊗(0,2)),

Table 5': (D12⊗(0,1,2))∪(D12'⊗(0,2)),

Table 6': (D21⊗(0,1,2))∪(D21'⊗(0,1)),

Table 7': (D22⊗(0,1,2))∪(D22'⊗(0,1)).

Step 3.3. Tables 4', 5', 6', and 7' are used to obtain Fig. 7.7, where F = F5 ∩ F6, F5 = F1 ∩ F2, F6 = F3 ∩ F4, and F1 is given by Table 4', F2 by Table 5', F3 by Table 6', and F4 by Table 7'.

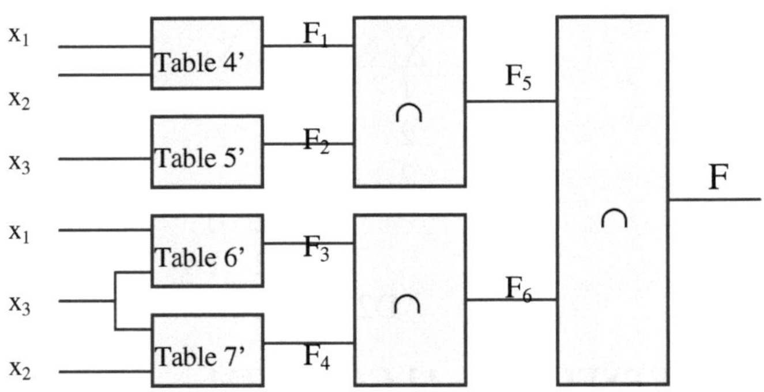

Fig. 7.7. The resulting decomposed structure by applying the multi-valued MRA.

The intermediate sub-functions, F5 and F6 are shown as follows.

x_1 \ x_2x_3	00	01	02	10	11	12	20	21	22
0	0,2	0,2	0,2	1	1	1	1	1	1
1	0,2	0,2	0,2	0,2	0,2	0,2	1	1	1
2	0,2	0,2	0,2	1	1	1	0,2	0,2	0,2

$F_5 = F_1 \cap F_2$

x_1 \ x_2x_3	00	01	02	10	11	12	20	21	22
0	0,1	0,1	0,1	0,1	0,1	0,1	0,1	0,1	0,1
1	0,1	0,1	2	0,1	0,1	2	0,1	0,1	0,1
2	0,1	2	0,1	0,1	0,1	0,1	2	2	0,1

$F_6 = F_3 \cap F_4$

THE UNION ALGORITHM

Steps 1 and 2 are the same as in the intersection algorithm.

Step 3.1. Using the decomposition model D01:D02:D11:D12:D21:D22 obtain D0, D1, and D2 by standard methods as follows:

D0 = (D01⊗x3)∩(D02⊗x1)∩(D03⊗x2),

D1 = (D11⊗x3)∩(D12⊗x1x2),

D2 = (D21⊗x2)∩(D22⊗x1),

where D0 is the decomposition for function value "0", D1 is for value "1", D2 for value "2", and x1, x2, and x3 ∈ {0,1,2}.

Step 3.2. Perform the set-theoretic operations to obtain the total function from the decomposed sub-functions. This can be done using only two of the three decompositions as in Step (3.2) of the union algorithm in Example 7.3, or alternatively, one can use all three decompositions as follows:

x1x2x3F = (D0⊗0)∪(D1⊗1)∪(D2⊗2).

The function value of (x1,x2,x3) is determined by the following block diagram, where G performs the following operation: F = 0 if (x1x2x3) ∈ D0, F = 1 if (x1x2x3) ∈ D1, and F = 2 if (x1x2x3) ∈ D2.

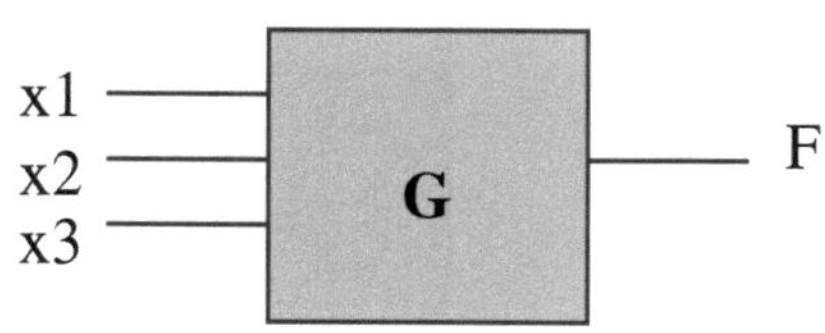

The logic function in Example 7.4 is decomposable using CRA with the lossless CRA model x_1x_2:x_2x_3:x_1x_3. Consequently, unlike the previous example, both many-valued MRA and CRA decompose losslessly. Since both CRA and MRA decompose this function, we would like to be able to compare the complexities of the two decompositions. The complexity measure used in Appendix G could be used, but needs to be extended to many-valued functions.

From the previous discussion, it follows that the extension of many-valued MRA from functions to relations is trivial. One just performs the union algorithm using all n decompositions, e.g., for three values (D0⊗0)∪(D1⊗1)∪(D2⊗2). One can observe that the set-theoretic formulation of multiple-valued MRA as the union of the individual sub-tables (e.g., $x1x2x3F = (D0 \otimes 0) \cup (D1 \otimes 1) \cup (D2 \otimes 2)$) is analogous to the algebraic formulation of multiple-valued Shannon expansion as the disjunction of the individual values of a logic function (e.g., Eq. (2.6): $f = {}^0x\, f_0 + {}^1x\, f_1 + {}^2x\, f_2$).

7.3 Reversible MRA

Reversible (3,3) gates, that are universal in two arguments, can be used for the construction of reversible MRA (RMRA) circuits. Figure 7.8 illustrates one example of a binary (3,3) reversible gate from [126], which is universal in two arguments.

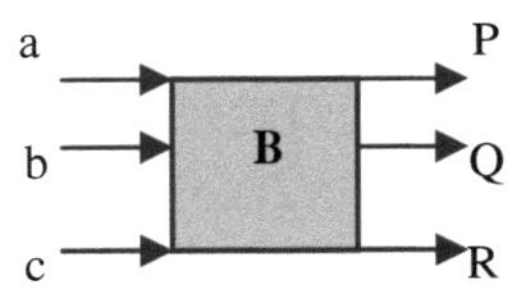

a

a	b	c	P	Q	R
0	0	0	1	1	0
0	0	1	1	0	1
0	1	0	1	0	0
0	1	1	0	1	1
1	0	0	0	1	0
1	0	1	0	0	0
1	1	0	1	1	1
1	1	1	0	0	1

b

Inputs	Outputs
a=0	**R=c, Q=(b⊕c)', P=b'+c'**
a=1	R=b, P=bc'
b=0	P=a'
b=1	R=a+c, Q=a⊕c, P=c'
c=0	R=ab, Q=a+b', P=a'+b
c=1	Q=a'b, P=a'b'
a=1, b=0	P=0
a=1,b=1	R=1

c

Fig. 7.8. **a** Diagram of the reversible (3,3) Boolean logic circuit, **b** truth table of this gate, and **c** proof of universality of the gate in two arguments.

The following example illustrates the use of the reversible gate in Fig. 7.8 for the synthesis of 1-MRA circuit for class 5. The 1-MRA decomposed Boolean circuit of class 5 can be realized using the binary (3,3) reversible circuit in Fig. 7.8b. This is done with the reversible circuit in Fig. 7.9, where blocks B1 and B2 are the reversible (3,3) gate from Fig. 7.8b, and block B3 is the reversible (3,3) Toffoli gate. For B3, c = 0 and thus B3 is a reversible logic AND gate.

Utilizing Fig. 7.8c, the Boolean reversible circuit in Fig. 7.9 implements the 1-MRA circuit of class 5 using the following input settings: $a = 0 \Rightarrow Q1 = f1 = (x1 \oplus x2)'$, $a = 0 \Rightarrow Q2 = f2 = (x1 \oplus x3)'$, and $F = Q1 \wedge Q2 = f1 \wedge f2 = (x1 \oplus x2)' \wedge (x1 \oplus x3)' = x1x2x3 + x1'x2'x3'$. Using similar substitutions with appropriate input values according to Fig. 7.8b, the reversible circuit in Fig. 7.9 can realize all 1-MRA circuits from classes 8 and 10, respectively.

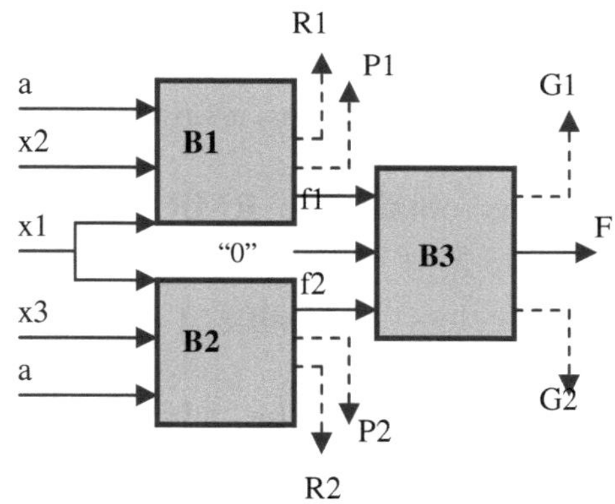

Fig. 7.9. Reversible (7,7) Boolean circuit that implements the 1-MRA circuit from class 5. Input {a} in B1 and B2 and the set of outputs {R1, P1, R2, P2, G1, G2} are needed for reversibility. Input {a} also selects the appropriate function value of which the universal B gate (Fig. 7.8b) is to implement.

The remaining classes can be realized using analogous techniques, by adding one more block from Fig. 7.8b to the first level of Fig. 7.9 in the case of class 1, and removing one block from the first level of Fig. 7.9 in the case of classes 4 and 7, respectively. Yet, as one can observe, the RMRA produces garbage in the outputs, and this output garbage has to be eliminated when the quantum counterpart of RMRA is created. To eliminate such garbage one needs to use the reversible inverse RMRA circuit as in Fig. 7.10.

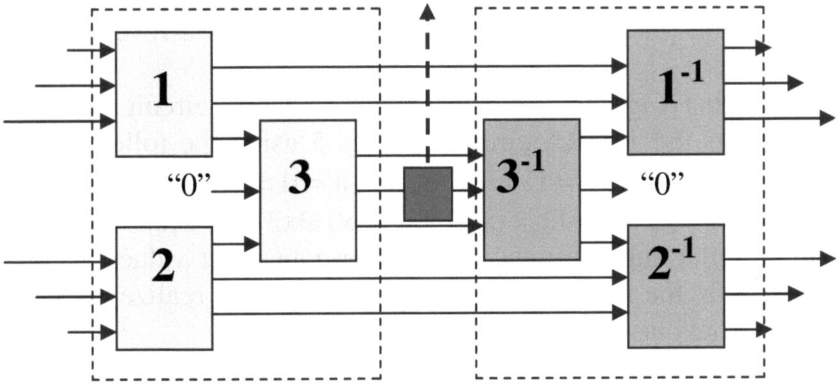

Fig. 7.10. Garbageless reversible MRA circuit.

Example.7.5. Let us obtain the RMRA circuit for the Boolean function in NPN class 5: $f = x_1x_2x_3 + \bar{x}_1\bar{x}_2\bar{x}_3$. Figure 7.11 illustrates the complete garbageless reversible 1-MRA for the function in Example 7.5.

The new MRA posseses the advantage of producing four distinct structures as shown in the middle column of Table 7.2, when decomposition occurs. Thus using one unified theory different topological structures are created, where such structures have to be created using separate decomposition methods in the case of other type of decompositions (such as in the case of Ashenhurst-Curtis (AC) based decompositions in Appendix H). As stated previously, this point can be of high importance when synthesizing logic circuits as it gives the designer more design options (i.e., larger design space) to choose from to meet specific design criteria, and this is an issue that needs further exploration.

The disadvantage of reversible MRA is that it creates big garbage, and thus the elimination of garbage in reversible MRA structures is done by using the reversible mirror image circuit of the forward reversible circuit as shown in Example 7.5. Another disadvantage of the current 1-MRA and 0-MRA decompositions is that MRA does not yet decompose for the ESOP-like Boolean function from NPN class 2, which is a very common and useful form in many applications in logic synthesis [217].

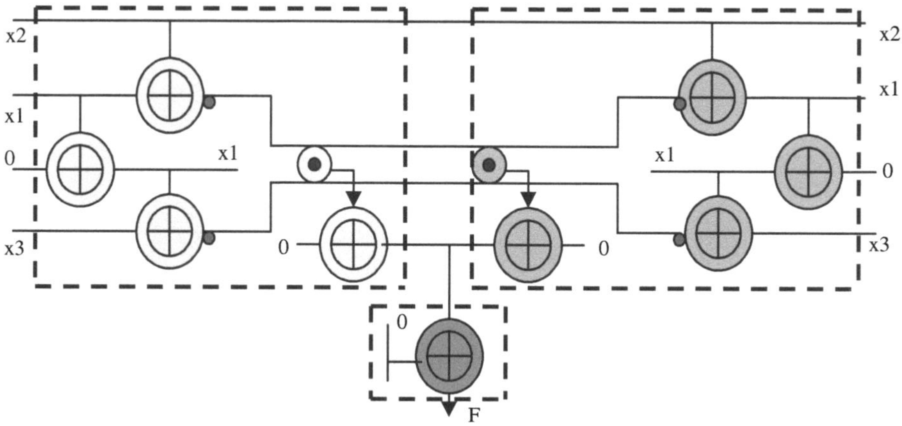

Fig. 7.11. Garbageless reversible 1-MRA circuit for NPN class 5 Boolean function.

7.4 Summary

A novel RA-based decomposition is introduced. The new decomposition is called the Modified Reconstructability Analysis (MRA). It is shown that in 4 out of 10 NPN classes while 3-variable NPN-classified Boolean functions are not decomposable using the Conventional Reconstructability Analysis (CRA) decomposition, they are decomposable using the Modified Reconstructability Analysis (MRA) decomposition. For the purpose of binary circuit design, it has been shown also that, by counting the total number of two-input gates, MRA is superior to CRA for both cases when including the cost of the inverters and when not including the cost of the inverters. The multiple-valued MRA has been also introduced. The reversible realization of the MRA has been introduced as a first step towards the quantum computation of such reversible structures.

8 New Reversible Structures: Reversible Nets, Reversible Decision Diagarams, and Reversible Cascades

In Chapts. 6 and 7 two reversible decompositions were created: reversible lattice structures and reversible Modified Reconstructability Analysis. While these reversible structures exhibit certain regularities, yet they generate a big “garbage” at the output which requires the use of the mirror image (i.e., inverse) reversible circuit to eliminate such “garbage”. This Chapt. will provide a variety of new reversible structures, which can possess advantages that the previous reversible structures did not have like the use of minimal garbage or no garbage at all in some cases. Since garbage is a big issue in reversible logic synthesis, search heuristics for reversible logic synthesis should include the following: (1) do not create many outputs of gates and sub-circuits, (2) re-use these outputs as inputs in other gates or sub-circuits, (3) apply re-usability properties of these common sub-functions and symmetry is one of such properties, (4) the method must be general, and (5) use regularity and algebraic properties (e.g., group, field, or linear properties) to create more powerful reversible structures. The new contributions of this Chapt. are:

- The creation of a new reversible structure, that uses the symmetry indices that were presented in Chapt. 4, called reversible Nets.
- The creation of new reversible Decision Trees and Diagrams that use the form of trees and diagrams to realize reversibly the corresponding logic functions.
- The invention of a multiple-valued reversible Cascade structures that show efficiency in the reversible synthesis of logic functions.

The remainder of this Chapt. is organized as follows. Reversible Nets are presented in Sect. 8.1. Reversible Decision Diagrams are presented in Sect. 8.2. Binary and multiple-valued reversible Cascades are presented in Sect. 8.3. Chapter Summary is presented in Sect. 8.4.

8.1 Reversible Nets

The basic idea of reversible Net structures [183,185] made up of (2,2) reversible gates is based on regular planes as shown in Fig. 8.1.

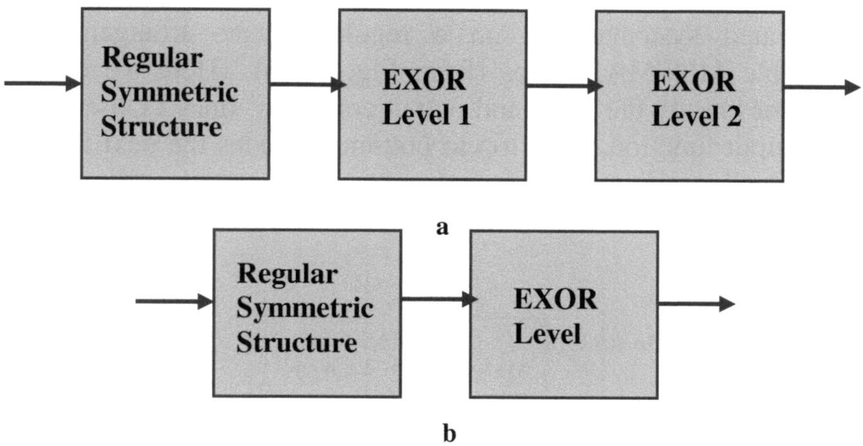

Fig. 8.1. a Three plane regular structure to realize arbitrary multi-input, multi-output Boolean functions using MIN/MAX reversible gates, and **b** an equivalent two plane regular structure that realizes arbitrary multi-input multi-output Boolean functions using MIN/MAX reversible gates.

The first plane from left in Fig. 8.1b is a levelized triangular structure in which the input variables correspond to the columns (we will call this plane the triangular plane). In contrast to lattices however, this structure, when realizing an arbitrary multi-output function with geometrically adjacent output signal, does not require variable repetition. The structure of the first plane is planar, regular, and algorithmically created. It realizes all positive unate symmetric functions of its input variables. The second structure plane in Fig. 8.1b is a plane of Feynman gates that uses their internal EXOR gates to realize every output function as an EXOR of single index symmetric functions from plane two. This plane can be compared in its functionality to the OR plane in the standard AND/OR PLA that is used to realize a Sum-Of-Product (SOP; DNF) expression. Because the functions on the output of the second plane are disjoint as single index functions, the OR of them is the same as the EXOR of them (i.e., this is based on the Boolean law: $A + B = A \oplus B \oplus$

AB, and thus when A and B are disjoint functions we obtain: AB = 0 and thus A + B = A ⊕ B).

The whole idea is thus based on the well known fact that every symmetric function can be realized as OR or EXOR of its single value symmetric functions (as was illustrated in Sect. 4.1). Figure 8.2 illustrates how positive polarity unate symmetric functions can be created systematically in a regular planar arrangement of reversible MIN/MAX gates (from Fig. 5.19a). (Here we don not show for brevity the input and output constants' lines.) Observe that each ouput function, from top to bottom, includes the next function and are all positive polarity and unate. The sets of indices of the adjacent functions differ by one.

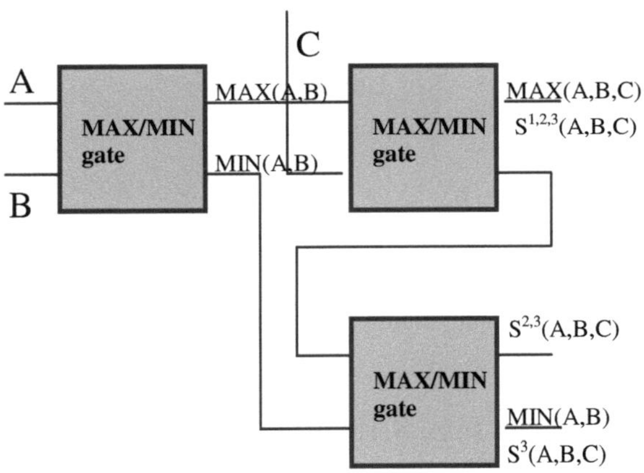

AB \ C	0	1
00	S^0	S^1
01	S^1	S^2
11	S^2	S^3
10	S^1	S^2

b

Fig. 8.2. Example of the realization of three-variable symmetric functions in the left plane from Fig. 8.1: **a** regular structure from reversible MAX/MIN gates that realizes positive polarity unate symmetric functions, and **b** indices of a symmetric function of three variables A, B, and C, where each cell includes an index of a symmetric function that corresponds to it, for instance function $S^{2,3}$(A,B,C) will have ones in cells with indices 2 and 3 and zeros in cells with indices 0 and 1.

Let us observe that positive unate symmetric functions that are generated on the outputs of the triangular plane have a very nice property: the EXOR of the neighbor functions creates a single index symmetric function. This is illustrated in Fig. 8.3. However, because the EXOR gate is not reversible, we have to complete it with a Feynman gate by repeating one of its inputs to the output. Because our structure is regular, this does not complicate the structure. In result, we obtain a structure such as the one shown in Fig. 8.4 for four variables.

As one observes, in this regular reversible Net structure, we obtain not only the single index symmetric functions, but also some interval functions whose parameters are highly correlated to the neighboring single index functions. We have also fan-out gates in the second plane as shown in Fig. 8.4. These fan-out gates (in the second plane of Fig. 8.4) uses Feynman gates (cf. Fig. 5.4d) with value "0" at the control line (input) to generate copies of the desired output signals from the first triangular plane.

Counts to characterize the complexity for this regular reversible Net structure are presented in Appendix I. Similarly to the argument presented for the case of lattice structures in Sect. 4.2 of Chapt. 4, evaluating the worst case for non-symmetric functions is difficult because we do not know yet how many times variables should be repeated in the process of symmetrization that is used to transform a non-symmetric function to a symmetric counterpart. Symmetrization however is a difficult problem for which no efficient algorithms have been created so far. When the input variables (in this particular case the inputs to the Net structure) have different polarities, the problem becomes even more complex, especially that repeated variables can have various distinct polarities.

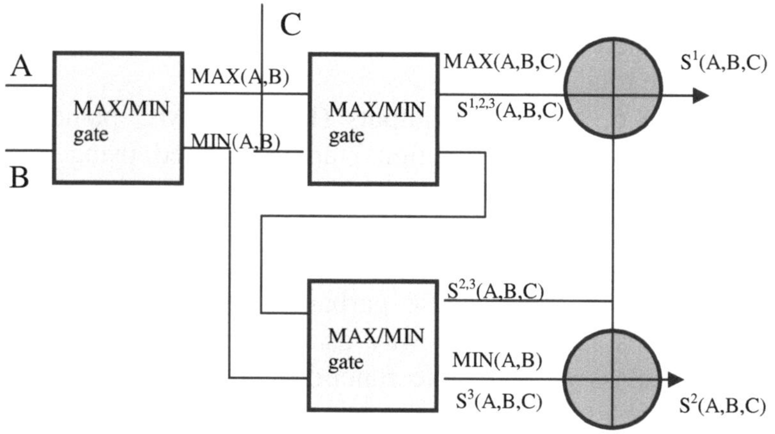

Fig. 8.3. Illustration of how a single index symmetric functions can be realized by an additional column of EXORs.

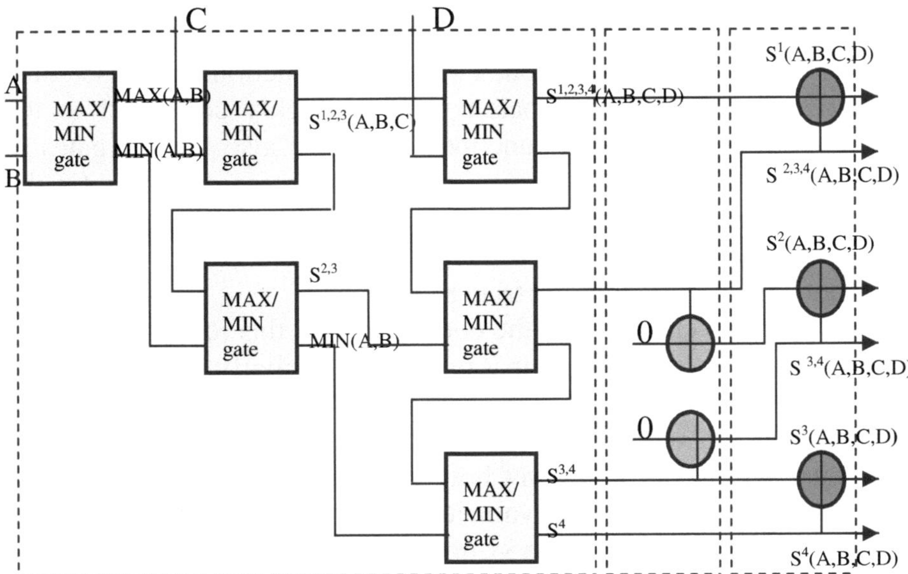

Fig. 8.4. Realization of all single index symmetric functions using only reversible gates.

It can be observed from Figs. 8.3 and 8.4 that the reversible Net structures produce garbage in their outputs. To eliminate this garbage, the reversible forward Net structure has to be serially interconnected with the reversible inverse Net structure as illustrated in Fig. 8.5, where the spy circuit is used to measure the intermediate outputs in the total network.

In Fig. 8.6, horizontal outputs (from the MC plane positive unate symmetric (PUS) functions) are EXOR-ed using Feynman gates in the right plane to create arbitrary symmetric functions at the bottom. Additional garbage outputs of MC gates must be forwarded to the primary outputs, shown in Fig. 8.6 as bold arrows from cells in the upper row only. These garbage outputs can be inputs to Feynman gates in the same way as the horizontal outputs. This extends the class of realizable functions in the structure using no repeated variables.

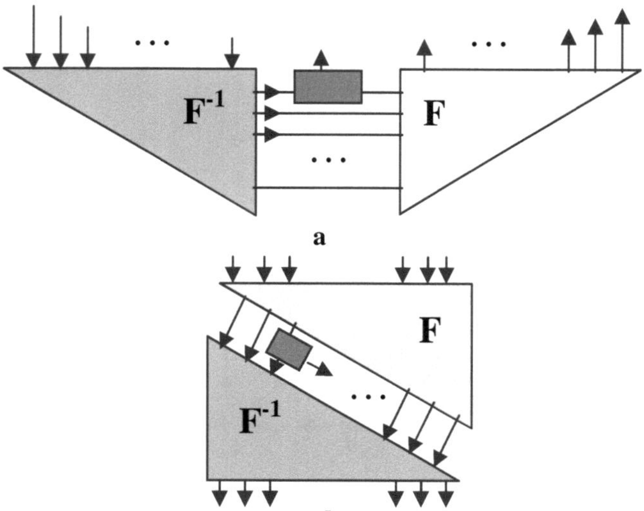

Fig. 8.5. Garbageless reversible net structure for 4-variable functions: **a** initial circuit, and **b** rectangular reversible net made up of one triangular reversible forward Net and one triangular reversible inverse Net.

Figure 8.6 illustrates the use of MC gates in contrast to the use of the reversible MIN/MAX gates (from Fig. 5.19a) to reversibly realize symmetric functions. The control signals for the first triangular plane that contains MC gates and the second plane that contains Feynman gates come from memory. These signals can take values of “0” or “1”. Feynman gates in the second plane can be alternatively replaced with Toffoli gates with control signals coming from memory with values “0” or “1”. This type of structure is semi-reversible because the memory from which the control signals come is not reversible, while the data path is reversible. The programmability of the Reversible Programmable Gate Array (RPGA) is done in the second plane in the form of interconnecting or disconnecting the gates in the columns in the second rectangular plane to achieve certain symmetric functions at the output of the second plane from which any other function can be synthesized.

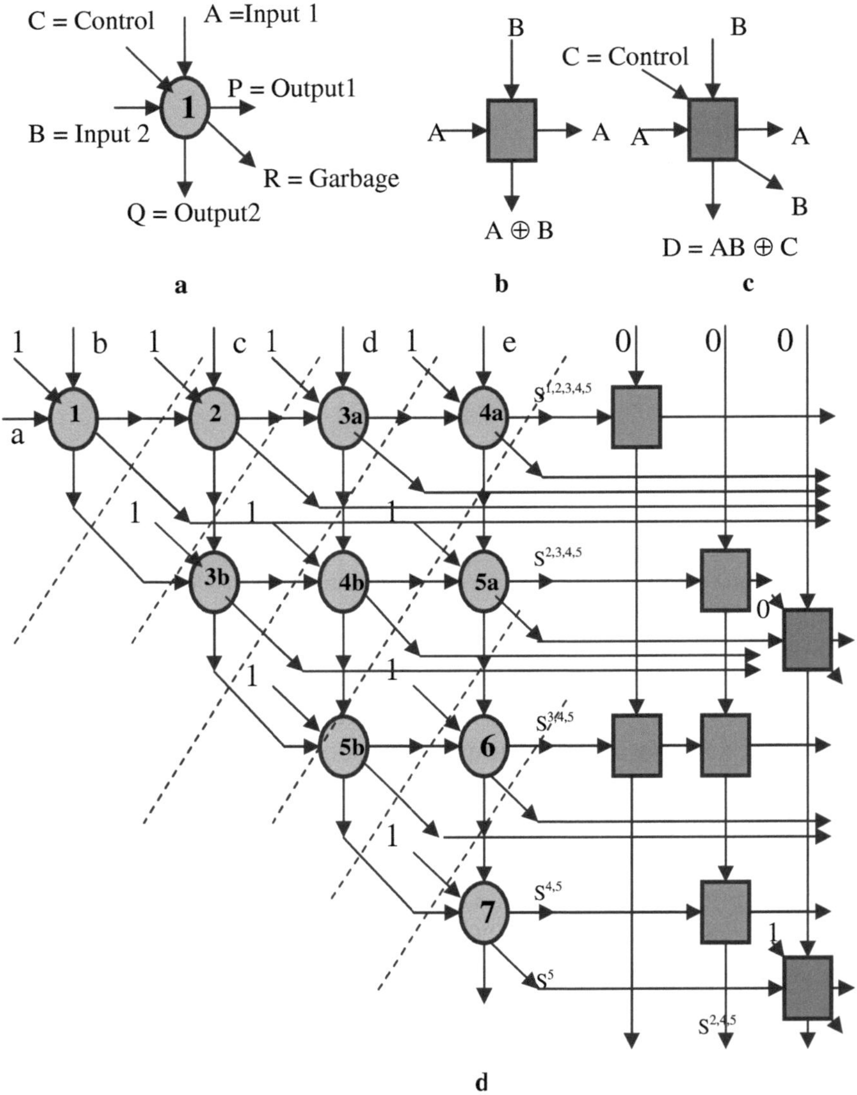

Fig. 8.6. Reversible Programmable Gate Array (RPGA): **a** the notation for MC gate used as a cell in the first plane, **b** the notation for Feynman gate used as a cell in the second plane, **c** symbol for Toffoli gate, and **d** an example of the realization of a 5-input, 2-output function ($S^{1,2}$(a,b,c,d,e), $S^{2,4,5}$(a,b,c,d,,e)) in RPGA.

8.2 Reversible Decision Diagrams

Because some decompositions require complex data processing to find a high quality solution, large multi-output functions should be first partitioned to smaller functions. This is based on their representations such as Binary Decision Diagrams, Pseudo-Kronecker Decision Diagrams (PKDD), Pseudo-Kronecker Diagrams with Complemented edges, Linearly Transformed Binary Decision Diagrams, Function-Driven Decision Diagrams, or other similar diagrams [180,181,184,189].

The goal of representing functions in such a representation is to find the "natural" structure of the function helpful for its subsequent partitioning to blocks of logic and subsets of variables. In PKDD a control variable goes through an entire level of a diagram. Therefore, a fast natural mapping from a PKDD into a reversible netlist with Toffoli gates exists. First a PKDD, or other similar diagram is mapped to Feynman, Fredkin, MC, and Toffoli gates and inverters, in such a way that there are no feedback loops and no fan-out larger than one from primary inputs and gates. In this Sect. the synthesis of reversible PKDDs with Toffoli gates are discussed, but similar approaches can be done with other types of diagrams as well.

After the mapping, every Toffoli gate in the mapped circuit can be in one of the following states: (1) no garbage outputs, (2) one garbage output, and (3) two garbage outputs. Thus a group of Toffoli gates that are controlled by the same control variable can have some percentage p of garbage outputs. T_j are different assigned values for various gate types. If this value p is higher than a certain threshold value T_j, the group is redesigned, otherwise it is retained. This is illustrated in Figs. 8.7 and 8.8.

The groups of Toffoli gates that are generated in this way create a partitioning of the initial circuit based on PKDD into blocks. Similar synthesis methods for various reversible Decision Diagrams have been shown in [184] in which one utilizes the signals from previous levels and inputs them into the next levels in a way that minimizes garbage and maximizes the use of the out functions.

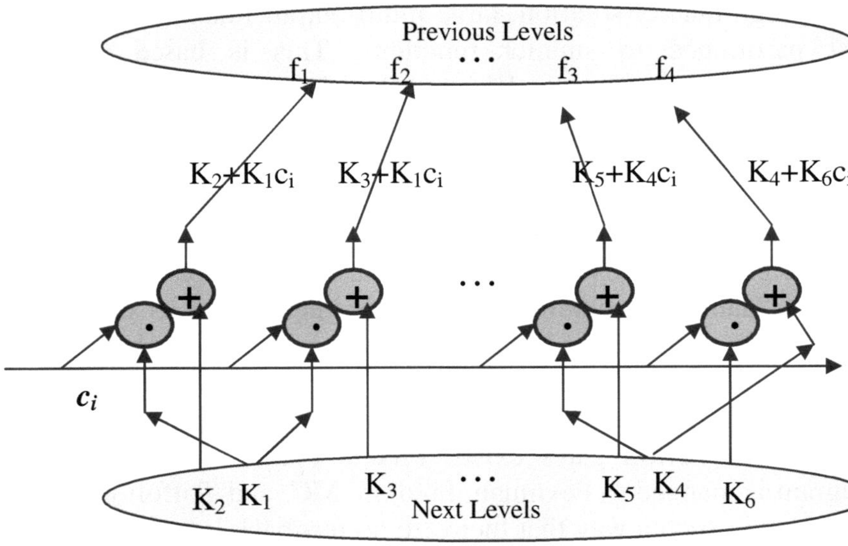

Fig. 8.7. A part of PKDD with positive Davio expansions in a level.

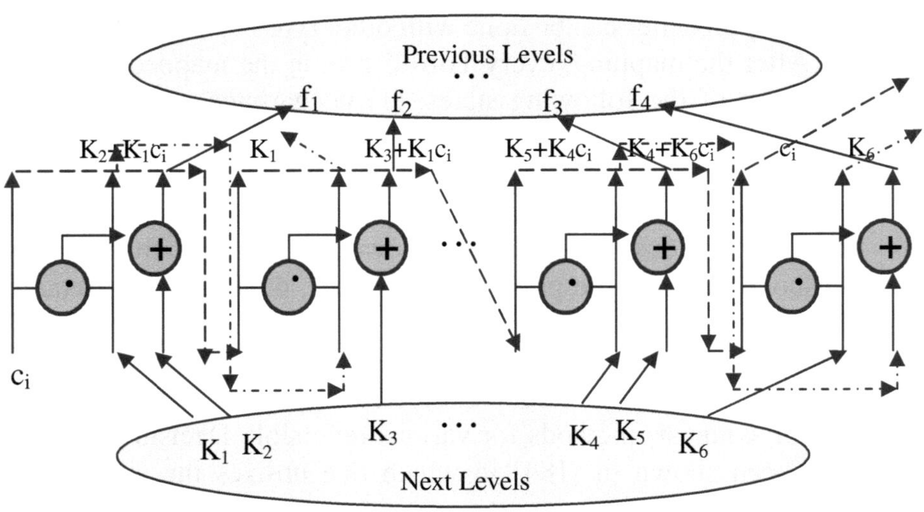

Fig. 8.8. A part of PKDD from Fig. 8.7 after the mapping into Toffoli gates. This mapping determined natural partitionaing to previous levels, next levels, and other gates of this level. Garbage outputs are drawn in interrupted lines.

This is done by a good selection of reversible gates such as Feynman, Inverters, Fredkin, Toffoli, or MC primitives. This type of realization produces garbage, thus the forward reversible circuit should be cascaded with the inverse (mirror image) reversible circuit to eliminate the resulting garbage outputs.

Example 8.1. This example shows the realization using reversible BDD (RBDD), for the Boolean function: F = ab ⊕ bc ⊕ ac, as shown in Fig. 8.9.

The reversible decision diagrams as shown in this Sect. produces garbage in the outputs. Consequently, to eliminate such garbage one needs to synthesize the reversible inverse decision diagram and then interconnect it to the reversible forward decision diagram. This is shown in Fig. 8.10.

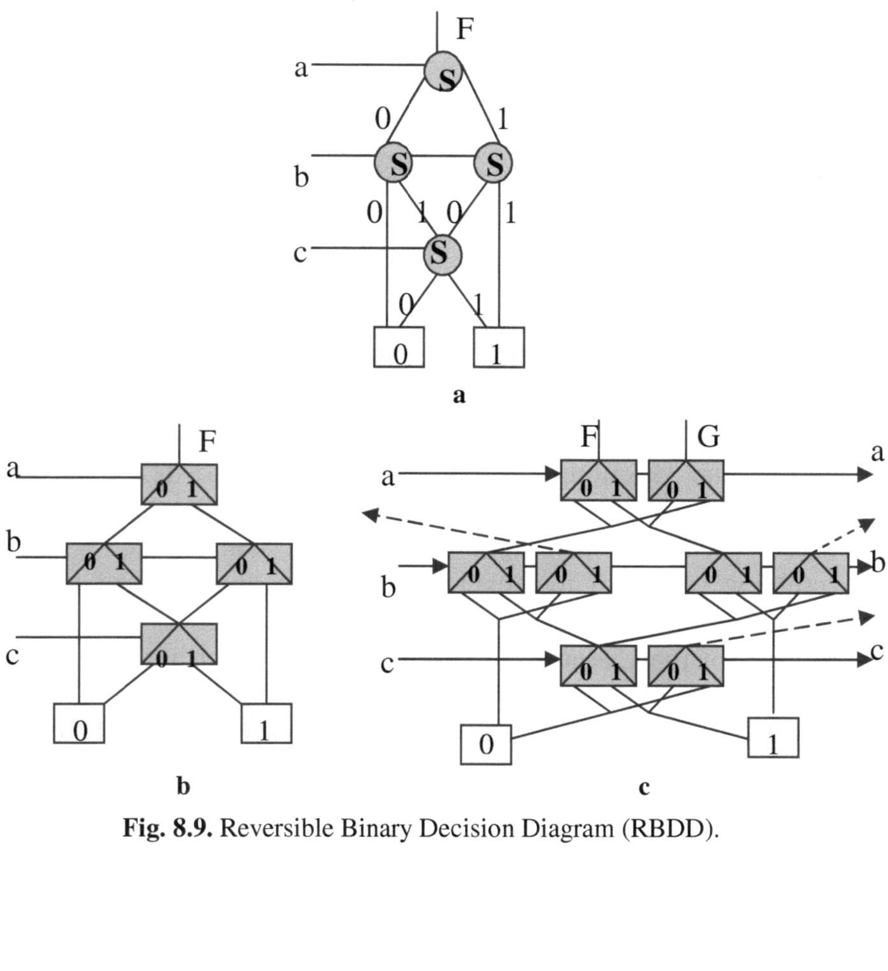

Fig. 8.9. Reversible Binary Decision Diagram (RBDD).

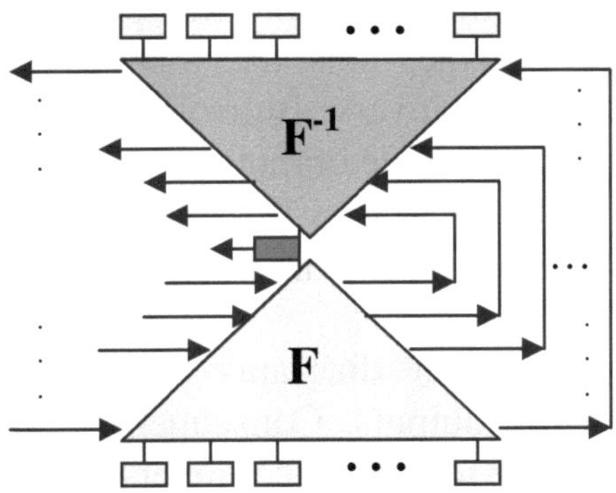

Fig. 8.10. The elimination of garbage in Reversible Decision Diagrams (RDDs).

8.3 Reversible Cascades

The following introduces a method to generate the Cascade of reversible complex Maitra terms [148]. The new structure is called a reversible Cascade [121,158,224]. The general structure for the reversible Cascade is shown in Fig. 8.11. The input variables (a_1, a_2, …, a_n) are the primiary inputs of function F. In the direct computation flow, they propagate from left to right and feed the two-input gates that form the individual stages of the Cascade. It is assumed, without the loss of the expressive power of the reversible Cascade, that one of the inputs of the topmost gate is the constant "0" Boolean function. The outputs of the Maitra terms feed the inputs of the EXOR gates at the bottom. The EXOR gates form the Cascade producing the output of function F. Without the loss of the expressive power of the reversible Cascade, the input of the first EXOR gate is set to the constant "1" Boolean function. The constant "1" input of the Cascade is the only garbage input in the reversible representation of the Cascade. In this implementation, the individual Cascades that are enclosed in the dashed lines in Fig. 8.11 can be viewed as (n+1) reversible gates belonging to Toffoli family of reversible gates.

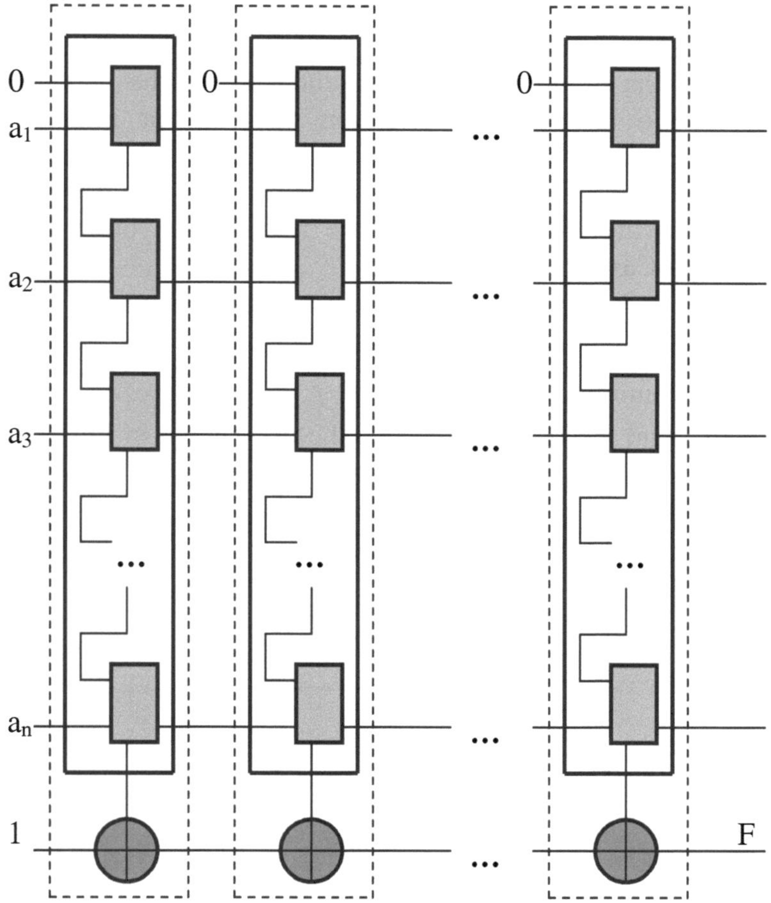

Fig. 8.11. Reversible wave cascades.

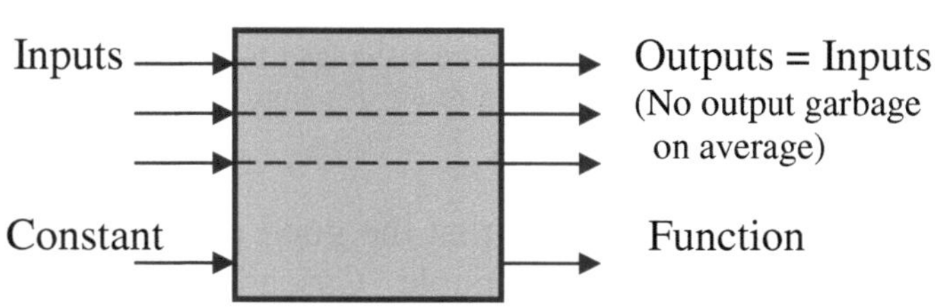

Fig. 8.12. Illustration of the fact that reversible cascades are naturally garbageless.

Note that the value "0" can be used as an input to the EXORs instead of the value "1". This should be determined by the minimum expression that a function can be expressed with. Note that in Fig. 8.11 that the input variables are produced for the output and thus there is no need for the creation of a mirror image (inverse) reversible circuit to be cascaded with the forward reversible circuit as was the case in the new Nets and reversible Decision Diagrams. The following example shows the reversible synthesis using the reversible Cascades.

Example 8.2. Figure 8.13a synthesizes using reversible Cascades of the logic function: $f = 1 \oplus \overline{a} \oplus b \oplus a\overline{b}$. Yet, if a minimizer was used to minimize the expression $f = 1 \oplus \overline{a} \oplus b \oplus a\overline{b}$ one obtains $f = \overline{a}b$, and the circuit would be the one shown in Fig. 8.13b. This example shows clearly the need for a minimizer to minimize the expression before it is realized in a reversible Cascade circuit. This is the basic idea the motivates the introduction of the algorithm in Sect. 8.3.1.

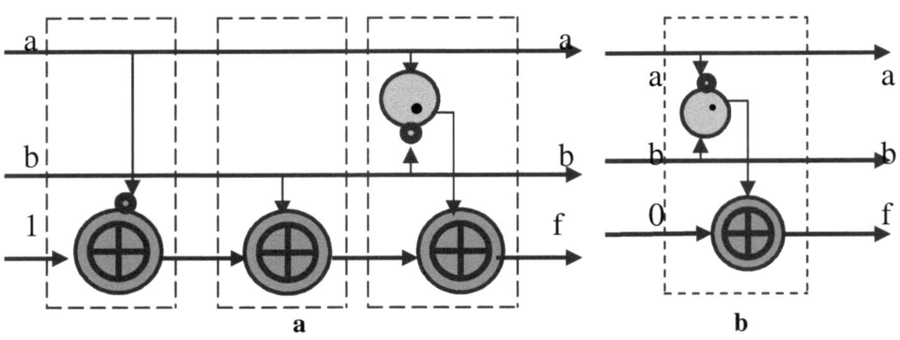

Fig. 8.13. a Reversible cascade composed of three stages of two Feynman gates in the first two stages and a Toffoli gate at the third stage, and **b** an equivalent reversible cascade circuit made only of Toffoli gate as a result of minimization. Here a bubble means an inverter.

It has been shown [158] that the good property of the reversible logic synthesis using reversible Cascades, compared to other previous reversible logic synthesis methods, is that it creates on average at most one constant input and no additional garbage outputs. Table 8.1 from [158] represents experimental results for the upper bound for the number of stages needed in reversible Cascades.

Table 8.1. Upper bound on the number of stages in two-valued reversible cascades.

Benchmark			Upper Bound
Name	# Inputs	# Outputs	
5xp1	7	10	31
9sym	9	1	51
add6	12	7	127
addm4	9	8	89
b12	15	9	28
clip	9	5	63
ex7	16	5	81
f51m	8	8	31
in7	26	10	35
intb	15	7	268
life1	9	1	48
m181	15	9	29
m4	8	16	76
max512	9	6	82
rd53	5	3	14
rd73	7	3	36
rd84	8	4	58
ryy6	16	1	40
sao2	10	4	28
seq	41	35	246
sym10	10	1	79
t3	12	8	24
t481	16	1	13
vg2	25	8	184
z4	7	4	29
Average	13.0	7.0	71.6

Although the reversible Cascade from Fig. 8.11 is for Boolean logic over a Galois field of radix two, the same topological structure can be straightforward generalized to a Galois field of any radix. The only difference is that reversible multiple-valued gates have to be used and the constant in the bottom layer can be either of any value within the radix instead of just being of values “0” or “1” for radix two. Example 8.3 illustrates this generalization.

Example 8.3. Let us synthesize the ternary input ternary output function using a multi-valued reversible Cascade:

$f = 2 +_{GF(3)} ab' +_{GF(3)} b'' +_{GF(3)} a''c$.

Figure 8.14 shows such a synthesis using the multi-valued Feynman and Toffoli gates that were presented in Chapt. 5.

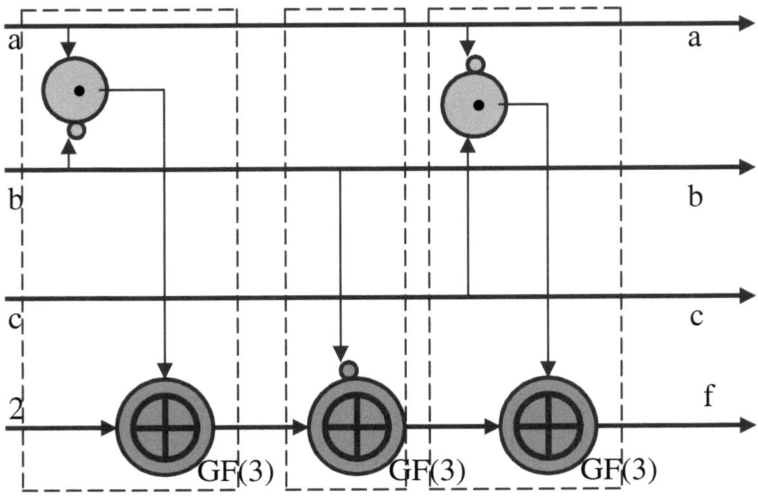

Fig. 8.14. Reversible multiple-valued cascade for the realization of the function in Example 8.3, where a bubble produces the corresponding multiple-valued literal.

Yet, and analogous to Example 8.2, a minimizer is needed to produce reversible Cascade circuits of minimal size.

Example 8.4. Figure 8.15a synthesizes using reversible cascades the logic function: $f = \bar{b} +_{GF(3)} \bar{c} +_{GF(3)} \bar{b}c$. Yet, if a minimizer was used to minimize the expression, one obtains the minimal form as follows:

$$f = \bar{b} +_{GF(3)} \bar{c} +_{GF(3)} \bar{b}c = \bar{b}(1 +_{GF(3)} c) +_{GF(3)} \bar{c},$$

$$= \bar{b}\bar{c} +_{GF(3)} \bar{c} = \bar{c}(1 +_{GF(3)} \bar{b}) = \bar{\bar{b}}\,\bar{c},$$

and the minimized circuit would be the one shown in Fig. 8.15b. Example 8.4 shows clearly the need for a multiple-valued minimizer to minimize the expression before it is realized in a reversible Cascade circuit. This is the basic idea that motivates the introduction of the algorithm in Sect. 8.3.1 for the synthesis of multiple-valued functions using reversible Cascade logic circuits.

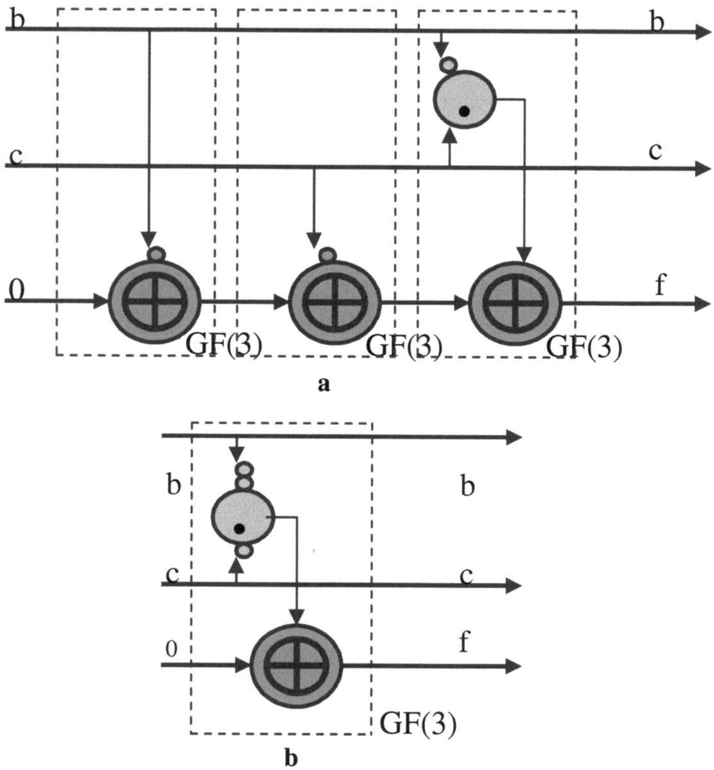

Fig. 8.15. The realization of multiple-valued expressions using reversible cascades: **a** reversible cascade for the original function $f = \bar{b} +_{GF(3)} \bar{c} +_{GF(3)} \bar{b}c$, and **b** reversible cascade circuit for the minimized function $f = \bar{\bar{b}}\,\bar{c}$.

8.3.1 The Realization of GFSOP Expressions Using Reversible Cascades

From Sect. 8.3 and Examples 8.2 and 8.4 one notes that having the function represented in a certain flattened form (i.e., certain polarity) can produce minimum size expression in terms of count of the number of literals and/or terms, and thus this can lead to a realization in a reversible Cascade with a minimal number of stages that are needed. Consequently, the issue of functional minimization becomes very important. The minimizer from [233,235] works for functions with few percent of don't cares, yet this minimizer does not work for functions with a high number of don't cares. Multiple-valued S/D trees developed previously in Chapt. 3 provide a general polarity of Inclusive Form (IF) polarity. For this reason we use the GFSOP evolutionary functional minimizer that was proposed in Chapt. 3, which uses the Inclusive Form (IF) polarity that is produced by the corresponding S/D tree. This is important in order to realize smallest functional forms using the reversible Cascades. Thus, the pre-processing step for the realization of functions using reversible Cascade is to minimize this function, and then using a search heuristic routine to search a library for an optimal synthesis of the function using a reversible Cascade in terms of minimizing the number of stages that are needed.

Figure 8.16 illustrates such pre-processing for the realization of logic functions in reversible Cascades. In Fig. 8.16, the reversible logic circuit synthesizer has four inputs. The first input is the cost of the final reversible circuit. This can be in general a combination (i.e., a linear superposition) of (1) the total number of two-input gates, (2) the total number of garbages, and (3) delay which is characterized by the critical path delay where the signal propagation from the inputs to the outputs require the most delay time. The second input is a library that contains (k,k) two-valued and multiple-valued reversible gates, from which the final reversible circuit will be designed. The third input is the minimal form of the function which is needed to be reversibly realized.

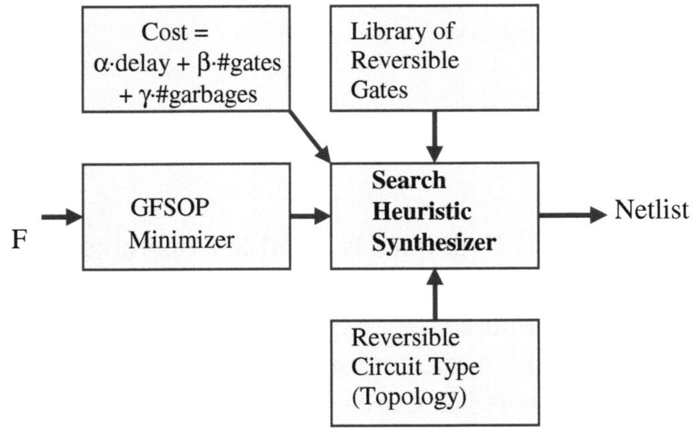

Fig. 8.16. General block diagram for the synthesis of logic functions using reversible logic.

This minimal size expression can be the output of the GFSOP evolutionary minimizer that was introduced in Sect. 3.7 using evolutionary methods from Fig. 3.15, or the output of other appropriate minimizers as well. For example, if one has functions with few percent of don't cares, one could use alternatively the minimizer from [233,235]. The fourth input is the internal specification of the type (topology) of the reversible structure that is used to realize reversibly the minimal size expression from the output of the GFSOP minimizer. This can be, in this case for instance, the general reversible Cascade structure. The synthesizer can be of any type of search-based synthesizers. This can include Evolutionary Algorithms type (as discussed in Sect. 3.7 and in Appendix E) or any other search heuristic. The synthesizer can use an exact search algorithm for functions with low dimensions (i.e., composed of small number of input variables), but must be heuristic for logic functions with high dimensions (i.e., high number of input variables). Consequently, "good" search heuristics in the synthesizer has yet to be found to minimize the time needed in order to design the final reversible netlist at the output of the synthesizer.

8.4 Summary

Novel methods for the synthesis of Boolean and multiple-valued logic functions have been presented in this Chapt. These methods are reversible Nets, reversible Decision Diagrams, and reversible Cascades.

Reversible Cascades show a big advantage over the rest of these methods, since reversible Cascades do not produce on average garbage at the outputs while the other methods do produce such garbage in the outputs. The garbage in the output of reversible Nets and Decision Diagrams is eliminated by interconnecting the forward reversible circuits to their inverse reversible circuits. This is important since these structures will be used in quantum computing, such as the quantum Cascades (including two-valued quantum Cascades and multiple-valued quantum Cascades), in Chapts. 10 and 11, where no garbage is expected at the output.

The type of reversible Cascades that was introduced in Sect. 8.3 of this Chapt. belongs to one specific family of circuit topology for the reversible synthesis of logic functions. Other families of various circuit topologies for Cascade-based reversible synthesis can also be created [121,224], and in general multiple-output multiple-valued reversible Cascades that use serial, parallel, or mix of serial and parallel interconnects of multiple-valued reversible *n-ary* operators (e.g., reversible 3-valued unary shift operators from the formalisms presented in Sect. 2.1 in Chapt. 2 such as: (1) Wire (Buffer; zero shift): x = x, (2) first shift: x' = x + 1, (3) second shift: x" = x + 2, and (4) other shifts of x by: 2·x, 2·x + 1, 2·x + 2, etc) can be also synthesized.

The following Chapt. introduces an initial evaluation of the implementation of logic functions using the various types of reversible logic structures and methods that have been introduced in Chapts. 6, 7, and 8, respectively.

9 Initial Evaluation of the New Reversible Logic Synthesis Methodologies

This Chapt. introduces an initial evaluation of the implementation of the reversible structures that have been presented in Chapts. 6, 7, and 8 to realize logic functions. Although this evaluation is for functions with relatively small number of arguments, it still gives an important first look to some of the weaknesses, strengths, and new properties of the previously introduced reversible structures.

The remainder of this Chapt. is organized as follows. Section 9.1 introduces complete examples for the synthesis of functions using the previously introduced reversible structures. Initial comparison between the various reversible realizations is introduced in Sect. 9.2. Summary of the Chapt. is presented in Sect. 9.3.

9.1 Complete Examples

In this Sect. complete examples are presented for the synthesis of symmetric and non-symmetric logic functions using different reversible logic synthesis methodologies in order to produce the corresponding reversible structures. In the following evaluations, we will consider the following criterion to characterize garbage in the outputs: If there is a need to create the reversible inverse circuit then the outputs contain garbage, otherwise the outputs do not contain garbage. This criterion is reflected in the following conclusion: If the structure produces only the output function and the inputs as outputs, then there is no need to create the reversible inverse circuit and we will consider that there is no garbage in the output of the structure. Otherwise, if the structure produces at least one garbage in the output of the forward reversible circuit then there is a need to create the reversible inverse circuit to eliminate the garbage. Consequently, outputs of a reversible circuit, that are generated by propagating the inputs through the reversible circuit, are not considered as garbage.

Example 9.1. Let us synthesize the following symmetric Boolean function, which is the representative of class 1 for the NPN classification of Boolean functions (See Table G.1 in Appendix G) that encompass eight Boolean functions, using reversible synthesis methods introduced in previous Chapts.: $f = ab + bc + ac$.

(1) Reversible Lattice Structure: Fig. 9.1 represents the realization of the Boolean function in Example 9.1 using the reversible lattice structure from Chapt. 6.

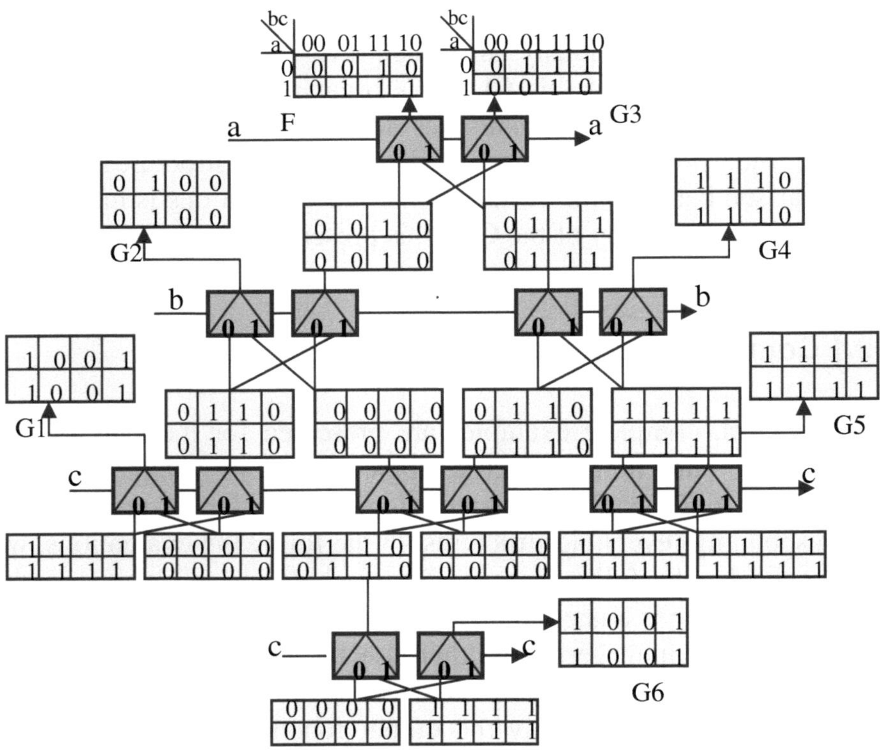

Fig. 9.1. Synthesis of a reversible lattice structure for the function from NPN class 1.

It is interesting to observe from this example that although the function in Example 9.1 is totally symmetric, variable c has to be repeated two times in order to realize the symmetric Boolean function in the reversible lattice structure in Fig. 9.1. This is totally different from the non-reversible lattice structures from Chapt. 4, and implies that reversible lattice structures require different type of symmetry than the regular lattice structures. The structure in Fig. 9.1 has a total of 6 garbage outputs, and 7 Fredkin gates.
(2) Reversible MRA: Fig. 9.2 represents the realization of the Boolean function in Example 9.1 using the reversible MRA structure from Chapt. 7.

As observed the circuit in Fig. 9.2 is fully regular since it uses one type of Toffoli primitives. Figure 9.2 has a total of 4 garbage outputs, and 5 Toffoli gates.

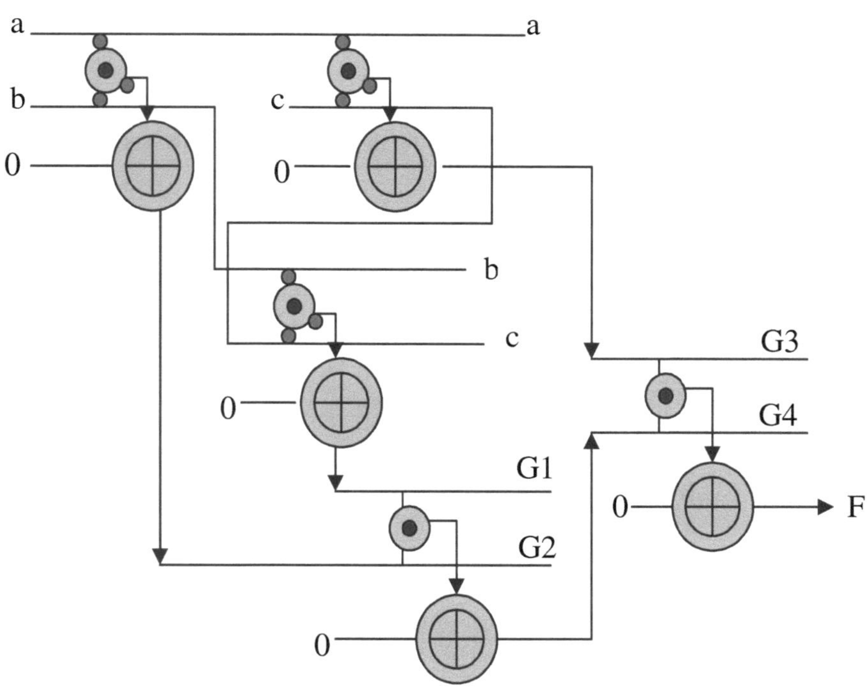

Fig. 9.2. Reversible Boolean circuit to implement 1-MRA circuit for the function from NPN class 1.

(3) Reversible Cascade: Fig. 9.3 presents the reversible Cascade realization, from Chapt. 8, for the Boolean function in Example 9.1. To realize the function from Example 9.1 which is in the SOP form, one has to transform it to the ESOP form. This is performed using the Boolean (i.e., two-valued) rule: $a+b=1\oplus\overline{a}\overline{b}$ as follows: $f=ab+bc+ac=1\oplus\overline{a}\overline{b}\oplus\overline{a}\overline{c}\oplus\overline{b}\overline{c}$.

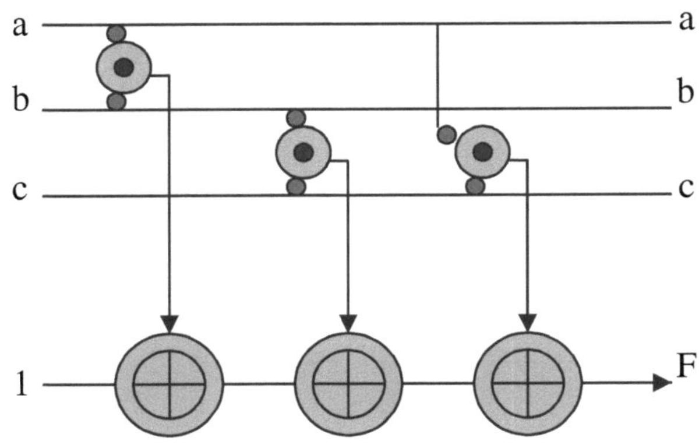

Fig. 9.3. Reversible cascade circuit for the function from NPN class 1.

One can note that the circuit in Fig. 9.3 produces no garbage at the output, and uses a total of 3 Toffoli gates.

(4) Reversible Net: Fig. 9.4 represents the realization of the Boolean function in Example 9.1 using the reversible Nets from Chapt. 8.

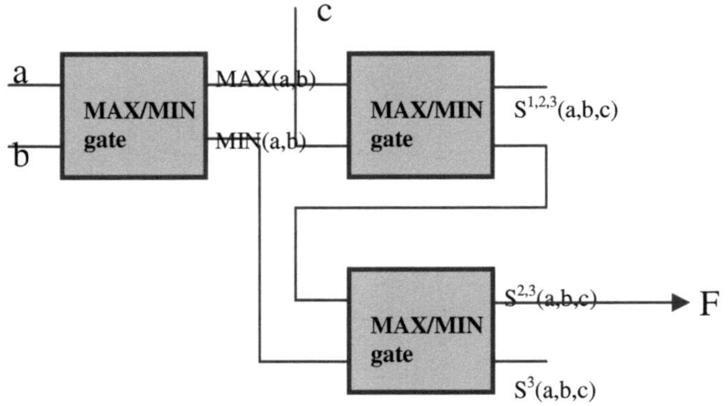

Fig. 9.4. Reversible Net for the function from NPN class 1.

One notes that the circuit in Fig. 9.4 produces two garbage outputs, and uses 3 reversible MAX/MIN gates (from Fig. 5.19a). (The garbage count performed here includes only the variable outputs and does not include the constant outputs.)

(5) Reversible Decision Diagram: Let us implement the function in Example 9.1 using reversible Positive Davio Decision Diagram (PDDD). Positive Davio DD is shown in Fig. 9.5. The reversible PDDD is shown in Fig. 9.6.

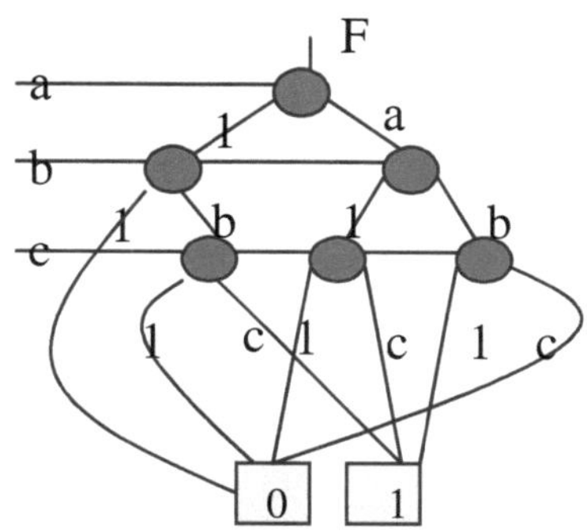

Fig. 9.5. Positive Davio DD for the function from NPN class1.

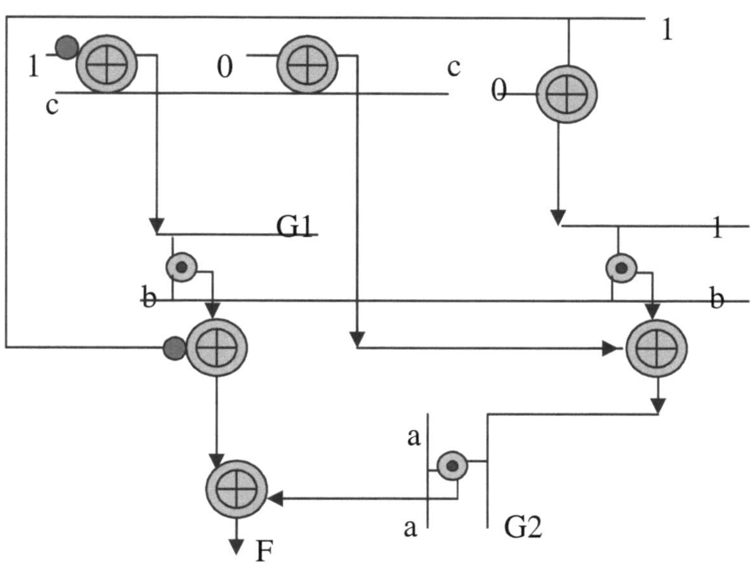

Fig. 9.6. Reversible positive Davio DD for the function in NPN class 1.

One notes that the total number of garbage outputs in Fig. 9.6 is 2, and that the reversible structure uses 6 reversible primitives.

Let us now realize a non-symmetric Boolean function using the same methods used in Example 9.1.

Example 9.2. Let us synthesize the following non-symmetric Boolean function, which is the representative of class 4 for the NPN classification of Boolean functions (See Table G.1 in Appendix G) that encompass 48 Boolean functions, using reversible synthesis methods that were introduced in the previous Chapts.: $f = ab + ac$.

(1) Reversible Lattice Structure: Fig. 9.7 represents the realization of the Boolean function in Example 9.2 using the reversible lattice structure from Chapt. 6.

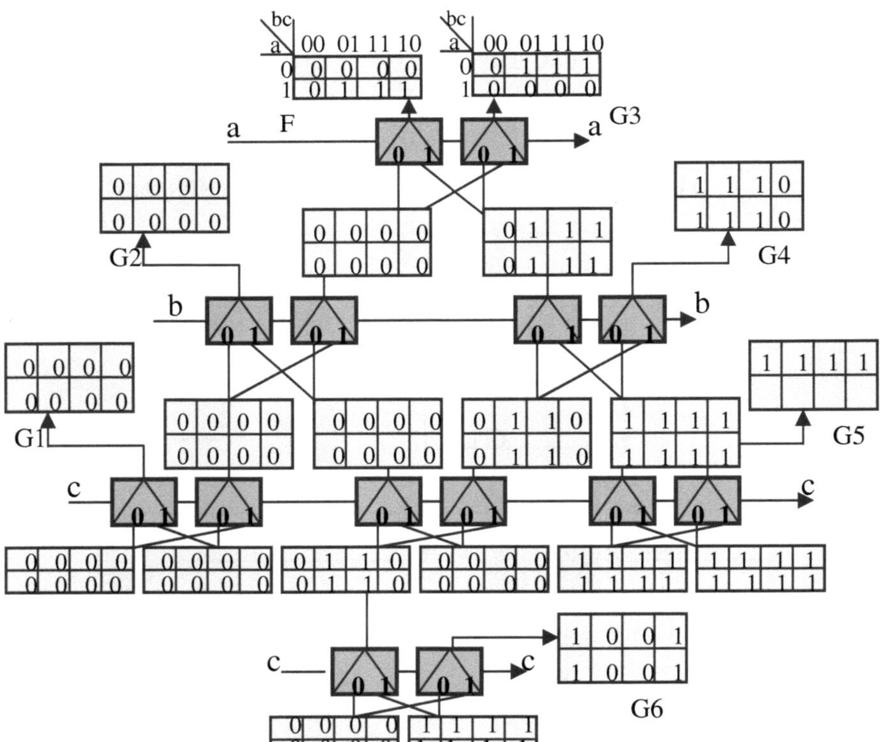

Fig. 9.7. Synthesis of a reversible lattice structure for the function from NPN class 4.

It is interesting to observe from this example that although the function in Example 9.2 is non-symmetric, variable c has to be repeated two times in order to realize the non-symmetric Boolean function in the reversible lattice structure in Fig. 9.7. This is the same structure from Fig. 9.1 with different leaf values. The structure in Fig. 9.7 has a total of 6 garbage outputs, and uses 7 Fredkin gates.
(2) Reversible MRA: Fig. 9.8 represents the realization of the Boolean function from NPN class 4 using the reversible MRA structure from Chapt. 7.

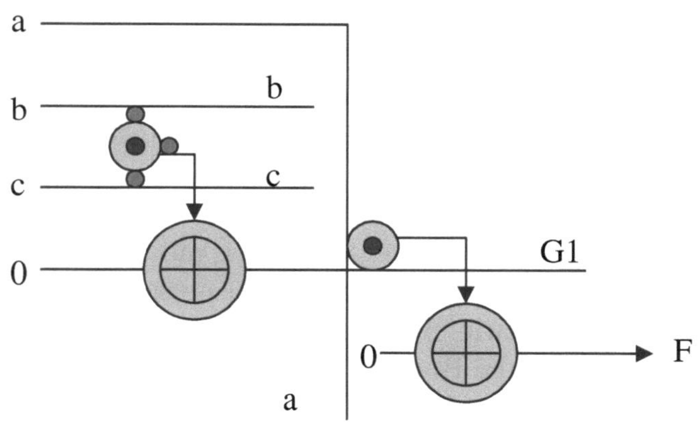

Fig. 9.8. Reversible Boolean circuit that implements the 1-MRA circuit for the function from NPN class 4.

As can be observed, the circuit in Fig. 9.8 is fully regular since it uses one type of Toffoli primitives. The circuit from Fig. 9.8 has a total of 1 garbage output, and uses two Toffoli gates.
(3) Reversible Cascade: Fig. 9.9 presents the reversible Cascade realization, from Chapt. 8, for the non-symmetric Boolean function from Example 9.2. To realize the function from Example 9.2 which is in the SOP form, one has to transform it to the ESOP form. This is done using the Boolean rule $a+b=1\oplus\overline{a}\overline{b}$ as follows:
$f=ab+ac=1\oplus\overline{a}\oplus a\overline{b}\overline{c}$.

One can note that the circuit in Fig. 9.9 produces 1 garbage at the output, and uses three reversible gates.

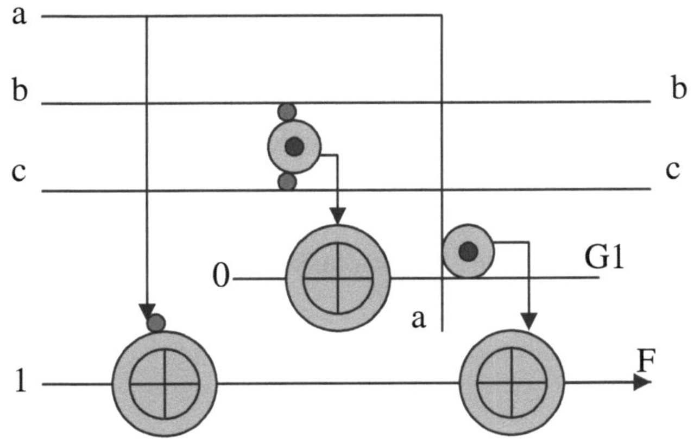

Fig. 9.9. Reversible cascade circuit for the function from NPN class 4.

(4) Reversible Net: The Boolean function in Example 9.2 is non-symmetric. Consequently, repetition of variables has to be made to symmetrize the function. Figure 9.10a presents the symmetrization of the non-symmetric Boolean function, and Fig. 9.10b illustrates the realization of the Boolean function from Example 9.2 using the reversible Nets from Chapt. 8.

Although Fig. 9.10b implements the circuit from Fig. 8.4, one notes that Fig. 9.10b implements the function $S^{3,4}(a,b,c,a)$ by repeating variable {a} two times, and thus there is no need for Feynman planes at the output (analogously to the symmetric function F in Fig. 9.4). Consequently, one observes that the circuit in Fig. 9.10b produces three (variable) garbage outputs, and uses 6 reversible MAX/MIN gates (from Fig. 5.19a).

(5) Reversible Decision Diagram: Let us implement the function in Example 9.2 using reversible Binary Decision Diagram (i.e., reversible Shannon Decision Diagram (SDD)). BDD is shown in Fig. 9.11, and the reversible BDD is shown in Fig. 9.12.

One notes that the total number of garbage outputs in Fig. 9.12 is 3, and that the reversible structure uses 3 Fredkin gates.

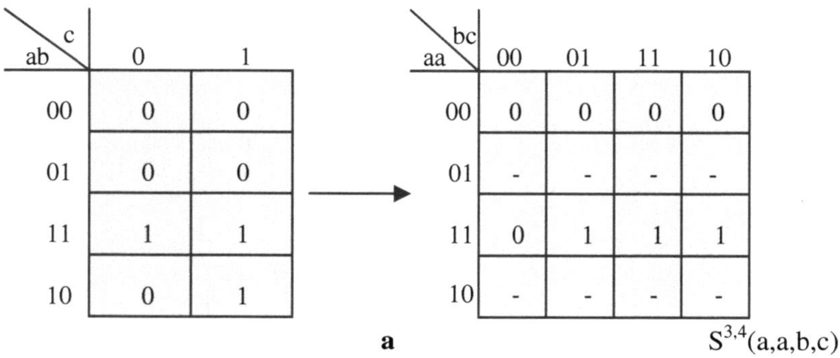

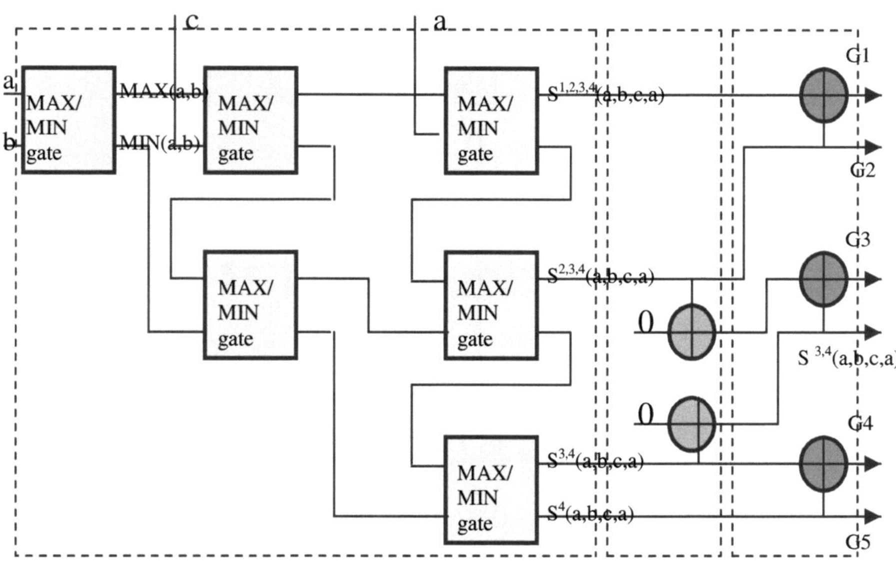

Fig. 9.10. Reversible Net for the function from NPN class 4: **a** symmetrization of the non-symmetric Boolean function, and **b** reversible Net structure.

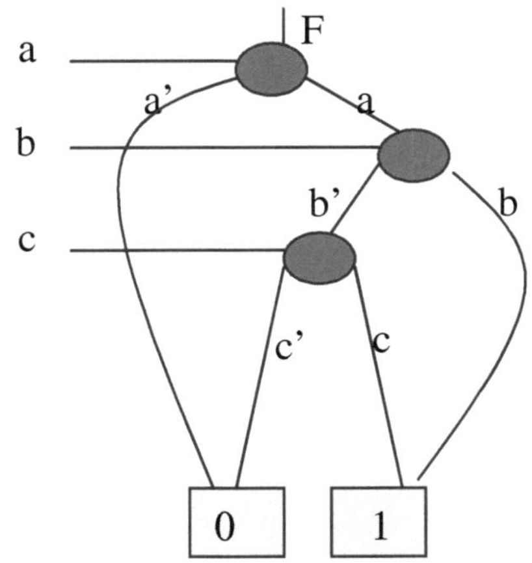

Fig. 9.11. BDD for the Boolean function from NPN class 4.

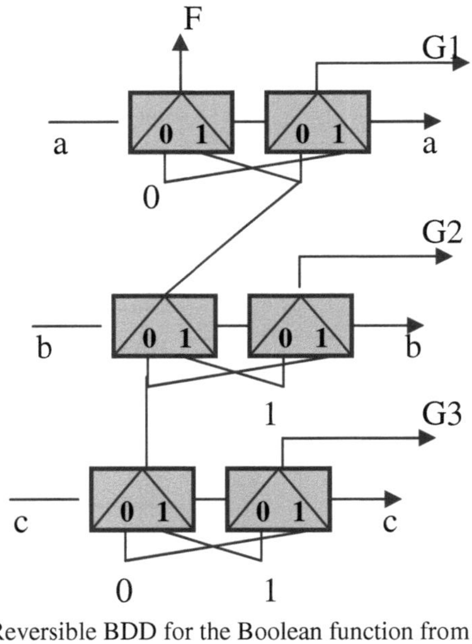

Fig. 9.12. Reversible BDD for the Boolean function from NPN class 4.

Other reversible decompositions have been implemented such as the compositional method where search heuristics are applied to compose a reversible circuit level-by-level by using backtracking search algorithm [184], and the reversible form of classical Ashenhurst-Curtis decomposition (from Appendix H).

It has been presented in Chapts. 6, 7, and 8 various new methods for the reversible synthesis of logic functions. Yet, advantages and disadvantages occur for the use of certain reversible decompositions over the others. Table 9.1 in the next Sect. illustrates some of the advantages and disadvantages that we see so far for the previously mentioned reversible decompositions.

9.2 Initial Comparison

While the methodologies for the reversible synthesis of logic functions that have been introduced in Chapts. 6, 7, and 8 are new, some of the introduced structures possess certain advantages over the other structures, and vice versa, which should be investigated in a further analysis. This was initially observed when symmetric and non-symmetric Boolean functions were synthesized in Examples 9.1 and 9.2, respectively. Table 9.1 shows some properties that have been observed so far when synthesizing reversibly the logic functions. Although this analysis is presented based on few previously presented examples, it produces some important first look at the property-based cost-benefit analysis of using the new reversible structures.

For example, one can observe (from Table 9.1) that reversible Nets are very suitable for single-output and multi-output symmetric functions. On the other hand, if the functions are not symmetric, then reversible Cascades can produce reversible circuits with small garbage on average. Also, as the reversible synthesis using reversible Cascades does not produce garbage in the outputs on average, it has been shown that this type of synthesis is very well suited for quantum computing [121] as will be shown in the quantum logic circuits that will be introduced in Chapt. 10. Although some initial insights and conclusions were made regarding reversible logic synthesis of logic functions using the reversible structures from Chapts. 6, 7, and 8, the various methods in Table 9.1 still need more extensive experimental results (evaluations) to have numerical comparisons on well-known binary and multiple-valued benchmarks. In future CAD tools, these methods must be combined.

Table 9.1. Advantages and disadvantages of various reversible logic synthesis methods.

Type	Advantages	Disadvantages
Reversible Lattices	(1) Regularity which is useful for testing, (2) methodological, (3) generally applied for incompletely specified functions.	(1) Produces garbage, (2) inefficient for multi-output functions.
Reversible Nets	(1) Good for multi-output functions, (2) uses symmetry, (3) can be extended to incompletely specified functions.	(1) Produces garbage, (2) inefficient for strongly non-symmetric functions.
Reversible Cascades	(1) No garbage is produced on average, (2) good for quantum circuits, (3) can be extended to incompletely specified functions, (4) can be created for efficient realization of multiple-input multiple-output (MIMO) functions (circuits).	(1) Not methodological and depends on search, (2) no efficient minimizer exists yet for this method, (3) has a single constant garbage at the input, (4) for functions with many inputs, cascades can be exponential in size that results in long circuits.
Reversible Decision Diagrams	(1) Exhibit certain regularities, (2) many types of decision diagrams exist from which minimal size reversible structure could be found.	(1) Produces garbage.
Reversible MRA	(1) Can produce a comparative small size reversible circuits.	(1) Produces garbage, (2) currently does not realize ESOP forms.

Also, one important factor for the evaluation of such reversible methods is the final total cost of the physical quantum circuits that will implement such reversible structures. While in conventional circuit design the cost of the design is measured by the total number of two-input gates that are used (i.e., $C_{\#}$ from Appendix H), in quantum circuits this is not the case, since in quantum circuits physical processes implement the quantum operations rather than simple hardware gates (e.g., CMOS) as in the case of the classical logic design. Quantum cost characterizes the physical process complexity that is needed to realize physically the corresponding reversible structures. Since little, if none, has been published on this quantum cost for the realization of the reversible structures in Table 9.1, one very important question is still open on how much complex the quantum realization of the structures in Table 9.1 will be, and the answer to this question may very well lead to new cost-benefit conclusions.

9.3 Summary

This Chapt. has introduced an initial evaluation of the various implementations of the reversible structures that have been presented in Chapts. 6, 7, and 8, respectively. Although this evaluation is for functions with relatively small number of arguments (variables), it still produces an initial important first look to some of the weaknesses, strengths, and new properties of the previously introduced reversible structures. Also one important issue still to be considered is the quantum cost of the physical processes that implement the new reversible structures, a subject that can lead to a new cost-benefit evaluations for the reversible methods introduced in Table 9.1.

Next Chapt. will introduce the physical quantum operational notation and quantum circuits that would be used to construct the counterparts of the reversible structures from Chapts. 6, 7, and 8. Two-valued and multiple-valued quantum computations of such circuits will be implemented in Chapt. 11.

10 Quantum Logic Circuits for Reversible Structures

The reversible structures that are introduced in previous Chapts. can be implemented, utilizing the garbageless circuits of such structures, using many possible technologies, including optical [17,24,62,63,64,65,222], and CMOS [35,68,70,206,262,263]. We are interested in this Chapt. with the implementation of the reversible structures using quantum logic circuits. This will be a second step, after the reversibility step from Chapts. 6, 7, and 8, towards quantum computing of such structures (that will be introduced in Chapt. 11). The main contribution of this Chapt. is the introduction of the quantum circuits for two-valued and multiple-valued reversible structures using the corresponding quantum notation.

The fact that the tendency of current technologies is towards the nano-scale (i.e., dimensions of a single atom in the order of 10^{-10} m = 1 Angestrom) will have, and already having, disaster effects on the signal integrity in classical designs of processing and transmitting information bits. The higher packing of devices in increasingly smaller and smaller areas will have, and is already having, tremendous power consumption effects. Thus, one possible solution for both problems is the implementation of logic functionalities using quantum circuits and the associated quantum technology. This Chapt. uses the quantum notation from [95,167] that emphasizes the use of control lines to activate the classical logical operations on other inputs. This is important since this can realize physical processes where there are specific physical control variables which one would like to use to obtain logical operations form other wires using such processes. Another important advantage of using the operational quantum notation is that the quantum notation shows the one-dimensional left-to-right time-based evolution of the physical process which is realized by the reversible circuits, and also shows the composition of the circuit as number of inputs that are equal to the number of outputs, and the propagation of the input signals from input lines to output lines throughout the gates. The main contributions of this Chapt. are:

- Multiple-valued quantum gates and their operational quantum notation. This includes for example the multiple-valued quantum Swap and controlled-Swap gates.
- Showing that minimum two-valued and multiple-valued reversible logic structures lead to minimum size two-valued and multiple-valued quantum logic circuits, respectively.

The remainder of this Chapt. is organized as follows. Section 10.1 presents the important quantum notation that is used for quantum circuits. Section 10.2 introduces the quantum circuits that correspond to the previously introduced reversible structures in Chapts. 6, 7, and 8. Summary is provided in Sect. 10.3.

The content of this Chapt. will be computationally implemented using the two-valued and multiple-valued quantum computing methods that will be introduced in Chapt. 11.

10.1 Notation for Two-Valued and Multiple-Valued Quantum Circuits

In quantum computing, usually a set of "quantum" notations are used for the corresponding reversible gates. The quantum notation for some basic logic primitives are shown in Fig. 10.1. This notation is very useful to explain the two-valued and multiple-valued quantum evolution (i.e., time-based) processes. We will follow here the two-valued quantum notation introduced in [95] and [167].

The primitive in Fig. 10.1a is a two way AND which is a dot on an intersection. Two-way OR gate in Fig. 10.1b is a box on an intersection. NOT gate is represented as an "X" on a wire. No connection is represented with a simple crossing in Fig. 10.1d, and control wire is an "O" on a wire in Fig. 10.1e. Using the quantum notation in Fig. 10.1, one can obtain two-valued and multiple-valued quantum gates using such notation as demonstrated in Fig. 10.2.

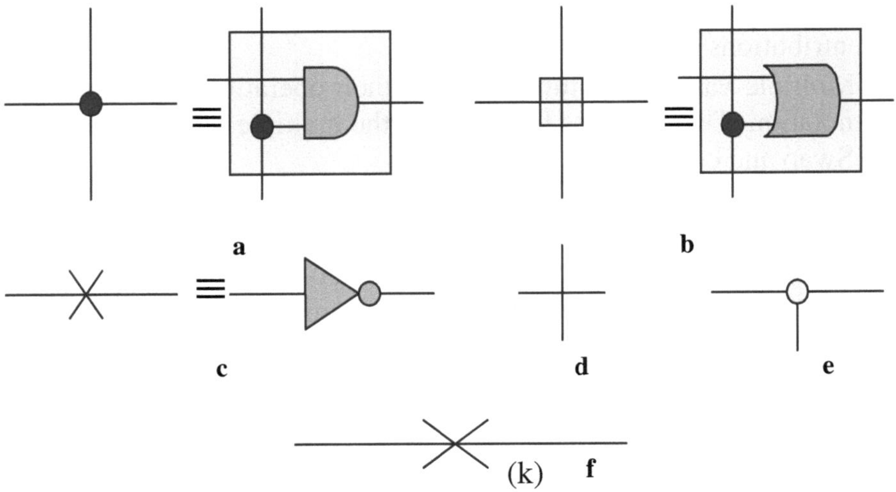

Fig. 10.1. Quantum notation of quantum circuits for: **a** two way AND which is a dot on an intersection, **b** two way OR which is a box on an intersection, **c** Not which is an "X" on a wire, **d** no connection which is simple crossing, **e** control wire which is an "O" on a wire, and **f** multiple-valued shift by (k).

This quantum notation is easier to use and conveys the important information regarding the "control lines" which control the operations of C-NOT gate, C-C-NOT gate, and C-Swap gate. The "X" (in Fig. 10.2) denotes a NOT operation, however, this NOT operation is not a conventional one; it is controlled by the input to the O-wire. Specifically, if the input to the O-wire is 1, then the input to the X-wire is inverted; if the O-input is zero, then the NOT gate does not work, and the signal on the X-wire goes through unchanged. In other words, the input to the O-line activates the NOT gate on the lower line. The O-output, however, is always the same as the O-input (i.e., the upper line is identity). In CCN gate, we have two control lines A and B, each marked by an O, and as with the CN gate, the signals on this line are unchanged on passage through the gate: A' = A, and B' = B. The remaining line, again, has a NOT on it, but this is only activated if both A = 1 and B = 1: then: C' = NOT C.

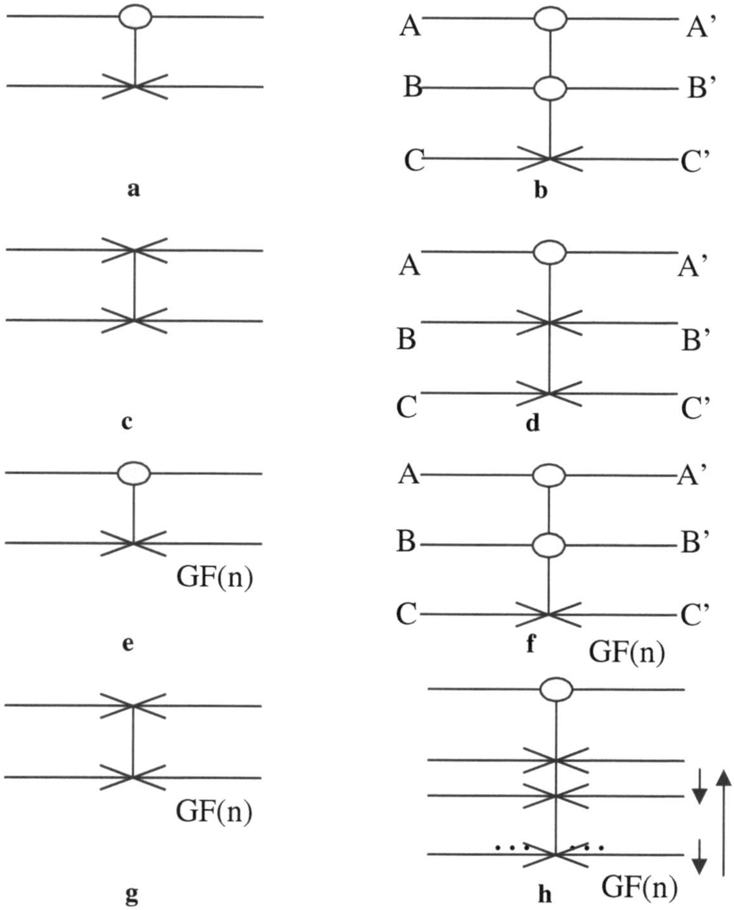

Fig. 10.2. Quantum notation for: **a** two-valued Controlled-NOT gate (Feynman gate), **b** two-valued Controlled-Controlled-NOT gate (Toffoli gate), **c** two-valued Swap gate, **d** two-valued Controlled-Swap gate (Fredkin gate), **e** multiple-valued Galois Controlled-NOT gate (Feynman gate), **f** multiple-valued Galois Controlled-Controlled-NOT gate (Toffoli gate), **g** multiple-valued Swap gate, and **h** multiple-valued Galois Controlled-Swap gate (Fredkin gate).

Notice that this single gate is very powerful. If we keep both A and B equal to one then the CCN gate is just a NOT gate. If we keep just A = 1, then the gate is just a CN gate with B and C as inputs. So, if we have a CCN gate and a source of 1's and 0's, we can implement both N (NOT) and CN gates, besides the CCN gate which is a universal gate. Thus the control lines just activate a more conventional operation on other inputs. In the case of Fredkin gate, the operation that is been controlled is exchange (swap): if A = 0, B and C are not exchanged. However, if A = 1 then B and C are exchanged.

Many of the two-valued and multiple-valued quantum circuit implementations in this Chapt. use two-valued and multiple-valued quantum *Swap-based* and *NOT-based* gates. This can be important, since the Swap and NOT gates are basic primitives in quantum computing, from which many other gates are built, such as (1) two-valued NOT gate, (2) two-valued Controlled-NOT gate (Feynman gate), (3) two-valued Controlled-Controlled-NOT gate (Toffoli gate), (4) two-valued Swap gate, (5) two-valued Controlled-Swap gate (Fredkin gate), (6) multiple-valued NOT gate, (7) multiple-valued Controlled-NOT gate (multiple-valued Feynman gate), (8) multiple-valued Controlled-Controlled-NOT gate (multiple-valued Toffoli gate), (9) multiple-valued Swap gate, and (10) multiple-valued Controlled-Swap gate (multiple-valued Fredkin gate).

10.2 Quantum Logic Circuits

Using the upper quantum notation one can obtain a half adder and full adder as shown in Fig. 10.3. The half adder and full adder are described by the following Boolean Eqs.: (1) Half Adder (HA): Sum = S = $a \oplus b$ and Carry = C = ab, and (2) Full Adder (FA): Sum = $a \oplus b \oplus c$ and Carry = C = $(a\oplus b)c \oplus ab$. Note that the dashed CN gate (in Fig. 10.3d) is used to produce the inputs a and b from the output garbage.

The quantum notation introduced in Sect. 10.1 emphasizes the use of control lines to activate conventional operations on other inputs. This is important since this can realize physical processes where there are specific physical control variables which we would like to use to obtain logic operations from such processes.

Using such operational quantum notation, one can realize the reversible structures that were introduced previously (in Chapts. 6, 7, and 8) using the form of quantum circuits.

a	b	C	S
0	0	0	0
0	1	0	1
1	0	0	1
1	1	1	0

a

a	b	c	C	S
0	0	0	0	0
0	0	1	0	1
0	1	0	0	1
0	1	1	1	0
1	0	0	0	1
1	0	1	1	0
1	1	0	1	0
1	1	1	1	1

b

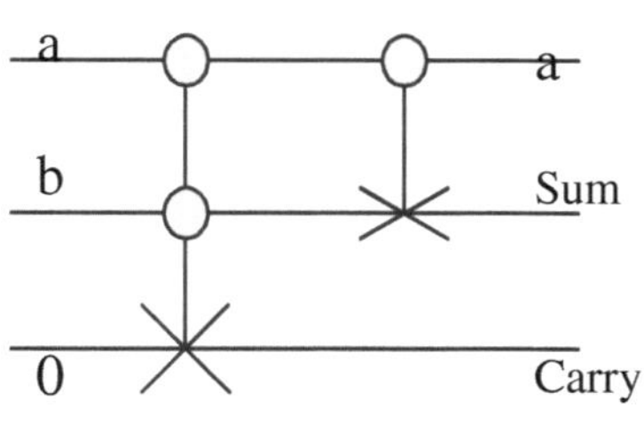

c

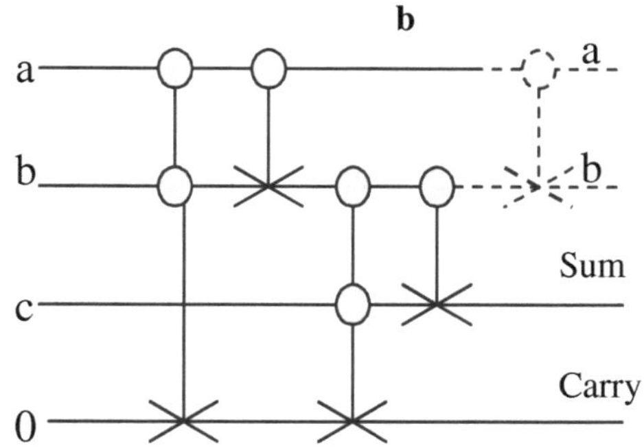

d

Fig. 10.3. Quantum circuits of adders: **a** a truth table for half adder, **b** truth table for full adder, **c** quantum circuit for half adder, and **d** quantum circuit for full adder.

For instance, Fig. 10.4 illustrates the quantum notation for the reversible lattice structure from Fig. 6.2. Here, another importance of the quantum notation is readily apparent: (1) the quantum notation shows the one-dimensional left-to-right time-based evolution of the process realized by the lattice structures, (2) it shows the composition of the circuit as number of inputs that are equal to the number of outputs, and (3) the propagation of the input signals from input lines to output lines throughout the gates.

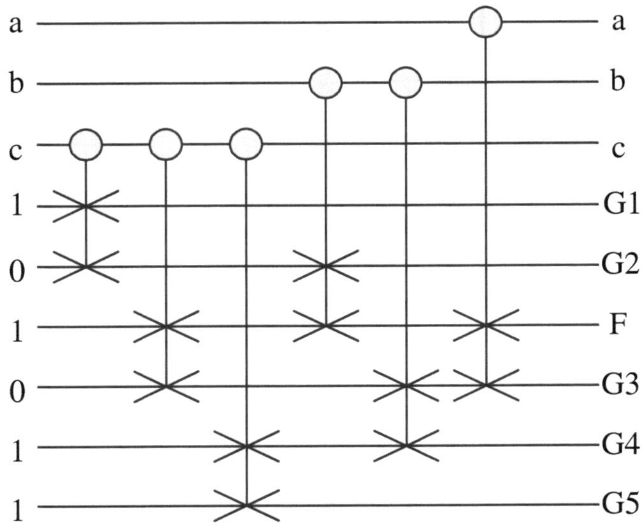

Fig. 10.4. One-dimensional left-to-right time-based evolution quantum circuit that realizes the process implemented by the 2-D lattice structure in Fig. 6.2.

Using the same quantum notation, one obtains the one-dimensional left-to-right time-based evolution quantum circuit in Fig. 10.5 that realizes the process which is implemented by the multiple-valued reversible lattice structure from Fig. 6.4.

It is observed that when the quantum lattice structure is of a big size, one can utilize the ISID decomposition from Chapt. 4 to decompose the total structure into multiple parts and thus potentially reducing the total size of the quantum lattice structure (and consequently reducing the total number of quantum operations that are needed to perform quantum computations using such structures as will be shown in Chapt. 11).

Using the two-valued operational quantum notation, one obtains the quantum logic circuits in Figs. 10.6a and 10.6b for the reversible Cascade circuits in Figs. 8.13a and 8.13b, respectively.

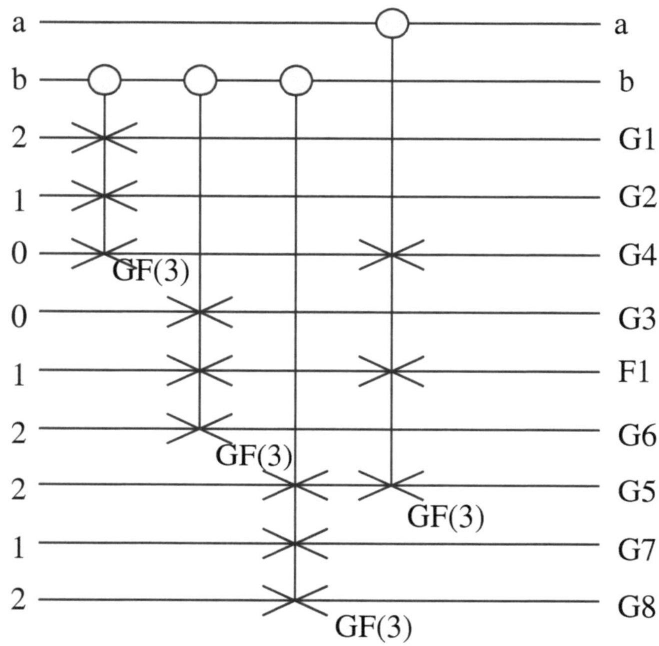

Fig. 10.5. One-dimensional left-to-right time-based evolution quantum circuit that realizes the process implemented by the lattice structure in Fig. 6.4.

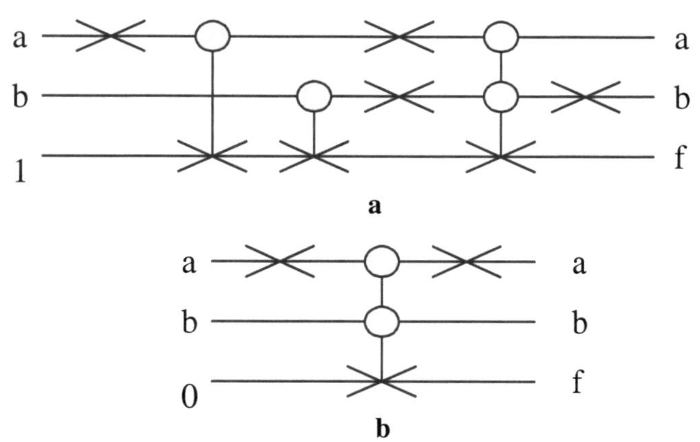

Fig. 10.6. One-dimensional left-to-right time-based evolution two-valued quantum logic circuits: **a** the implementation of the reversible cascade circuit in Fig. 8.13a, and **b** the implementation of the reversible cascade circuit in Fig. 8.13b.

From Fig. 10.6, one notes that while the quantum logic circuit in Fig. 10.6a requires a total of 7 quantum primitives (4 N gates, 2 CN gates, and 1 CCN gate), the quantum logic circuit in Fig. 10.6b requires a total of three quantum primitives (2 N gates, and 1 CCN gate).

Using the multiple-valued quantum notation, one obtains the following quantum logic circuits in Figs. 10.7a and 10.7b for the reversible Cascade circuits in Figs. 8.15a and 8.15b, respectively.

From Fig. 10.7, one notes that while the quantum logic circuit in Fig. 10.7a requires a total of 7 ternary quantum primitives (2 $N_{(1)}$ gates, 2 $N_{(2)}$ gates, 2 CN gates, and 1 CCN gate), the quantum logic circuit in Fig. 10.7b requires a total of five ternary quantum primitives (2 $N_{(1)}$ gates, 2 $N_{(2)}$ gates, and 1 CCN gate).

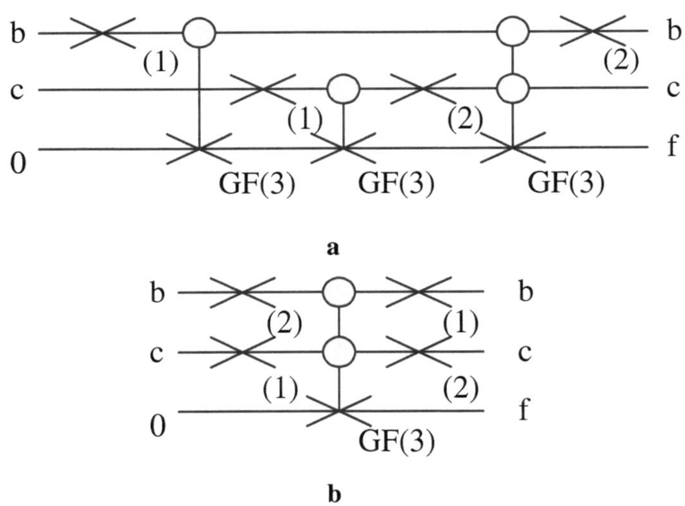

Fig. 10.7. One-dimensional left-to-right time-based evolution ternary quantum logic circuits: **a** the implementation of the reversible cascade circuit in Fig. 8.15a, and **b** the implementation of the reversible cascade circuit in Fig. 8.15b.

The examples from Figs. 10.6 and 10.7 show very clearly that the result of minimum reversible logic circuit will lead to a minimum quantum logic circuit, in terms of minimizing the length (number of levels) and the width (number of gates per level) of the corresponding quantum logic circuit. (The size of the quantum logic circuit will be reflected in the size of the quantum (scratchpad) register, as will be shown in the next Chapt.) This is illustrated in Fig. 10.8.

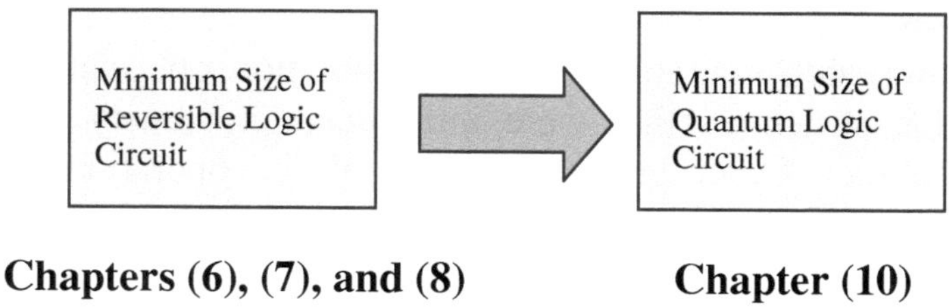

Fig. 10.8. Relationship between the size of the reversible circuit and the corresponding size of the quantum logic circuit.

Also, one can note the difference between the reversible notation in Figs. 8.13 and 8.15 and the corresponding operational quantum notation from Figs. 10.6 and 10.7, respectively. The difference is not only in the fact that the quantum notation reflects the 1-D quantum evolution from left-to-right, but also the quantum notation will lead to different number of primitives than the corresponding reversible notation, which are needed to show the implementation of the reversible circuit using garbage-free quantum logic circuit.

Although the conclusion in Fig. 10.8 has a common sense ground, it can be of limited importance. This is because another very important factor besides the size of the quantum logic circuit is the cost of the physical quantum processes that implement the corresponding quantum logic circuits, and thus smaller size quantum logic circuit is not necessarily implemented by a less complex quantum physical processes, and vice versa. This issue is an open subject with very little, if none, literature available, and needs much further future investigations.

10.3 Summary

This Chapt. has inroduced the quantum notation and the corresponding quantum logic circuits for the previously invented reversible circuits. The quantum notation introduced emphasizes the use of control lines to activate conventional operations on other inputs. This is important since this can realize physical processes where there are specific physical control variables which one would like to use to obtain logical operations from other wires within such processes.

Using such operational quantum notation one can realize the reversible structures that were introduced previously in quantum logic circuits. Other good properties of the operational quantum notation can be observed when implemented upon reversible structures from Chapts. 6, 7, and 8 as follows: (1) the quantum notation shows the one-dimensional left-to-right time-based evolution of the process realized by the reversible circuits, (2) it shows the composition of the circuit as number of inputs that are equal to the number of outputs, and (3) it shows the propagation of the input signals from input lines to output lines throughout the quantum gates (primitives).

Besides the circuits shown in Sect. 10.2 of this Chapt., other quantum circuits could be created as well, such as for the general synthesis of multiple-output multiple-valued quantum Cascades that use serial, parallel, or mix of serial and parallel interconnects of multiple-valued quantum *n-ary* operators (e.g., the equivalent quantum counterparts of the reversible 3-valued unary shift operators from the mathematical formalisms that were introduced in Sect. 2.1 in Chapt. 2 such as: (1) Wire (Buffer; zero shift): x = x, (2) first shift: x' = x + 1, (3) second shift: x" = x + 2, and (4) other shifts of x by: 2·x, 2·x + 1, 2·x + 2, etc). (Methods of quantum computations using similar Cascades will be shown in the next Chapt.)

Next Chapt. will introduce the two-valued and multiple-valued quantum computing methodologies for the two-valued and multiple-valued quantum circuits that were introduced in this Chapt.

11 Quantum Computing: Basics and New Results

Trends in computer hardware are leading toward higher density and lower energy dissipation. Ultimately, some approaches should result in packing extremely high densities in excess of 10^{17} logic devices in a cubiccentimeter. The trend towards higher packing density strongly influences energy dissipation. Conventional devices must dissipate more than K·T·ln(2) Joules in switching (cf. Fig. 1.1), and thus enormous amounts of power will be needed for computing using classical methods of computations [116]. Even an idealized device, which uses a one Volt power supply and dissipatively discharges a single electron to ground during the switching operation, would dissipate one electron Volt per switching operation. At T = 300 Kelvins, this is 40·K·T per switching operation or about 160,000,000 Watts for a computer with 10^{17} logic elements operating at 10 GHz. If each switching operation involves hundreds of electrons then the energy dissipation enters the multi gigaWatt range.

New thermodynamically reversible circuits (e.g., CMOS, nMOS, CCD-based logic circuits, etc) would be far better [116], but these circuits still have some amount of dissipative losses that are caused by the unavoidable parasetic resistance that exists in the circuit. Thus quantum computing technology, which is naturally reversible, offers a very promising solution for this big problem.

Logic circuits within the quantum barrier (nano-scale) have already been fabricated using nanotechnology [111,112,196]. Figures 11.1 and 11.2 from [111] show the electron microscope images of a two-dimensional lattice-like structure which was fabricated using nanotechnology.

The nano circuits in Figs. 11.1 and 11.2 are very similar to the (2,2) two-dimensional lattice structures from Chapt. 4. Figures 11.1 and 11.2 implement at the nano scale a binary adder, where s_1 and s_0 represent the sum outputs, and c_1 represents the output carry.

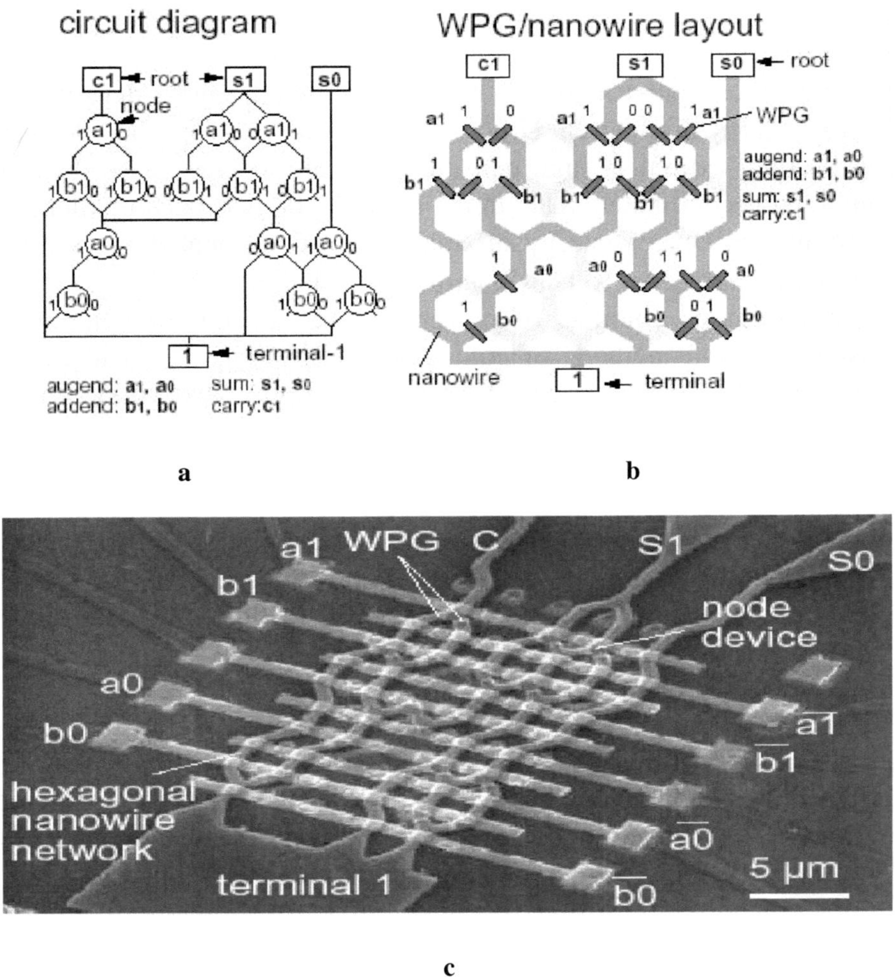

Fig. 11.1. Circuit fabricated using nanotechnology: **a** logic circuit diagram, **b** nanowire layout, and **c** image of the physical nano circuit.

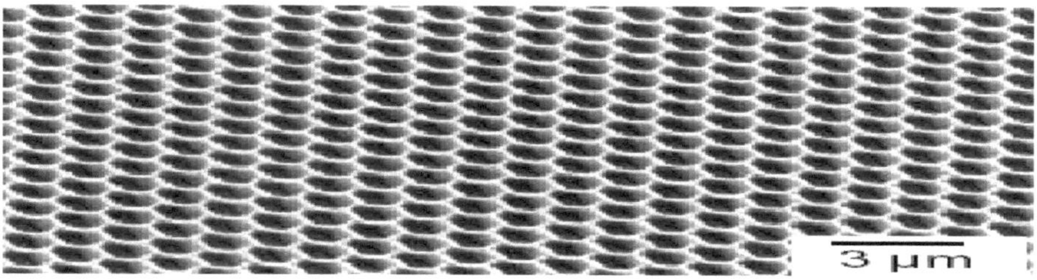

a

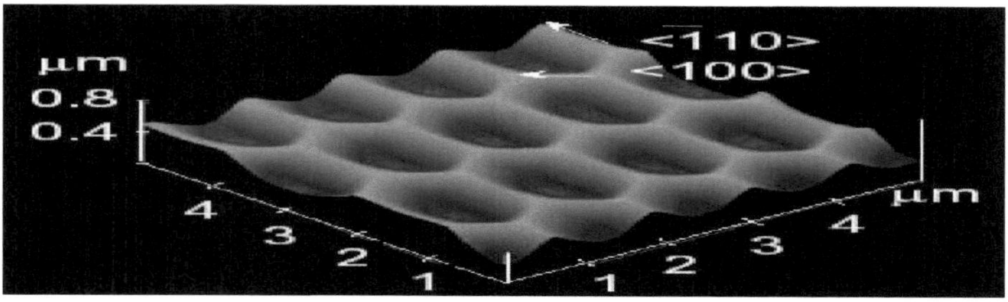

b

Fig. 11.2. The (2,2) 2-D lattice-like structure from Fig. 9.1: **a** image, and **b** visualization.

Due to the anticipated failure of Moore's law around the year 2020, quantum computing will play an important role in building smaller size and less power-consuming computers [40,107,115,167,168,170]. Because all quantum computer gates (i.e., building blocks) must be reversible, reversible computing will also be increasingly important in the future design of *regular*, *minimal size*, and *universal (complete)* systems. Due to the fact that higher power consumption occurs at higher frequencies of operation, and due to the fact that minimal power consumption is needed for mobile and remote tele-communications (as cellular phones) and computing (as laptops), two major solutions have been proposed for this increasing power consumption problem: (1) asynchronous design of sequential machines, and (2) adiabatic design of circuits and systems, which is related to reversible logic [206]. Reversible computing can be considered a necessary but not sufficient step towards quantum computing, where no power is consumed for internal information processing, and the power is only consumed when reading and writing quantum bits (quantum data) from and into the quantum computer, respectively [93,122,160,167,205].

While some logic systems are non-linear (like fuzzy logic), quantum logic (QL) is linear, since all the evolutions (operations) of input quantum bits are performed using linear unitary transformations (evolution processes). Quantum computing requires the following main constraints that distinguish it from classical computing (The notation that is used in this Chapt. follows the well-known "Dirac notation" of quantum mechanics [81]):

(1) Quantum operations are done on vectors of bits called qubits (quantum bits).

(2) Quantum functions are complex-weighted probability-based linear combinations (superpositions) of orthonormal basis states [167] as follows: $|\psi\rangle = \sum_i c_i |\varphi_i\rangle$, where the coefficients c_i are called probability amplitudes, and $|c_i|^2$ produces the probability of the quantum state $|\psi\rangle$ collapsing into the state $|\varphi_i\rangle$, and the unitarity condition $\sum_i |c_i|^2 = 1$. Some of the basis states that are used in a 1-qubit binary quantum systems include [95,107,115,167,253,254]: (a) the computational basis states $\{|0\rangle, |1\rangle\}$, and (b) the composite basis states $\left\{ |+\rangle = \frac{|0\rangle + |1\rangle}{\sqrt{2}}, |-\rangle = \frac{|0\rangle - |1\rangle}{\sqrt{2}} \right\}$. Some of the basis states that are used in a 2-qubit binary quantum systems include: (a) Einstein-Podolsky-Rosen (EPR) basis states $\left\{ \frac{|00\rangle + |11\rangle}{\sqrt{2}}, \frac{|00\rangle - |11\rangle}{\sqrt{2}}, \frac{|01\rangle + |10\rangle}{\sqrt{2}}, \frac{|01\rangle - |10\rangle}{\sqrt{2}} \right\}$, and (b) the computational basis states $\{|00\rangle, |01\rangle, |10\rangle, |11\rangle\}$.

(3) Quantum computations, algorithms, and circuits must be reversible.

(4) Linear and Unitary operations (evolutions; transformations). These operations are performed in n-dimensional Hilbert space, which is in general a linear complex vector space. The states |0> and |1> are the computational basis states that form an orthonormal basis set in Hilbert space. Two-valued single qubits exist in 2-D Hilbert

space, and three-valued single qubits exist in 3-D Hilbert space. The quantum operators in Hilbert space describe how one wave function is changed into another wave function. In the formalism as an eigenvalue Eq. (problem), one obtains the following eigenvalue Eq.: $\hat{B}|\varphi_i\rangle = b_i|\varphi_i\rangle$, where b_i is the eigenvalue, and the solutions to the eigenvalue Eq. are $|\varphi_i\rangle$, which are called eigenstates that can be used to construct the basis states of a Hilbert space. Consequently, the Hilbert space has a set of basis states $|\varphi_i\rangle$ and the quantum system is described by a quantum state $|\psi\rangle$ as: $|\psi\rangle = \sum_i c_i|\varphi_i\rangle$, where in general the coefficients c_i may be complex. In Dirac notation [81], the probability that a quantum state $|\psi\rangle$ will collapse into an eigenstate $|\varphi_i\rangle$ is equal to $|\langle\varphi_i|\psi\rangle|^2$ which is the dot (inner; scalar) product (projection) of the two vectors $|\varphi_i\rangle$ and $|\psi\rangle$. The *unitarity* of operators (matrices) (cf. Sect. 11.1) implies *reversibility*. (5) The quantum register, which is an array of qubits, can be in any of the individual states of its qubits at any instant of time or at all of the states at the same time, thus allowing for parallelism at the quantum level. When the quantum states exhibit correlations that cannot be accounted for classically, then the quantum state is said to be in *entanglement*.

Quantum computing has important advantages in comparison to classical computing as follows:
(1) Transforming highly complex problems from the real domain to other domains (like Fourier domain, Walsh domain, etc) does not reduce the problem complexity, but transforming such complex problems to the quantum domain *does* reduce the problem complexity. Due to this fact, some problems that are not solvable in polynomial time in classical domains can be solvable in polynomial time in the quantum domain, and a well-known example is the factoring problem [226,228]. Solving the factoring problem results in the ability to penetrate encrypted messages utilizing any communication channels like the internet for instance, which makes it a vital issue for national security of any country [162,163], and thus quantum cryptography becomes very important. Fast quantum algorithms for database search have been also created [105,106].

(2) Quantum logic (QL) permits intensive parallel computations.

(3) QL executes on 2-valued as well as many-valued logics, as was shown in [54,165]. Thus by utilizing the physical properties of atoms, one can use the same aperture to perform two-valued or multiple-valued quantum computing. This is different from classical computing where different complex devices have to be designed for two-valued and multiple-valued logics. This can be performed using the polarization of light for binary and multiple-valued computing, spin of particles for two-valued computing [79,167], the energy (eigen) state transitions of cold trapped ions for two-valued computing [54] and multiple-valued computing [165]. (This is illustrated in Fig. 11.3.)

In Fig. 11.3a, for the equally distributed quantum states (E_i), the eigenstates that are obtained by solving Schrodinger Eq. for the simple harmonic oscillator (SHO), can be considered as the positions where an electron can be found purely in any of the eigenstates or as entangled states where the electron exists in intermediate (superimposed) eigenstates and not in any individual eigenstate. Figure 11.3b illustrates the famous Stern-Gerlach experiment [167] where the quantum number, which is associated with the spin of particles, was discovered. Such quantum spin number, in the case of an electron, proton, or neutron for example, has two possible values: $(+\frac{1}{2})$ for spin-up and $(-\frac{1}{2})$ for spin-down, and thus two-valued quantum computations can be achieved using such unique spins [107,167]. (Quantum spin numbers for other particles such as a photon (+1 and -1) can be different). Figure 11.3c illustrates the potential use of light polarization ($\vec{E}$) as quantum states for two-valued and multiple-valued quantum computations. Figure 11.3d illustrates the multiple-valued quantum computing using linear ion trap scheme from [165], where by using d-level ions in this scheme, (d-1) neighboring transitions occur by illuminating (d-1) distinct laser beams on the linearly trapped ions, and thus multiple-valued quantum computing can be accomplished.

(4) The requirement of reversible algorithms/codes on quantum machines, thus very little power is needed (theoretically zero). This issue is important in mobile and remote communications (as in cellular phones) and computing (such as laptops).

(5) The size of a future quantum computer (QC) will be extremely small (nano-scale), since all the quantum operations are performed on the level of atoms.
(6) In teleportation (e.g., teleporting quantum state of an object to a distant object and thus teleporting the encoding (qubit)) the quantum effects are vital, thus the need for quantum computers [44,107,167].
(7) A quantum computer is a *true* random number generator [253].

As was discussed in Chapt. 5, while the physical constraints allow for some reversible gates to have the number of inputs to be not equal to the number of outputs like in the case of optical computing using the Interaction and Switch gates [62], other physical constraints allow for reversible gates only to have the number of inputs equal to the number of outputs like in the case of quantum computing.

Three layers of abstraction are distinguished in investigating quantum logic design. The first layer is the *quantum algorithmic (mathematical)* level where algorithms (procedures) are developed that use fundamental concepts from the quantum domain such as the entanglement and quantum transforms (e.g., the quantum Fourier transforms (QFT) on dyadic and p-adic groups) to efficiently solve problems that were previously thought to be unsolvable in the classical domain. The second layer is the *quantum logic synthesis* level where methodologies for the synthesis of circuits using quantum primitives are created. This includes synthesis methods such as various mathematical decompositions, factorizations, etc. The third layer is the *quantum physical (device)* level which consists of device modeling, simulation, and fabrication in the nano scale (i.e., less or equal to 10^{-10} m = 1 Angestrom).

In general, two methodologies exist to synthesize quantum computers: (1) quantum Turing machine [167] which requires: (a) "infinite" memory, (b) read/write head, (c) information encoding (0,1,-), and (d) set of instructions (i.e., program or code) that specify: (1) the output, (2) next state of the head as a function of the present state of the head and inputs, and (3) shift-left or shift-right operation of the head, or (2) a combination of interconnections between (universal) logic primitives [95,107,167]. In this Chapt. the focus will be on synthesizing quantum circuits using the second approach of interconnections between quantum logic primitives.

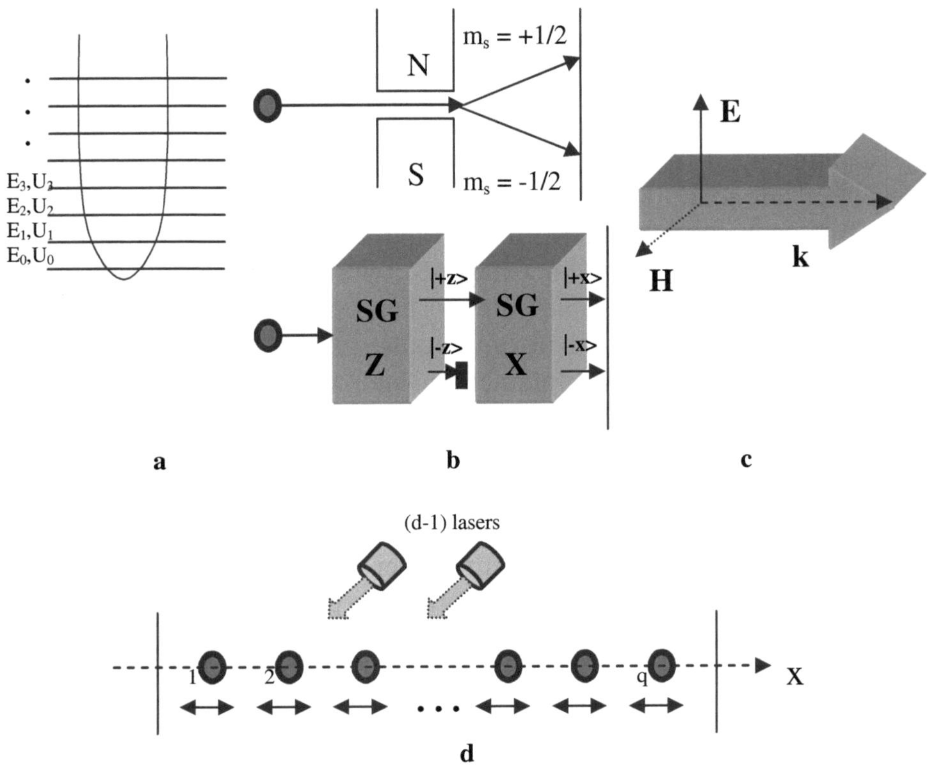

Fig. 11.3. Various physical realization methodologies for the implementation of two-valued and multiple-valued quantum computing: **a** energy states for Simple Harmonic Oscillator (SHO) potential for two-valued and multiple-valued quantum computing, **b** particle spin angular momentum in Stern-Gerlach experiment for two-valued quantum computing, **c** light polarization for two-valued and multiple-valued quantum computing, and **d** cold trapped ions for two-valued and multiple-valued quantum computing.

Many types of quantum computers have been proposed that utilize a combination of interconnections between quantum logic primitives: Feynman quantum computer [93,253], Deutsch quantum computer [69], Czerny quantum computer, Benhoff quantum computer, Chuang-Yamamoto quantum computer [53], and others in [41,169]. Physical devices have been proposed for quantum computing like the ½ spin particles [248], and Nuclear Magnetic Resonance (NMR) machines [167]. NMR is one of the main quantum engines that have been proposed and can currently operate on up to seven qubits.

Since little has been developed towards the analysis and synthesis of multiple-valued quantum logic circuits, this Chapt. serves as a first step to fill this gap. This Chapt. provides the theoretical background for multiple-valued quantum logic, especially that multiple-valued quantum primitives have been constructed [165], and quantum hardware devices have been built [54,162,163,165] that can utilize such theories. Figure 11.4 illustrates the analogy graph that motivates the developments for the various quantum circuits in this Chapt.

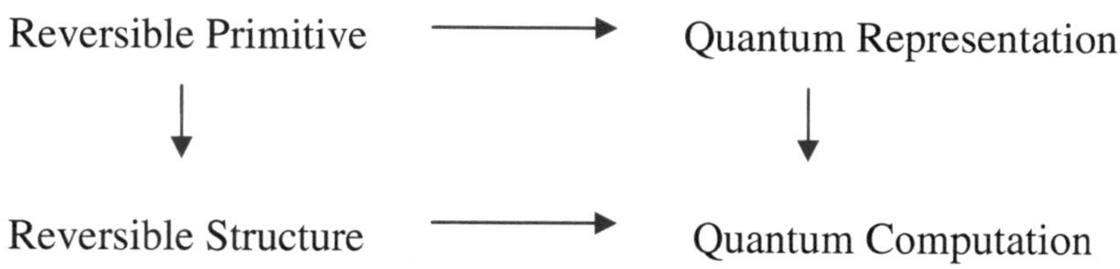

Fig. 11.4. Analogy graph that motivates the developments in this Chapt.

So far, not much has been published on multiple-valued quantum logic gates and especially their characterization and representation formalisms. It is the main goal of this Chapt. to start building a systematic theory of multiple-valued quantum gates, structures, and synthesis methods. This Chapt. introduces the following new results:

- New two-valued and multiple-valued quantum primitives and evolution processes.
- New multiple-valued composite basis states and Einstein-Podolsky-Rosen (EPR) basis states.
- Generalized multiple-valued quantum permuters.
- Two-valued and multiple-valued canonical quantum decision trees (QDTs) and decision diagrams (QDDs).
- Introducing the mathematical operations for the analysis and synthesis of serial, parallel, and mixture of serial and parallel multiple-valued quantum circuits.
- Showing that minimum two-valued and multiple-valued quantum logic circuits, that result from minimum size two-valued and multiple-valued reversible logic structures, lead to minimum complexity of two-valued and multiple-valued quantum computing, respectively.

Items 1 and 3 are necessary for the automated analysis and verification of netlists of the corresponding quantum gates. They are also necessary for the automated synthesis of a netlist described as an evolution matrix from quantum gates, especially for try-and-check (i.e., try-and-error) methods such as evolutionary computations [100,110,207,236,237,249,252,255,258]. Since decision diagrams allow for efficient representation of large sparse matrices, they found applications in many computer aided design (CAD) algorithms, and we believe that their quantum counterparts will be useful for quantum logic synthesis and analysis. Finally, item 2 is important because new forms (representations) of quantum decision trees and diagrams can be produced for the new quantum multiple-valued Einstein-Podolsky-Rosen (EPR) basis states, and thus allowing for further possible optimizations in the design of quantum circuits, analogous to the classical (non-quantum) case where different forms of decision trees and diagrams lead to different scales of optimizations in the design of logic circuits [213,217,219].

The remainder of this Chapt. is organized as follows: Background and preliminaries on two-valued quantum evolution processes and synthesis are included in Sect. 11.1. Different types of mathematical decompositions that can be used for quantum computing and for the synthesis of quantum logic circuits using basic quantum primitives are presented in Sect. 11.1.1. New two-valued quantum evolution processes are introduced in Sect. 11.2. Two-valued quantum decision trees and diagrams are presented in Sect. 11.3. Fundamentals of multiple-valued quantum computing are presented in Sect. 11.4. New multiple-valued composite and EPR basis states and quantum Chrestenson evolution process are presented in Sect. 11.5. New multiple-valued generalized quantum permuters and evolution processes are presented in Sect. 11.6. Novel representations of multiple-valued quantum decision trees and diagrams for multiple-valued quantum computing are introduced in Sect. 11.7. A methodology of the automatic synthesis of quantum circuits based on evolutionary algorithms is presented in Sect. 11.8. Quantum computations using reversible structures (from Chapts. 5, 6, 7, and 8) and initial comparison for such synthesis methods are presented in Sect. 11.9. Chapter Summary is presented in Sect. 11.10.

11.1 Fundamentals of Two-Valued Quantum Evolution Processes and Synthesis

In general, the dynamical behavior of quantum systems is governed by the solution of $|\Psi\rangle$ in the time-dependent Shrodinger Eq. (TDSE) [81,208]. The following is the 1-D TDSE [81,208]:

$$-\frac{(h/2\pi)^2}{2m}\frac{\partial^2|\Psi\rangle}{\partial x^2}+V|\Psi\rangle=i(h/2\pi)\frac{\partial|\Psi\rangle}{\partial t},$$

$$H|\Psi\rangle=i(h/2\pi)\frac{d|\Psi\rangle}{dt}, \tag{11.1}$$

where h is Planck constant ($6.626 \cdot 10^{-34}$ J·s), V(x,t) is the potential, m is particle mass, i is the imaginary number ($i=\sqrt{-1}$), $|\Psi(x,t)\rangle$ is the time-dependent quantum state, and H is the Hamiltonian operator (e.g., for 1-D: $H=-\frac{(h/2\pi)^2}{2m}\frac{\partial^2}{\partial x^2}+V$).

A general solution to TDSE is the expansion of a stationary (time-independent; spatial) basis functions (eigen states) $U_e(\vec{r})$ using a time-dependent (i.e., temporal) expansion coefficients $C_e(t)$, as follows:

$$\Psi(\vec{r},t)=\sum_{e=0}^{n} c_e(t)u_e(\vec{r}),$$
$$=c_0(t)u_0(\vec{r})+c_1(t)u_1(\vec{r})+c_2(t)u_2(\vec{r})+\ldots+c_n(t)u_n(\vec{r}).$$

The expansion coefficients $C_e(t)$ are a scaled complex exponentials as follows:

$$C_e(t)=k_e e^{-i\frac{E_e}{(h/2\pi)}t}.$$

Most of the proposed quantum computers [95,167], and the proposed quantum algorithms to solve (time-independent) optimization problems (e.g., Traveling Slaseman problem, Graph Coloring problem, Maximum Clique problem, etc) are systems that evolve according to the time-independent Shrodinger Eq. (TISE):

$$E|\Psi\rangle=H|\Psi\rangle, \tag{11.2}$$

where E is the system energy, H is the system Hamiltonian operator (which is related to the total energy of the quantum system, and can be thought of as the quantum computer hardware that evolves the input quantum state vector or quantum data), and $|\psi|^2$ is the wave Eq. (probability density). (For a complete example for obtaining ψ and $|\psi|^2 = \psi\psi^*$ please refer to Example K.1 and Fig. K.1 in Appendix K.)

Thus, and according to Fig. 11.3a for instance, if one considers the eigen states $U_e(\vec{r})$ to be the qubits $|e\rangle$ of the quantum system, then the expansion coefficients C_e are the square roots of the probability density functions (PDFs) as follows:

$$
\begin{aligned}
|\Psi\rangle = \sum_{e=0}^{n} c_e u_e &= \sum_{e=0}^{n} \alpha_e |e\rangle, \\
&= \sum_{e=0}^{n} \sqrt{\alpha_e \alpha_e^*}\,|e\rangle = \sum_{e=0}^{n} \sqrt{|\alpha_e|^2}\,|e\rangle, \\
&= \sum_{e=0}^{n} \sqrt{PDF_e}\,|e\rangle.
\end{aligned}
$$

Where PDF_e is the probability of finding the state $|\Psi\rangle$ in the state U_e, and:

$$\sum_e |c_e|^2 = \sum_e |\alpha_e|^2 = 1.$$

The evolution operator (transformation; process) U is a unitary operator like the Walsh-Hadamard operator [167]. This operator is a function of the Hamiltonian (H), which is obtained by solving Shrodinger Eq., and thus:

$$\therefore\ U = f(H). \tag{11.3}$$

If the initial quantum state of a quantum system is:

$$|\Psi\rangle = \sum_k \mu_k |\Psi_k\rangle, \tag{11.4}$$

then the quantum system evolves over time, using the *linear* and *unitary* operator U, to the following state:

$$|\Psi'\rangle = U|\Psi\rangle = \sum_k \mu'_k |\Psi_k\rangle,$$

$$\therefore \vec{\mu}' = U\vec{\mu}.$$

Where the unitarity condition holds: $U^H U = I$, $(\vec{\mu}')^H \vec{\mu}' = 1$, and H is the Hermitian (i.e., transpose of the conjugate).

As stated previously, quantum logic incorporates physical axioms into the abstract mathematical axioms in order to make the resulting logic a true representation of the natural physical reality, an ability the classical irreversible logics lack. Thus, in the quantum domain, the issues of: *interference, entanglement, decoherence, and measurement (observation)* are essential. These quantum-based phenomena have been experimentally verified.

The *quantum interference* exists to explain the 2-slot experiment: when a beam is projected into a plane through two "tiny" slots, a detector on the projected plane detects a varying amplitude wave-like beam [81,208]. This varying amplitude wave-like beam can be constructed by super-imposing two waves, called "probability waves". Mathematically, this quantum interference corresponds to the fact that when a unitary operator is applied upon an input, some "amplitudes" increase while others decrease (i.e., they interfere with each other). Therefore, when the linear unitary operator is applied upon the wave function, the wave function interferes with itself, and consequently different parts of the wave function interfere constructively or destructively according to their relative phases, and thus the interference pattern occurs, where for the total amplitudes c_j some amplitudes c_i increase and other amplitudes c_{j-i} decrease. The physical phenomenon of interference is illustrated as shown in Fig. 11.5.

The quantum *entanglement* means that a quantum state cannot be written as the tensor product of individual states (i.e., non-decomposable). When the quantum states $|\varphi_i\rangle$ exhibit correlations that cannot be accounted for classically, then the quantum state $|\psi\rangle$ is said to be in entanglement. (Further explanation of two-valued and multiple-valued entanglement will be shown in Examples 11.1 and 11.5, respectively.)

The *quantum measurement (observation)* means that while the state of the quantum register can be in general in any of the individual states at any instant of time or at all of the states at the same time, the state of the binary quantum register "collapses" into either the quantum state $|0\rangle$ or quantum state $|1\rangle$ when measurement is conducted. Measurement in the quantum domain is governed by Heisenberg's "*uncertainty principle*" [81] which states that one cannot measure with absolute certainty specific multiple quantum quantities (i.e., observable variables) at the same time since a measurement of one variable affects the measurement of the other variable (e.g., the measurement of the pairs: (position, momentum) and (time, energy)). Because in measuring the state of a quantum system $|\psi\rangle$ the superposition, in which quantum state exists, collapses into a single state $|\varphi_i\rangle$, and because this measurement is governed by the uncertainty principle, the collapsing of the quantum state $|\psi\rangle$ into a single state $|\varphi_i\rangle$, through measurement in the quantum domain, changes the state of the quantum register according to the uncertainty principle.

The physical process of *decoherence* (dephasing) [117] (or the loss of phase coherence) indicates the tendency of a quantum system (such as a quantum computer) to decay from a given quantum state into an incoherent state as it interacts, or entangles, with the state of the environment in which that specific quantum system exists. A quantum system is said to be *coherent* if it exists in a linear superposition of its basis states. Yet, as a result of quantum mechanics, if the quantum system interacts in any means with its environment (i.e., the surrounding or outer system), then the linear superposition of that specific quantum system will be destroyed. When the quantum system $|\psi\rangle$ is coherent, one cannot decide the state $|\varphi_i\rangle$ in which the quantum system exists, since the state of such quantum system is the probabilistic superposition of basic orthogonal quantum basis states. On the other hand, when the quantum system $|\psi\rangle$ decoheres, then one can decide which state $|\varphi_i\rangle$ the system $|\psi\rangle$ will be in with a probability according to the scalar projection $|\langle\varphi_i|\psi\rangle|^2$.

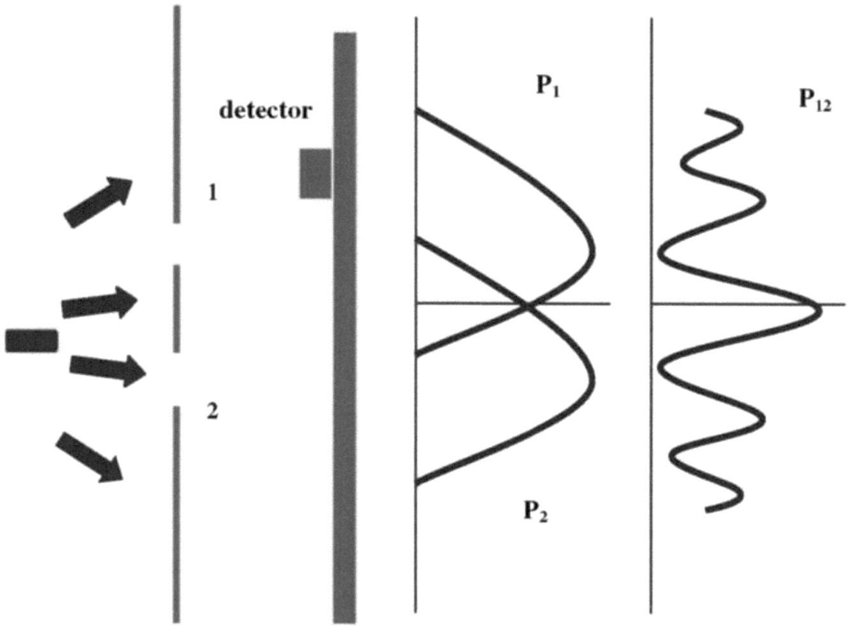

Fig. 11.5. The quantum interference experiment, which by analogy to quantum computing implies that when a unitary quantum operator is applied, some amplitudes increase while others decrease (i.e., they *interfere* with each other). The probability density function PDF $= P_{12} = |\beta|^2 = \beta\beta^*$. Thus, $P_{12} = \gamma_1 P_1 + \gamma_2 P_2$.

These interactions between the environment and the states of the quantum system are unavoidable, and induce the breakdown of information stored in the quantum system (computer), and thus errors in computation. Other types of quantum noise also appear in the nano scale besides the noise from decoherence [261]. Consequently, error correction methods have to be used to obtain the phase coherence and to counteract the effects of noise [48,227,241]. The phenomena of decoherence is one of the biggest obstacles that prevent us today from building a quantum computer, or more precisely, building a quantum computer that can rival today's modern digital computer, and still need to be solved to obtain reliable quantum computations.

Conservation law states that matter cannot be created or vanished but can be transformed from one form to another. Conservative logic implements the physical law of conservation by having the number of ones in the input vector equal to the number of ones in the output vector for binary reversible logic, and number of ones and twos in the input vector equal to the number of ones and twos in the output vector for ternary reversible logic, etc.

Thus, in general, the axiomatic algebraic systems, which are used in quantum logic, incorporate the following set of physical phenomena: *uncertainty principle, interference, entanglement, decoherence, linear evolution, conservativeness, and reversibility.*

In binary quantum logic, two qubits can be represented by the vector corresponding to the spin of atomic particles as follows:

$$|0\rangle = \begin{bmatrix} 1 \\ 0 \end{bmatrix} \Rightarrow spin-up(\uparrow 90^0), |1\rangle = \begin{bmatrix} 0 \\ 1 \end{bmatrix} \Rightarrow spin-down(\downarrow 270^0).$$

Figure 11.6 illustrates the spin-up and spin-down for particles to perform two-valued quantum computing. A qubit can be mathematically modeled in the spherical coordinate system using the Bloch sphere [167,270], where a vector of fixed length rotates with angles inside the sphere and each of these rotations correspond to a specific quantum gate (primitive). For example, the core of quantum Fourier transform (QFT) (i.e., n = 1) that will be presented in Sect. 11.1.1 is performed by the composition of two vectors (qubits) in 2-D (vector or function) Hilbert space: (1) quantum state $|0\rangle$ and (2) a specific rotation equal to $e^{\pi i j}$ of quantum state $|1\rangle$. (In the case of p-adic groups the core of QFT is performed by the composition of p vectors each with different rotation in p-dimensional space. For example, in the case of ternary logic [268], the core of QFT is generated by the composition of three vectors (qubits) in 3-D space: (1) quantum state $|0\rangle$, (2) rotation1 of quantum state $|1\rangle$, and (3) rotation2 of quantum state $|2\rangle$.)

An n-qubit binary quantum register (also called a scratchpad register [95,98,167]) is an array of n binary qubits. For a quantum register that is composed of 2 binary qubits, one obtains 4 possible states of the register. These states are as follows:

$$|00\rangle = |0\rangle \otimes |0\rangle = \begin{bmatrix} 1 \\ 0 \end{bmatrix} \otimes \begin{bmatrix} 1 \\ 0 \end{bmatrix} = [1 \quad 0 \quad 0 \quad 0]^T, |01\rangle = |0\rangle \otimes |1\rangle = \begin{bmatrix} 1 \\ 0 \end{bmatrix} \otimes \begin{bmatrix} 0 \\ 1 \end{bmatrix} = [0 \quad 1 \quad 0 \quad 0]^T,$$

$$|10\rangle = |1\rangle \otimes |0\rangle = \begin{bmatrix} 0 \\ 1 \end{bmatrix} \otimes \begin{bmatrix} 1 \\ 0 \end{bmatrix} = [0 \quad 0 \quad 1 \quad 0]^T, |11\rangle = |1\rangle \otimes |1\rangle = \begin{bmatrix} 0 \\ 1 \end{bmatrix} \otimes \begin{bmatrix} 0 \\ 1 \end{bmatrix} = [0 \quad 0 \quad 0 \quad 1]^T.$$

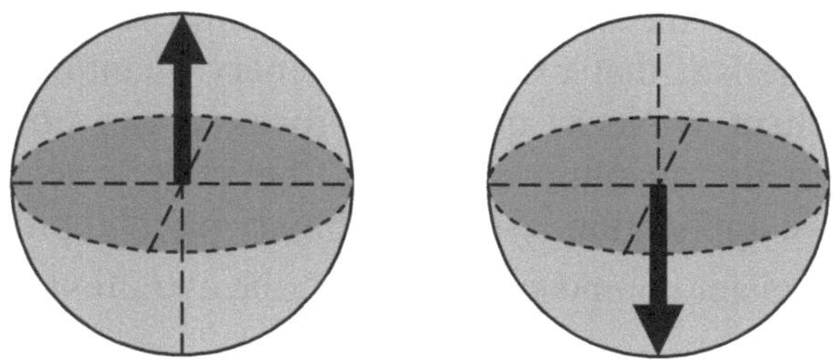

Fig. 11.6. a Atomic particle spin-up that represents qubit |0>, and **b** atomic particle spin-down that represents qubit |1>. An m-qubit binary quantum register consists of m of such spins, and thus can have up to 2^m distinct states. Any quantum state is a linear combination (superposition) of the orthonormal computational basis states |0> and |1>.

Where ⊗ is the tensor (Kronecker) product [217]. For a scratchpad register composed of three binary qubits, one obtains 8 possible states of the quantum register. These states are generated as follows:

$$|000\rangle = |0\rangle \otimes |0\rangle \otimes |0\rangle = \begin{vmatrix}1\\0\end{pmatrix} \otimes \begin{vmatrix}1\\0\end{pmatrix} \otimes \begin{vmatrix}1\\0\end{pmatrix} = [1 \quad 0 \quad 0 \quad 0 \quad 0 \quad 0 \quad 0 \quad 0]^T,$$

$$|001\rangle = |0\rangle \otimes |0\rangle \otimes |1\rangle = \begin{vmatrix}1\\0\end{pmatrix} \otimes \begin{vmatrix}1\\0\end{pmatrix} \otimes \begin{vmatrix}0\\1\end{pmatrix} = [0 \quad 1 \quad 0 \quad 0 \quad 0 \quad 0 \quad 0 \quad 0]^T,$$

$$|010\rangle = |0\rangle \otimes |1\rangle \otimes |0\rangle = \begin{vmatrix}1\\0\end{pmatrix} \otimes \begin{vmatrix}0\\1\end{pmatrix} \otimes \begin{vmatrix}1\\0\end{pmatrix} = [0 \quad 0 \quad 1 \quad 0 \quad 0 \quad 0 \quad 0 \quad 0]^T,$$

$$|011\rangle = |0\rangle \otimes |1\rangle \otimes |1\rangle = \begin{vmatrix}1\\0\end{pmatrix} \otimes \begin{vmatrix}0\\1\end{pmatrix} \otimes \begin{vmatrix}0\\1\end{pmatrix} = [0 \quad 0 \quad 0 \quad 1 \quad 0 \quad 0 \quad 0 \quad 0]^T,$$

$$|100\rangle = |1\rangle \otimes |0\rangle \otimes |0\rangle = \begin{vmatrix}0\\1\end{pmatrix} \otimes \begin{vmatrix}1\\0\end{pmatrix} \otimes \begin{vmatrix}1\\0\end{pmatrix} = [0 \quad 0 \quad 0 \quad 0 \quad 1 \quad 0 \quad 0 \quad 0]^T,$$

$$|101\rangle = |1\rangle \otimes |0\rangle \otimes |1\rangle = \begin{vmatrix}0\\1\end{pmatrix} \otimes \begin{vmatrix}1\\0\end{pmatrix} \otimes \begin{vmatrix}0\\1\end{pmatrix} = [0 \quad 0 \quad 0 \quad 0 \quad 0 \quad 1 \quad 0 \quad 0]^T,$$

$$|110\rangle = |1\rangle \otimes |1\rangle \otimes |0\rangle = \begin{vmatrix}0\\1\end{pmatrix} \otimes \begin{vmatrix}0\\1\end{pmatrix} \otimes \begin{vmatrix}1\\0\end{pmatrix} = [0 \quad 0 \quad 0 \quad 0 \quad 0 \quad 0 \quad 1 \quad 0]^T,$$

$$|111\rangle = |1\rangle \otimes |1\rangle \otimes |1\rangle = \begin{vmatrix}0\\1\end{pmatrix} \otimes \begin{vmatrix}0\\1\end{pmatrix} \otimes \begin{vmatrix}0\\1\end{pmatrix} = [0 \quad 0 \quad 0 \quad 0 \quad 0 \quad 0 \quad 0 \quad 1]^T.$$

In general, a binary quantum register that is composed of k binary qubits can have up to 2^k possible states. The quantum register can be in any of the individual states at any instant of time or in all of the states at the same time. The fact that the scratchpad register can be in all of the states at the same time is the reason for the binary parallelism that exists at the binary quantum level. Due to this parallelism, a binary quantum processor can operate on all of the states of the quantum register at the same time (i.e., it can be modeled like application-specific 2^k binary parallel processors).

For a register composed of 1 qubit, the evolution state ($|\Psi\rangle$) is represented as follows:

$$|\Psi\rangle_{binary-qubit} = \sqrt{p_0}|0\rangle + \sqrt{p_1}|1\rangle, \tag{11.5}$$

where in general:

$$p(|0\rangle) = p_0 = \frac{|\alpha|^2}{|\alpha|^2 + |\beta|^2} \Rightarrow \sqrt{p_0} = \frac{|\alpha|}{\sqrt{|\alpha|^2 + |\beta|^2}},$$

$$p(|1\rangle) = p_1 = \frac{|\beta|^2}{|\alpha|^2 + |\beta|^2} \Rightarrow \sqrt{p_1} = \frac{|\beta|}{\sqrt{|\alpha|^2 + |\beta|^2}},$$

$1 \geq p_i \geq 0,\ i \in \{0,1\}$, *and* $p_0 + p_1 = 1$.

Here p_0 is the probability of the qubit being in state |0>, and p_1 is the probability of the qubit being in state |1>, α and β are complex numbers called "probability amplitudes", and in general $\alpha\alpha^* + \beta\beta^* = |\alpha|^2 + |\beta|^2 = 1$. Thus, if $\{|\alpha| = 0, |\beta| = 1\}$, $\{|\alpha| = 1, |\beta| = 0\}$, $\{|\alpha| = 1/\sqrt{2}, |\beta| = 1/\sqrt{2}\}$, or $\{|\alpha| = 1/\sqrt{4/3}, |\beta| = 1/\sqrt{4}\}$, etc, then $|\alpha|^2 + |\beta|^2 = p_0 + p_1 = 1$.

$$\therefore\ |\Psi\rangle_{binary-qubit} = \sqrt{p_0}|0\rangle + \sqrt{p_1}|1\rangle = \alpha|0\rangle + \beta|1\rangle. \tag{11.6}$$

Eq. (11.6) is the amplitude Eq. that describes two-valued quantum computing. The more general Eq. that includes the phase as well is as follows [167]:

$$|\Psi\rangle_{binary-qubit} = e^{i\gamma}\cos(\frac{\theta}{2})|0\rangle + e^{i(\gamma+\varphi)}\sin(\frac{\theta}{2})|1\rangle, \tag{11.7}$$

where $\{\gamma,\varphi,\theta\} \in R$. Equation (11.6) can be re-written as:

$$|\Psi\rangle_{binary-qubit} = [|0\rangle \quad |1\rangle][E]\begin{bmatrix}\alpha \\ \beta\end{bmatrix}, \tag{11.8}$$

where [E] is an evolution matrix.

In general, the size of the evolution matrix [E] for (k,k) reversible gates over any radix and for an arbitrary number of inputs is governed by the following Eq. [15]:

$$||E|| = radix^{\#inputs = \#outputs} \cdot radix^{\#inputs = \#outputs}, \tag{11.9}$$

where the radix indicates the radix of logic which is used for reversible and quantum computing.

For a two-valued quantum register that is composed of two binary qubits, the quantum state $|\Psi\rangle$ is represented using the computational basis states $\{|0\rangle, |1\rangle\}$, as follows:

$$|\Psi\rangle_{binary-qubit1} = \sqrt{p_0}|0\rangle + \sqrt{p_1}|1\rangle,$$

$$|\Psi\rangle_{binary-qubit2} = \sqrt{p_2}|0\rangle + \sqrt{p_3}|1\rangle,$$

$$|\Psi\rangle_{2-binary-qubit} = |\Psi\rangle_1 \otimes |\Psi\rangle_2$$

$$= (\sqrt{p_0}|0\rangle + \sqrt{p_1}|1\rangle) \otimes (\sqrt{p_2}|0\rangle + \sqrt{p_3}|1\rangle),$$

$$= \sqrt{p_0}\sqrt{p_2}|00\rangle + \sqrt{p_0}\sqrt{p_3}|01\rangle + \sqrt{p_1}\sqrt{p_2}|10\rangle + \sqrt{p_1}\sqrt{p_3}|11\rangle.$$

Given that:

$$\sqrt{p_0} = \frac{|\alpha_1|}{\sqrt{|\alpha_1|^2 + |\beta_1|^2}},$$

$$\sqrt{p_2} = \frac{|\alpha_2|}{\sqrt{|\alpha_2|^2 + |\beta_2|^2}},$$

$$\sqrt{p_1} = \frac{|\beta_1|}{\sqrt{|\alpha_1|^2 + |\beta_1|^2}},$$

$$\sqrt{p_3} = \frac{|\beta_2|}{\sqrt{|\alpha_2|^2 + |\beta_2|^2}}.$$

One obtains $|\Psi\rangle_{2-binary-qubit}$ equals to:

$$\left|\Psi\right\rangle_{2-binary-qubit} =$$

$$\frac{|\alpha_1|}{\sqrt{|\alpha_1|^2+|\beta_1|^2}}\frac{|\alpha_2|}{\sqrt{|\alpha_2|^2+|\beta_2|^2}}\left|00\right\rangle+\frac{|\alpha_1|}{\sqrt{|\alpha_1|^2+|\beta_1|^2}}\frac{|\beta_2|}{\sqrt{|\alpha_2|^2+|\beta_2|^2}}\left|01\right\rangle+$$

$$\frac{|\beta_1|}{\sqrt{|\alpha_1|^2+|\beta_1|^2}}\frac{|\alpha_2|}{\sqrt{|\alpha_2|^2+|\beta_2|^2}}\left|10\right\rangle+\frac{|\beta_1|}{\sqrt{|\alpha_1|^2+|\beta_1|^2}}\frac{|\beta_2|}{\sqrt{|\alpha_2|^2+|\beta_2|^2}}\left|11\right\rangle,$$

$$=$$

$$\frac{|\alpha_1||\alpha_2|}{\sqrt{(|\alpha_1|^2+|\beta_1|^2)(|\alpha_2|^2+|\beta_2|^2)}}\left|00\right\rangle+\frac{|\alpha_1||\beta_2|}{\sqrt{(|\alpha_1|^2+|\beta_1|^2)(|\alpha_2|^2+|\beta_2|^2)}}\left|01\right\rangle+$$

$$\frac{|\beta_1||\alpha_2|}{\sqrt{(|\alpha_1|^2+|\beta_1|^2)(|\alpha_2|^2+|\beta_2|^2)}}\left|10\right\rangle+\frac{|\beta_1||\beta_2|}{\sqrt{(|\alpha_1|^2+|\beta_1|^2)(|\alpha_2|^2+|\beta_2|^2)}}\left|11\right\rangle.$$

Specifically, given the orthonormalization conditions: $|\alpha_1|^2 + |\beta_1|^2 = 1$ and $|\alpha_2|^2 + |\beta_2|^2 = 1$, one obtains:

$(|\alpha_1|^2 + |\beta_1|^2)(|\alpha_2|^2 + |\beta_2|^2)= 1,$

$\alpha_1\alpha_2|^2 + |\alpha_1\beta_2|^2 +|\beta_1\alpha_2|^2 +|\beta_1\beta_2|^2 = 1.$

$$\therefore \left|\Psi\right\rangle_{2-binary-qubit} = \left|\Psi\right\rangle_1 \otimes \left|\Psi\right\rangle_2 = (\alpha_1\left|0\right\rangle+\beta_1\left|1\right\rangle)\otimes(\alpha_2\left|0\right\rangle+\beta_2\left|1\right\rangle),$$
$$=\alpha_1\alpha_2\left|0\right\rangle\left|0\right\rangle+\alpha_1\beta_2\left|0\right\rangle\left|1\right\rangle+\beta_1\alpha_2\left|1\right\rangle\left|0\right\rangle+\beta_1\beta_2\left|1\right\rangle\left|1\right\rangle,$$
$$=\alpha_1\alpha_2\left|00\right\rangle+\alpha_1\beta_2\left|01\right\rangle+\beta_1\alpha_2\left|10\right\rangle+\beta_1\beta_2\left|11\right\rangle.$$

$$\therefore \left|\Psi\right\rangle_{2-binary-qubit} = \left[\left|00\right\rangle \quad \left|01\right\rangle \quad \left|10\right\rangle \quad \left|11\right\rangle\right][E]\begin{bmatrix}\alpha_1\alpha_2\\ \alpha_1\beta_2\\ \beta_1\alpha_2\\ \beta_1\beta_2\end{bmatrix}. \qquad (11.10)$$

where [E] is the quantum evolution process.

It is interesting to note that the binary 2-qubit orthonormal computational basis states: $\{\left|00\right\rangle,\left|01\right\rangle,\left|10\right\rangle,\left|11\right\rangle\}$ in Eq. (11.10) is just one possible set of orthonormal basis states. Other possible orthonormal basis states for a binary 2-qubit quantum register include the Einstein-Podolsky-Rosen (EPR) basis states [167]: $\left\{\frac{\left|00\right\rangle+\left|11\right\rangle}{\sqrt{2}},\frac{\left|00\right\rangle-\left|11\right\rangle}{\sqrt{2}},\frac{\left|01\right\rangle+\left|10\right\rangle}{\sqrt{2}},\frac{\left|01\right\rangle-\left|10\right\rangle}{\sqrt{2}}\right\}$ (Example 11.1 will illustrate the quantum circuit that creates the EPR set of orthonormal basis states). The following represents the unitary evolution matrices for some quantum gates that are used in quantum computing [36,167,230,253] (quantum gates, such as the Pauli gates, can be obtained using quantum mechanical formalisms [81]):

(1,1) gates:

Walsh-Hadamard: $\frac{1}{\sqrt{2}}\begin{bmatrix}1 & 1\\1 & -1\end{bmatrix}$, *Pauli-X (NOT)*: $\begin{bmatrix}0 & 1\\1 & 0\end{bmatrix}$, *Pauli-Y*: $\begin{bmatrix}0 & -i\\i & 0\end{bmatrix}$, *Pauli-Z*: $\begin{bmatrix}1 & 0\\0 & -1\end{bmatrix}$, *Phase*: $\begin{bmatrix}1 & 0\\0 & i\end{bmatrix}$, $\frac{\Pi}{8}$: $\begin{bmatrix}1 & 0\\0 & e^{i\pi/4}\end{bmatrix}$,

R_k : $\begin{bmatrix}1 & 0\\0 & e^{\frac{2\pi i}{2^k}}\end{bmatrix}$, U_θ: $\begin{bmatrix}e^{i(\delta+\sigma+\tau)}\cos(\frac{\theta}{2}) & e^{-i(\delta+\sigma-\tau)}\sin(\frac{\theta}{2})\\-e^{i(\delta-\sigma+\tau)}\sin(\frac{\theta}{2}) & e^{i(\delta-\sigma-\tau)}\cos(\frac{\theta}{2})\end{bmatrix}$,

$\sqrt{NOT}$: $\begin{bmatrix}\frac{1+i}{2} & \frac{1-i}{2}\\\frac{1-i}{2} & \frac{1+i}{2}\end{bmatrix}$, *Wire (Buffer; Identity)*: $\begin{bmatrix}1 & 0\\0 & 1\end{bmatrix}$.

(2,2) gates:

Feynman (CN): $\begin{bmatrix}1 & 0 & 0 & 0\\0 & 1 & 0 & 0\\0 & 0 & 0 & 1\\0 & 0 & 1 & 0\end{bmatrix}$, *Swap*: $\begin{bmatrix}1 & 0 & 0 & 0\\0 & 0 & 1 & 0\\0 & 1 & 0 & 0\\0 & 0 & 0 & 1\end{bmatrix}$,

Controlled-Z: $\begin{bmatrix}1 & 0 & 0 & 0\\0 & 1 & 0 & 0\\0 & 0 & 1 & 0\\0 & 0 & 0 & -1\end{bmatrix}$, *Controlled-Phase*: $\begin{bmatrix}1 & 0 & 0 & 0\\0 & 1 & 0 & 0\\0 & 0 & 1 & 0\\0 & 0 & 0 & i\end{bmatrix}$,

Barenco-DiVincenzo: $\begin{bmatrix}1 & 0 & 0 & 0\\0 & 1 & 0 & 0\\0 & 0 & e^{i\alpha}\cos(\theta) & -ie^{i(\alpha-\phi)}\sin(\theta)\\0 & 0 & -ie^{i(\alpha+\phi)}\sin(\theta) & e^{i\alpha}\cos(\theta)\end{bmatrix}$.

where α, ϕ, and θ are constant irrational multiples of π and one another.

(3,3) gates:

$$Toffoli_{abc/def}: \begin{bmatrix} 1&0&0&0&0&0&0&0\\ 0&1&0&0&0&0&0&0\\ 0&0&1&0&0&0&0&0\\ 0&0&0&1&0&0&0&0\\ 0&0&0&0&1&0&0&0\\ 0&0&0&0&0&1&0&0\\ 0&0&0&0&0&0&0&1\\ 0&0&0&0&0&0&1&0 \end{bmatrix}, \quad Fredkin0_{abc/cf_{r0}f_{r1}}: \begin{bmatrix} 1&0&0&0&0&0&0&0\\ 0&0&1&0&0&0&0&0\\ 0&0&0&0&1&0&0&0\\ 0&0&0&0&0&0&1&0\\ 0&1&0&0&0&0&0&0\\ 0&0&0&0&0&1&0&0\\ 0&0&0&1&0&0&0&0\\ 0&0&0&0&0&0&0&1 \end{bmatrix}.$$

The size of each evolution matrix is governed by Eq. (11.9). For example, for (2,2) two-valued gate, the size of the evolution matrix is equal to $2^2 \cdot 2^2 = 4 \cdot 4 = 16$ elements. Note that the orthogonal Walsh-Hadamard transform $\begin{bmatrix} 1 & 1 \\ 1 & -1 \end{bmatrix}$ is normalized by $\frac{1}{\sqrt{2}}$ to make the matrix unitary (i.e., the columns are orthogonal to each other and the Euclidian length of each column is one).

It has been shown [167] that any quantum computer can be constructed using only the Barenco-DiVincenzo gates, quantum XOR (Feynman gate) and one generalized quantum inverter (i.e., single-input single-output gates), or only using Toffoli gate together with (1,1) gates. This stems from the fact that a complete (universal) system should consist of at least a linear part and a non-linear part, not only a linear part. Since a Feynman gate is only made of a linear part (XOR gate) then it cannot be a universal gate on its own. On the contrary is the Toffoli gate, which consists of a linear part (XOR gate) and a non-linear part (AND gate) that qualify the gate to be universal (complete) (i.e., all possible 256 3-variable binary functions are produced using the quantum Toffoli gate).

From the matrix representation of the quantum gates, the matrix representation is equivalent to the input-output (I/O) mapping representation of quantum gates, as follows: If one considers each *row* in the input side of the I/O map as an input vector represented by the natural binary code of 2^{index} with row index starting from 0, and similarly for the output row of the I/O map, then the matrix transforms the input vector to the corresponding output vector by transforming the code for the input to the code for the output. One notes from this example, that the Feynman gate is merely a permuter, i.e., it produces output vectors, which are permutations of the input vectors.

Although some quantum gates like Feynman gates are merely permuters, not all quantum gates do simple permutations. For example the Walsh-Hadamard quantum gate is a transformer and not only a permuter. The mapping of a set of inputs into any set of outputs can be obtained using quantum computing methods.

According to the principles of quantum mechanics, the combination of quantum state qubits can be in either (1) decomposable or (2) in entangled states [167]. While each individual state qubit can be observed in the former case, the same is impossible in the later. The combination of two systems with the bases $\{|x_1\rangle, |x_2\rangle, \ldots, |x_n\rangle\}$ and $\{|y_1\rangle, |y_2\rangle, \ldots, |y_m\rangle\}$ is described as a pair $(|x_i\rangle, |y_j\rangle)$, and the composite quantum state is expressed as:

$$\sum_{i=1}^{n}\sum_{j=1}^{m}\alpha_{ij}\left|x_i, y_j\right\rangle. \tag{11.11}$$

In quantum logic, one defines a state to be decomposable if it can be expressed as:

$$\begin{aligned}\sum_{i=1}^{n}\ldots\sum_{j=1}^{m}\alpha_{i\ldots j}\left|x_i,\ldots,x_j\right\rangle &= \sum_{i=1}^{n}\ldots\sum_{j=1}^{m}\alpha_i\ldots\beta_j\left|x_i\right\rangle\ldots\left|x_j\right\rangle,\\ &= \sum_{i=1}^{n}\alpha_i\left|x_i\right\rangle\otimes\ldots\otimes\sum_{j=1}^{m}\beta_j\left|x_j\right\rangle,\\ &= \sum_{i=1}^{n}\alpha_i\left|x_i\right\rangle\ldots\sum_{j=1}^{m}\beta_j\left|x_j\right\rangle.\end{aligned} \tag{11.12}$$

Otherwise, the state is entangled. The speedups in quantum computations seem to be due to the entanglement, by which many computations are performed in parallel. The following example illustrates the concept of two-valued quantum entanglement.

Example 11.1. This example demonstrates the concept of two-valued entanglement.

11.1a. Consider a two-valued quantum system of two qubits, given as:

$$\frac{1}{2}(|00\rangle+|01\rangle+|10\rangle+|11\rangle),$$

$$=\frac{1}{\sqrt{2}}(|0\rangle+|1\rangle)\otimes\frac{1}{\sqrt{2}}(|0\rangle+|1\rangle)=\frac{1}{\sqrt{2}}(|0\rangle+|1\rangle)\frac{1}{\sqrt{2}}(|0\rangle+|1\rangle).$$

This system is decomposable, as the functions of the first and second qubits are disentangled according to Eq. (11.12).

11.1b. Consider now the two-valued quantum system:

$$\frac{1}{\sqrt{2}}\left(|01\rangle + |10\rangle\right).$$

This system is entangled, as no decomposition according to Eq. (11.12) is possible.

Among the previous quantum gates, the Walsh-Hadamard gate and Feynman gate are the most important quantum gates that are often used in binary quantum computing and synthesis. Analogous to the classical non-quantum operators (e.g., optical systems [49,261]), each quantum evolution matrix represents a unique quantum gate. For an evolution matrix of the size given by Eq. (11.9), the input qubit to the quantum gate corresponds to the column index of the evolution matrix, and the output qubit of the quantum gate corresponds to the row index of the evolution matrix. The column and row indices of the evolution matrix are in the following order for 1-input 1-output gates, 2-input 2-output gates, and 3-input 3-output gates, respectively:

$$1-input \quad 1-output \qquad gate: |0\rangle, |1\rangle.$$

$$2-input \quad 2-output \qquad gate: |00\rangle, |01\rangle, |10\rangle, |11\rangle.$$

$$3-input \quad 3-output \qquad gate: |000\rangle, |001\rangle, |010\rangle, |011\rangle, |100\rangle, |101\rangle, |110\rangle, |111\rangle.$$

Quantum circuits are synthesized using interconnects between quantum primitives. Interconnects can be serial-like interconnects, parallel-like interconnects, or a mixture of serial-like and parallel-like inter-connects.

Example 11.2. This example illustrates a number representation and qubit transformation in binary quantum computing.

11.2a. Number representation:

$$|13\rangle_{decimal} = |1101\rangle_{Boolean} = |\downarrow\downarrow\uparrow\downarrow\rangle_{Boolean}.$$

11.2b. Pauli-Z transformation: $\begin{bmatrix} 1 & 0 \\ 0 & -1 \end{bmatrix}|0\rangle = |0\rangle, \begin{bmatrix} 1 & 0 \\ 0 & -1 \end{bmatrix}|1\rangle = -|1\rangle.$

11.2c. Normalized Walsh-Hadamard transformation:

$$\frac{1}{\sqrt{2}}\begin{bmatrix} 1 & 1 \\ 1 & -1 \end{bmatrix}|0\rangle = \frac{1}{\sqrt{2}}\begin{bmatrix} 1 \\ 1 \end{bmatrix},$$

$$\left(\frac{1}{\sqrt{2}}\begin{bmatrix}1 & 1\\ 1 & -1\end{bmatrix}\cdot\frac{1}{\sqrt{2}}\begin{bmatrix}1 & 1\\ 1 & -1\end{bmatrix}\right)|0\rangle = |0\rangle .$$

Thus, two serially inter-connected Walsh-Hadamard gates lead to the Identity transformation.

11.2d. Figure 11.7 illustrates the quantum logic circuit that creates the EPR basis states [167]:

$$\left\{\frac{|00\rangle+|11\rangle}{\sqrt{2}}, \frac{|00\rangle-|11\rangle}{\sqrt{2}}, \frac{|01\rangle+|10\rangle}{\sqrt{2}}, \frac{|01\rangle-|10\rangle}{\sqrt{2}}\right\}.$$

Input qubits	Output qubits
|00>	(|00> + |11>)/√2
|01>	(|01> + |10>)/√2
|10>	(|00> - |11>)/√2
|11>	(|01> - |10>)/√2

a

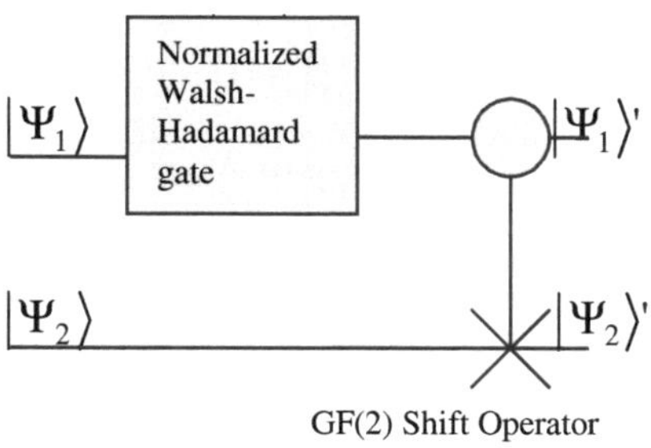

b

Fig. 11.7. a Input-Output quantum truth table, and **b** the corresponding quantum circuit that creates the orthonormal EPR basis states.

Utilizing Fig. 11.7.b, and the shift operator over GF(2): {0→1,1→0}, one obtains the following Eqs.:

$$|\Psi_1\rangle = \alpha_1|0\rangle + \beta_1|1\rangle,$$

$$|\Psi_2\rangle = \alpha_2|0\rangle + \beta_2|1\rangle,$$

$$|\Psi_1\rangle' = [|0\rangle \quad |1\rangle]\frac{1}{\sqrt{2}}\begin{bmatrix}1 & 1\\ 1 & -1\end{bmatrix}\begin{bmatrix}\alpha_1\\ \beta_1\end{bmatrix},$$

$$= \alpha_1\frac{|0\rangle+|1\rangle}{\sqrt{2}} + \beta_1\frac{|0\rangle-|1\rangle}{\sqrt{2}},$$

$$|\Psi_2\rangle' = Shift_{|\Psi_2\rangle}(|\Psi_1\rangle'),$$

where $|\Psi_2\rangle' = Shift_{|\Psi_2\rangle}(|\Psi_1\rangle')$ means to shift the value of the basis states of $|\Psi_1\rangle'$ by the amount $|\Psi_2\rangle$ over GF(2).

Then: For $\{\alpha_1 = 1, \beta_1 = 0\}$: $|\Psi_1\rangle' = \dfrac{|0\rangle + |1\rangle}{\sqrt{2}}$,

$\Rightarrow$ For$\{\alpha_2=1,\beta_2=0\}$:

$$|\Psi_2\rangle = |0\rangle, \quad \therefore |\Psi_2\rangle' = \frac{|0\rangle + |1\rangle}{\sqrt{2}} \Rightarrow |\Psi\rangle = |\Psi_1{}'\Psi_2{}'\rangle = \frac{|00\rangle + |11\rangle}{\sqrt{2}}.$$

$\Rightarrow$ For$\{\alpha_2=0,\beta_2=1\}$:

$$|\Psi_2\rangle = |1\rangle, \quad \therefore |\Psi_2\rangle' = \frac{|1\rangle + |0\rangle}{\sqrt{2}} \Rightarrow |\Psi\rangle = |\Psi_1{}'\Psi_2{}'\rangle = \frac{|01\rangle + |10\rangle}{\sqrt{2}}.$$

For $\{\alpha_1 = 0, \beta_1 = 1\}$: $|\Psi_1\rangle' = \dfrac{|0\rangle - |1\rangle}{\sqrt{2}}$,

$\Rightarrow$ For$\{\alpha_2=1,\beta_2=0\}$:

$$|\Psi_2\rangle = |0\rangle, \quad \therefore |\Psi_2\rangle' = \frac{|0\rangle - |1\rangle}{\sqrt{2}} \Rightarrow |\Psi\rangle = |\Psi_1{}'\Psi_2{}'\rangle = \frac{|00\rangle - |11\rangle}{\sqrt{2}}.$$

$\Rightarrow$ For$\{\alpha_2=0,\beta_2=1\}$:

$$|\Psi_2\rangle = |1\rangle, \quad \therefore |\Psi_2\rangle' = \frac{|1\rangle - |0\rangle}{\sqrt{2}} \Rightarrow |\Psi\rangle = |\Psi_1{}'\Psi_2{}'\rangle = \frac{|01\rangle - |10\rangle}{\sqrt{2}}.$$

11.2e. The following is the derivation of the orthonormal composite basis states: $\left\{|+\rangle = \dfrac{|0\rangle + |1\rangle}{\sqrt{2}}, |-\rangle = \dfrac{|0\rangle - |1\rangle}{\sqrt{2}}\right\}$.

Assuming the normalization of the "probability amplitudes", and using the quantum signal $|\Psi\rangle = \alpha|0\rangle + \beta|1\rangle$ as an input to the normalized Walsh-Hadamard circuit in Fig. 11.8.

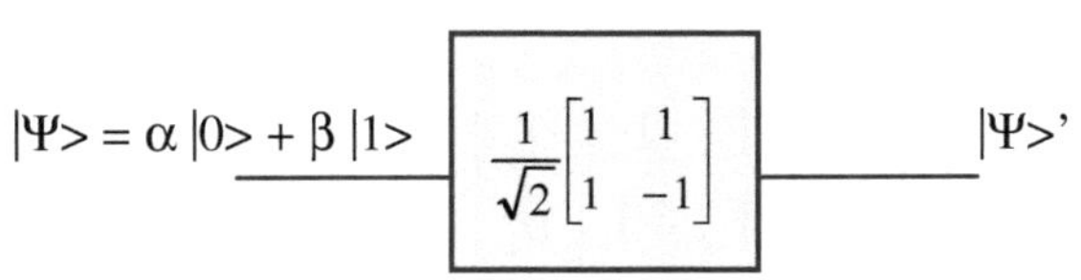

Fig. 11.8. Walsh-Hadamard logic circuit.

One obtains the following quantum signal at the output of the gate:

$$|\Psi\rangle' = [|0\rangle \quad |1\rangle]\frac{1}{\sqrt{2}}\begin{bmatrix}1 & 1\\ 1 & -1\end{bmatrix}\begin{bmatrix}\alpha\\ \beta\end{bmatrix} = [|0\rangle \quad |1\rangle]\begin{bmatrix}\frac{\alpha+\beta}{\sqrt{2}}\\ \frac{\alpha-\beta}{\sqrt{2}}\end{bmatrix},$$

$$= \frac{\alpha+\beta}{\sqrt{2}}|0\rangle + \frac{\alpha-\beta}{\sqrt{2}}|1\rangle.$$

$$|\Psi\rangle' = [|0\rangle \quad |1\rangle]\frac{1}{\sqrt{2}}\begin{bmatrix}1 & 1\\ 1 & -1\end{bmatrix}\begin{bmatrix}\alpha\\ \beta\end{bmatrix} = \begin{bmatrix}\frac{|0\rangle+|1\rangle}{\sqrt{2}} & \frac{|0\rangle-|1\rangle}{\sqrt{2}}\end{bmatrix}\begin{bmatrix}\alpha\\ \beta\end{bmatrix},$$

$$= [|+\rangle \quad |-\rangle]\begin{bmatrix}\alpha\\ \beta\end{bmatrix} = \alpha|+\rangle + \beta|-\rangle.$$

Where:

$$\left\{|+\rangle = \frac{|0\rangle+|1\rangle}{\sqrt{2}}, |-\rangle = \frac{|0\rangle-|1\rangle}{\sqrt{2}}\right\} \text{ and } \left\{|0\rangle = \frac{|+\rangle+|-\rangle}{\sqrt{2}}, |1\rangle = \frac{|+\rangle-|-\rangle}{\sqrt{2}}\right\}.$$

Therefore, one obtains, at the input side of the quantum Walsh-Hadamard gate, the following quantum signal:

$$\therefore |\Psi\rangle = \alpha|0\rangle + \beta|1\rangle = \alpha\frac{|+\rangle+|-\rangle}{\sqrt{2}} + \beta\frac{|+\rangle-|-\rangle}{\sqrt{2}},$$

$$= \frac{\alpha+\beta}{\sqrt{2}}|+\rangle + \frac{\alpha-\beta}{\sqrt{2}}|-\rangle.$$

Therefore: $\{\sqrt{p_{|+\rangle}} = \frac{\alpha+\beta}{\sqrt{2}}, \sqrt{p_{|-\rangle}} = \frac{\alpha-\beta}{\sqrt{2}}\}$.

Consequently, measuring $|\Psi\rangle$ with respect to the new basis $\{|+\rangle, |-\rangle\}$ will result in the state (basis) $\{|+\rangle\}$ with probability $\frac{|\alpha+\beta|^2}{2}$ and the state (basis) $\{|-\rangle\}$ with probability $\frac{|\alpha-\beta|^2}{2}$.

Example 11.3. The following circuits illustrate the process of evolving the input binary quantum bits using a composite of binary quantum primitives in serial-like, parallel-like, and a mixture of serial-like and parallel-like interconnects, respectively.

11.3a. Figure 11.9 illustrates the process of evolving the input binary qubits using the corresponding quantum circuits. Let us evolve the input binary qubit $|11\rangle = \begin{bmatrix}0\\1\end{bmatrix} \otimes \begin{bmatrix}0\\1\end{bmatrix} = [0 \;\; 0 \;\; 0 \;\; 1]^T$.

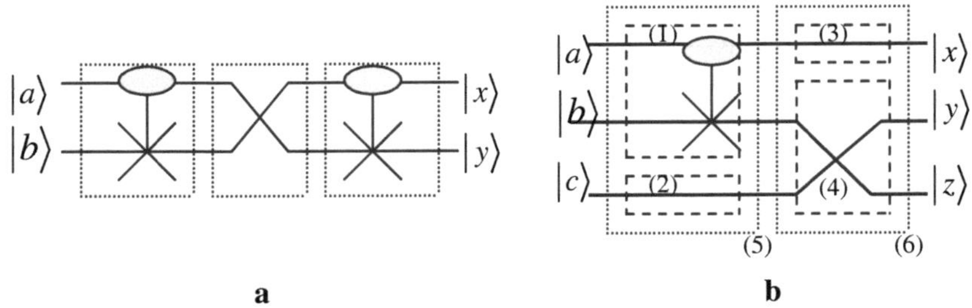

Fig. 11.9. a Quantum circuit composed of a serial interconnect of two Feynman gates and a Swap gate, and **b** quantum circuit composed of serial and parallel interconnects of a single Feynman gate, two Wires (Buffers), and a single Swap gate.

The evolution of the input qubit using cascaded (i.e., serially-interconnected) quantum gates can be viewed in two equivalent perspectives. The first perspective is to evolve the input qubit step by step using the serially interconnected gates. The second perspective, is to evolve the input qubit using the total quantum circuit at once, since the total evolution transformation $[M_{net}]$ is equal to the multiplication of the individual evolution matrices $[M_q]$ that correspond to the individual quantum primitives: $\therefore [M_{net}]_{serial} = \prod_q [M_q]$.

<u>Perspective #1:</u>

$$\begin{bmatrix}1&0&0&0\\0&1&0&0\\0&0&0&1\\0&0&1&0\end{bmatrix}\begin{bmatrix}0\\0\\0\\1\end{bmatrix} = \begin{bmatrix}0\\0\\1\\0\end{bmatrix} \Rightarrow \begin{bmatrix}1&0&0&0\\0&0&1&0\\0&1&0&0\\0&0&0&1\end{bmatrix}\begin{bmatrix}0\\0\\1\\0\end{bmatrix} = \begin{bmatrix}0\\1\\0\\0\end{bmatrix},$$

$$\Rightarrow \begin{bmatrix}1&0&0&0\\0&1&0&0\\0&0&0&1\\0&0&1&0\end{bmatrix}\begin{bmatrix}0\\1\\0\\0\end{bmatrix} = \begin{bmatrix}0\\1\\0\\0\end{bmatrix}.$$

Perspective #2:

$$\left(\begin{bmatrix}1&0&0&0\\0&1&0&0\\0&0&0&1\\0&0&1&0\end{bmatrix}\begin{bmatrix}1&0&0&0\\0&0&1&0\\0&1&0&0\\0&0&0&1\end{bmatrix}\begin{bmatrix}1&0&0&0\\0&1&0&0\\0&0&0&1\\0&0&1&0\end{bmatrix}\right)\begin{bmatrix}0\\0\\0\\1\end{bmatrix}=\begin{bmatrix}0\\1\\0\\0\end{bmatrix}.$$

Thus, the quantum circuit shown in Fig. 11.9a evolves the qubit $|11\rangle$ into the qubit $|01\rangle$.

The quantum circuit in Fig. 11.9b is composed of a serial interconnect of two parallel circuits as follows: dashed boxes ((1),(2)) and ((3),(4)) are parallel interconnected, and dotted boxes (5) and (6) are serially interconnected. The total evolution transformation $[M_{net}]$ of the total parallel-interconnected quantum circuit is equal to the tensor (Kronecker) product of the individual evolution matrices $[M_q]$ that correspond to the individual quantum primitives: $\therefore [M_{net}]_{parallel} = \otimes [M_q]$. Thus, analogously to the operations of the circuit in Fig. 11.9a, the evolution of the input qubit, in Fig. 11.9b, can be viewed in two equivalent perspectives. The first perspective is to evolve the input qubit stage by stage. The second perspective is to evolve the input qubit using the total quantum circuit at once. Let us evolve the input binary qubit $|111\rangle$ using the quantum circuit in Fig. 11.9b. The evolution matrices of the parallel-interconnected dashed boxes in (5) and (6), are as follows (where the symbol || means parallel connection):

$$\text{input} = |1\rangle \otimes |1\rangle \otimes |1\rangle = \begin{pmatrix}0\\1\end{pmatrix} \otimes \begin{pmatrix}0\\1\end{pmatrix} \otimes \begin{pmatrix}0\\1\end{pmatrix} = [00000001]^T.$$

The evolution matrix for (5) = (1) || (2) is:

$$\text{Feynman}\otimes\text{Wire} = \begin{bmatrix}1&0&0&0\\0&1&0&0\\0&0&0&1\\0&0&1&0\end{bmatrix} \otimes \begin{bmatrix}1&0\\0&1\end{bmatrix},$$

$$= \begin{bmatrix} \begin{bmatrix} 1 & 0 \\ 0 & 1 \end{bmatrix} & & & \\ & \begin{bmatrix} 1 & 0 \\ 0 & 1 \end{bmatrix} & & \\ & & & \begin{bmatrix} 1 & 0 \\ 0 & 1 \end{bmatrix} \\ & & \begin{bmatrix} 1 & 0 \\ 0 & 1 \end{bmatrix} & \end{bmatrix}.$$

The evolution matrix for (6) = (3) || (4) is:

$$\text{Wire} \otimes \text{Swap} = \begin{bmatrix} 1 & 0 \\ 0 & 1 \end{bmatrix} \otimes \begin{bmatrix} 1 & 0 & 0 & 0 \\ 0 & 0 & 1 & 0 \\ 0 & 1 & 0 & 0 \\ 0 & 0 & 0 & 1 \end{bmatrix},$$

$$= \begin{bmatrix} \begin{bmatrix} 1 & 0 & 0 & 0 \\ 0 & 0 & 1 & 0 \\ 0 & 1 & 0 & 0 \\ 0 & 0 & 0 & 1 \end{bmatrix} & \\ & \begin{bmatrix} 1 & 0 & 0 & 0 \\ 0 & 0 & 1 & 0 \\ 0 & 1 & 0 & 0 \\ 0 & 0 & 0 & 1 \end{bmatrix} \end{bmatrix}.$$

<u>Perspective #1</u>: input$\Rightarrow$ (5) $\Rightarrow$ output$_1$, input$_2$ (= output$_1$)$\Rightarrow$ (6) $\Rightarrow$ output$_2$.

$$\begin{bmatrix} \begin{bmatrix} 1 & 0 \\ 0 & 1 \end{bmatrix} & & & \\ & \begin{bmatrix} 1 & 0 \\ 0 & 1 \end{bmatrix} & & \\ & & & \begin{bmatrix} 1 & 0 \\ 0 & 1 \end{bmatrix} \\ & & \begin{bmatrix} 1 & 0 \\ 0 & 1 \end{bmatrix} & \end{bmatrix} \begin{bmatrix} 0 \\ 0 \\ 0 \\ 0 \\ 0 \\ 0 \\ 0 \\ 1 \end{bmatrix} = \begin{bmatrix} 0 \\ 0 \\ 0 \\ 0 \\ 0 \\ 1 \\ 0 \\ 0 \end{bmatrix},$$

$$\begin{bmatrix} \begin{bmatrix} 1 & 0 & 0 & 0 \\ 0 & 0 & 1 & 0 \\ 0 & 1 & 0 & 0 \\ 0 & 0 & 0 & 1 \end{bmatrix} & \\ & \begin{bmatrix} 1 & 0 & 0 & 0 \\ 0 & 0 & 1 & 0 \\ 0 & 1 & 0 & 0 \\ 0 & 0 & 0 & 1 \end{bmatrix} \end{bmatrix} \begin{bmatrix} 0 \\ 0 \\ 0 \\ 0 \\ 0 \\ 1 \\ 0 \\ 0 \end{bmatrix} = \begin{bmatrix} 0 \\ 0 \\ 0 \\ 0 \\ 0 \\ 0 \\ 1 \\ 0 \end{bmatrix} = |110\rangle .$$

<u>Perspective #2:</u> input$\Rightarrow$ ((6)(5)) $\Rightarrow$ output$_2$.

$$\left(\begin{bmatrix} \begin{bmatrix} 1 & 0 & 0 & 0 \\ 0 & 0 & 1 & 0 \\ 0 & 1 & 0 & 0 \\ 0 & 0 & 0 & 1 \end{bmatrix} & \\ & \begin{bmatrix} 1 & 0 & 0 & 0 \\ 0 & 0 & 1 & 0 \\ 0 & 1 & 0 & 0 \\ 0 & 0 & 0 & 1 \end{bmatrix} \end{bmatrix} \begin{bmatrix} \begin{bmatrix} 1 & 0 \\ 0 & 1 \end{bmatrix} & & & \\ & \begin{bmatrix} 1 & 0 \\ 0 & 1 \end{bmatrix} & & \\ & & \begin{bmatrix} 1 & 0 \\ 0 & 1 \end{bmatrix} & \\ & & & \begin{bmatrix} 1 & 0 \\ 0 & 1 \end{bmatrix} \end{bmatrix} \right) \cdot$$

$$\begin{bmatrix} 0 \\ 0 \\ 0 \\ 0 \\ 0 \\ 0 \\ 0 \\ 1 \end{bmatrix} = \begin{bmatrix} 0 \\ 0 \\ 0 \\ 0 \\ 0 \\ 0 \\ 1 \\ 0 \end{bmatrix} = |110\rangle .$$

Thus, the quantum circuit shown in Fig. 11.9b evolves the qubit $|111\rangle$ into the qubit $|110\rangle$.

(Note here the reason for the name "tensor product"; the form of the matrix for a parallel interconnect is in the form of matrix of matrix elements (i.e., tensor) generated by this type of product.)

11.3b. Let us evolve the input qubit |1111> using the following parallel-like interconnected quantum circuit.

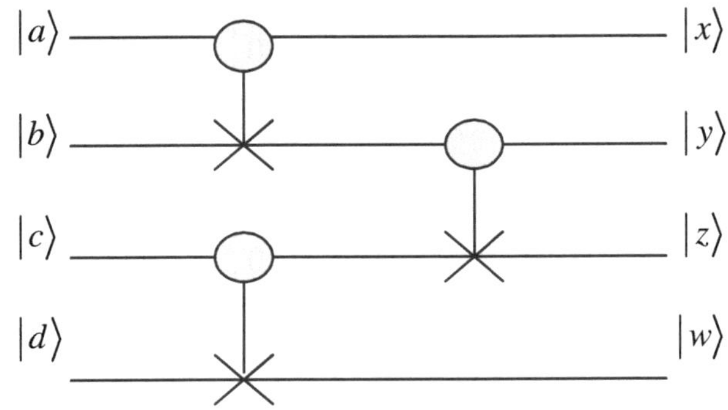

Fig. 11.10. Quantum circuit composed of parallel-like interconnect of three Feynman gates.

It can be shown the qubit |1111> evolves using the quantum circuit in Fig. 11.10 into the qubit |1010>.

11.3c. Let us evolve the input qubit |100> using the following mixture of interconnected quantum circuits:

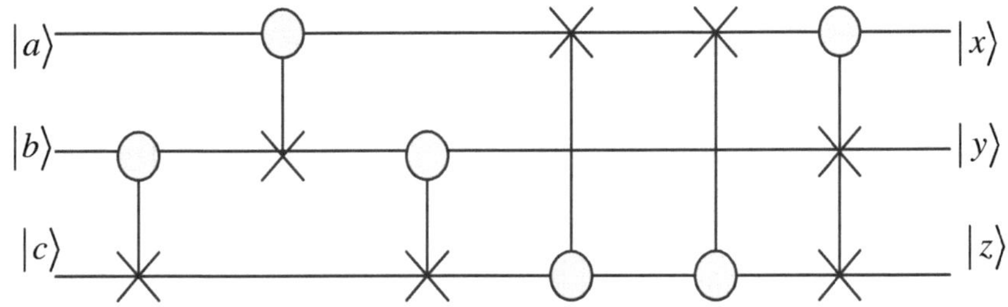

Fig. 11.11. Quantum circuit composed of parallel-like and serial-like interconnects of five Feynman gates and one controlled-Swap gate.

It can be shown the qubit |101> evolves using the quantum circuit in Fig. 11.11 into the qubit |101>.

11.3d. Since, in logic design, the replacement of one "large" circuit with an equivalent smaller size circuit, using components from library, is very important because it reduces the cost of synthesis, the same reasoning would hold in the synthesis of quantum logic circuits. The following shows an example of quantum circuit equivalence of two quantum logic circuits, where the first circuit is a binary quantum Swap gate and the second equivalent circuit is made of three serially-interconnected binary Feynman gates.

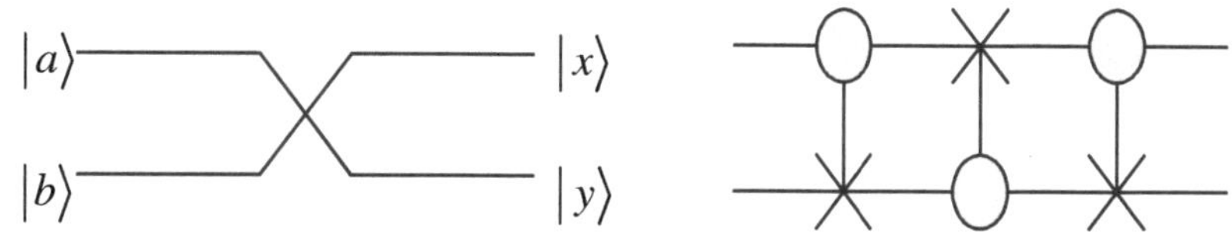

a **b**

Fig. 11.12. **a** Binary quantum Swap gate, and **b** its equivalence in terms of three serially-interconnected binary Feynman gates. This equivalence in structural transformation is important since Swap, as crossing wires, can be realizable in quantum logic using Feynman gates.

11.1.1 Mathematical Decompositions for Quantum Computing

Factorization of the evolution process leads to the serial decomposition of the total quantum circuit into serially interconnected quantum sub-circuits. Utilizing Example 11.3, it is interesting to solve for the following binary evolution process factorization (or equivalently quantum circuit decomposition) problem: given the output (evolved) qubit, factorize the total composite evolution process into known evolution sub-processes. This type of quantum decomposition can be very useful in the synthesis of quantum logic circuits.

Quantum Analysis means to take the total synthesized quantum circuit of interconnected quantum sub-circuits and produce the total evolution matrix from it. Quantum Synthesis is the opposite; by having the total evolution matrix we want to produce within specific design constraints certain topological quantum circuit made up of either totally serial interconnects (i.e., using only matrix product), totally parallel interconnects (i.e., using only Kronecker product), or a hybrid of serial and parallel interconnects (i.e., using both matrix product and Kronecker product). The definitions of quantum analysis and quantum synthesis are illustrated in Fig. 11.13.

Consequently, many decompositions that are commonly used in linear algebra have been proposed [15,167] for the decomposition of the unitary evolution matrices, like: spectral theorem, Z-Y decomposition, Polar decomposition, LDU decomposition, Jordan decomposition, fast Fourier-like decomposition [123], and Singular Value Decomposition (SVD) [15].

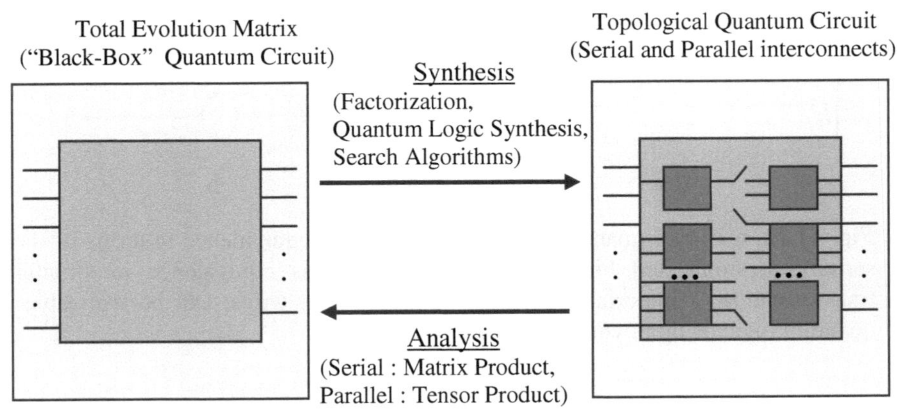

Fig. 11.13. Demonstration of Quantum Analysis versus Quantum Synthesis.

Example 11.4. This example illustrates the use of factorization to decompose the total evolution processes into quantum-realizable sub-processes, and to find quantum circuit equivalences.

11.4a. For a complex matrix [M], in general, M is said to be symmetric iff $[M] = [M]^H$, where H is the Hermitian (i.e., transpose of the conjugate). If [M] is symmetric, then the spectral theorem (in geometry and mechanics this is also known as "principal axis theorem") is as follows:

$$[M]=[Q][\Lambda][Q]^H=\begin{bmatrix} | & . & | \\ eve_1 & . & eve_n \\ | & . & | \end{bmatrix}_{nxn}\begin{bmatrix} eva_1 & 0 & 0 \\ 0 & . & 0 \\ 0 & 0 & eva_n \end{bmatrix}_{nxn}\begin{bmatrix} - & eve_1^H & - \\ . & . & . \\ - & eve_n^H & - \end{bmatrix}_{nxn}$$

where eve is the eigenvector of [M], and eva is the eigenvalue of [M], [Q] is an orthonormal matrix, and $[\Lambda]$ is a diagonal matrix. The following is the result of the application of spectral theorem on the evolution matrix $\begin{bmatrix} \cos\tau & -\sin\tau \\ \sin\tau & \cos\tau \end{bmatrix}$, which is unitary and symmetric.

$$\Rightarrow M=\begin{bmatrix} \cos\tau & -\sin\tau \\ \sin\tau & \cos\tau \end{bmatrix}=\frac{1}{\sqrt{2}}\begin{bmatrix} 1 & 1 \\ -i & i \end{bmatrix}\begin{bmatrix} e^{i\tau} & 0 \\ 0 & e^{-i\tau} \end{bmatrix}\frac{1}{\sqrt{2}}\begin{bmatrix} 1 & i \\ 1 & -i \end{bmatrix}.$$

But:

$$\begin{bmatrix}1 & i\\1 & -i\end{bmatrix}=\begin{bmatrix}1 & 1\\1 & -1\end{bmatrix}\begin{bmatrix}1 & 0\\0 & i\end{bmatrix},\begin{bmatrix}1 & 1\\-i & i\end{bmatrix}=\begin{bmatrix}1 & 0\\0 & -i\end{bmatrix}\begin{bmatrix}1 & 1\\1 & -1\end{bmatrix},$$
$$=\begin{bmatrix}1 & 0\\0 & -1\end{bmatrix}\begin{bmatrix}1 & 0\\0 & i\end{bmatrix}\begin{bmatrix}1 & 1\\1 & -1\end{bmatrix}.$$

Where: $\begin{bmatrix}1 & 1\\1 & -1\end{bmatrix}$ is Walsh-Hadamard gate, $\begin{bmatrix}1 & 0\\0 & i\end{bmatrix}$ is Phase gate, $\begin{bmatrix}1 & 0\\0 & -1\end{bmatrix}$ is Pauli-Z gate, and $\begin{bmatrix}e^{i\tau} & 0\\0 & e^{-i\tau}\end{bmatrix}$ is U_θ gate with $\theta = 0$, $\delta = 0$, $\sigma = 0$, and $\tau = 1$. Therefore:

$$\begin{bmatrix}\cos\tau & -\sin\tau\\\sin\tau & \cos\tau\end{bmatrix}=\begin{bmatrix}1 & 0\\0 & -1\end{bmatrix}\begin{bmatrix}1 & 0\\0 & i\end{bmatrix}\frac{1}{\sqrt{2}}\begin{bmatrix}1 & 1\\1 & -1\end{bmatrix}\begin{bmatrix}e^{i\tau} & 0\\0 & e^{-i\tau}\end{bmatrix}\frac{1}{\sqrt{2}}\begin{bmatrix}1 & 1\\1 & -1\end{bmatrix}\begin{bmatrix}1 & 0\\0 & i\end{bmatrix}.$$

This result is presented in Fig. 11.14.

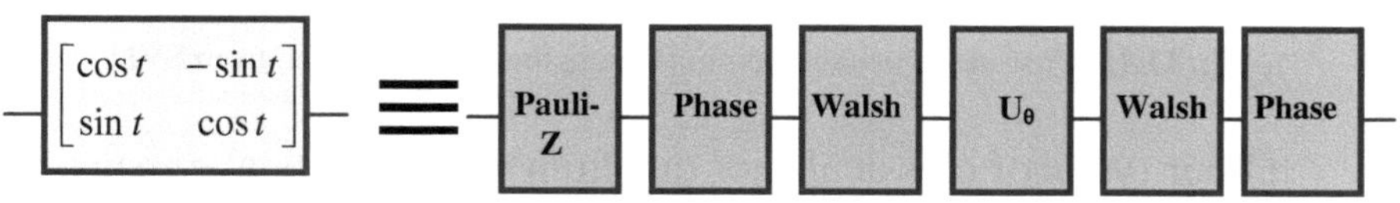

Fig. 11.14. Cascade of quantum circuits using the decomposition from the spectral theorem.

11.4b. The following illustrates finding serially-interconnected quantum sub-blocks which is equivalent to a single quantum evolution process using the SVD decomposition. For *any* matrix $[N]_{mxn}$, the Singular Value Decomposition (SVD) is as follows:

$$[N]=[Q_1][\Xi][Q_2]^T,$$
$$=\begin{bmatrix}| & . & |\\ eve_1(NN^H) & . & eve_m(NN^H)\\ | & . & |\end{bmatrix}_{mxm}\begin{bmatrix}\sqrt{eva_1} & . & 0\\ . & . & .\\ 0 & . & \sqrt{eva_r}\end{bmatrix}_{mxn}\begin{bmatrix}| & . & |\\ eve_1(N^HN) & . & eve_n(N^HN)\\ | & . & |\end{bmatrix}_{nxn}^T$$

where *eve* is the eigenvector, and *eva* is the eigenvalue of $[NN^H]$ and $[N^HN]$, $[Q_1]$ and $[Q_2]$ are orthogonal matrices, and $[\Xi]$ is a diagonal matrix of r singular-value elements. Let's use the SVD decomposition to decompose the normalized Walsh-Hadamard operator into an equivalent cascade of serially-interconnected sub-operators.

$$\Rightarrow \frac{1}{\sqrt{2}}\begin{bmatrix}1 & 1\\ 1 & -1\end{bmatrix} = \frac{1}{\sqrt{2}}\begin{bmatrix}1 & 1\\ -1 & 1\end{bmatrix}\begin{bmatrix}1 & 0\\ 0 & 1\end{bmatrix}\begin{bmatrix}0 & 1\\ 1 & 0\end{bmatrix}.$$

But:

$$\begin{bmatrix}1 & 1\\ -1 & 1\end{bmatrix} = \begin{bmatrix}1 & 0\\ 0 & -1\end{bmatrix}\begin{bmatrix}1 & 1\\ 1 & -1\end{bmatrix} \Rightarrow$$

$$\frac{1}{\sqrt{2}}\begin{bmatrix}1 & 1\\ 1 & -1\end{bmatrix} = \begin{bmatrix}1 & 0\\ 0 & -1\end{bmatrix}\frac{1}{\sqrt{2}}\begin{bmatrix}1 & 1\\ 1 & -1\end{bmatrix}\begin{bmatrix}1 & 0\\ 0 & 1\end{bmatrix}\begin{bmatrix}0 & 1\\ 1 & 0\end{bmatrix},$$

Where: $\begin{bmatrix}1 & 0\\ 0 & -1\end{bmatrix}$ is Pauli-Z, $\begin{bmatrix}1 & 0\\ 0 & 1\end{bmatrix}$ is a wire (Buffer), and $\begin{bmatrix}0 & 1\\ 1 & 0\end{bmatrix}$ is Pauli-X. This is presented in Fig. 11.15.

Fig. 11.15. Cascade of quantum circuits using the decomposition from SVD.

Other types of classical (not quantum) mathematical expansions have been also developed to fit the context of quantum computing. One important example is the quantum Fourier transform (QFT) [167,226,228,249]. Analogously to the classical discrete Fourier transform (DFT) with its crucial role in so many real-world data processing applications (e.g., in computer speech recognition), QFT has many important application such as in the attempt to solve the factoring problem [226,228].

The derivation of the quantum Fourier expansion (transform) is performed as follows [167]: having the mathematical notation for DFT as follows: $y_k \equiv \frac{1}{\sqrt{N}}\sum_{j=0}^{N-1} x_j e^{\frac{2\pi ijk}{N}}$, then the quantum Fourier transform is exactly the same as the classical discrete Fourier transform, except it is written in the quantum notation instead of the conventional notation. This is done in the quantum notation as a linear combination (operator, superposition) of the orthonormal computational basis set $\{|0\rangle, |1\rangle, \ldots, |N-1\rangle\}$ of an n qubit quantum computer, where $N = 2^n$, as follows:

$$|j\rangle \rightarrow \frac{1}{\sqrt{N}} \sum_{k=0}^{N-1} e^{\frac{2\pi ijk}{N}} |k\rangle,$$

$$= \frac{1}{\sqrt{2^n}} \sum_{k=0}^{2^n-1} e^{\frac{2\pi ijk}{2^n}} |k\rangle = \frac{1}{2^{\frac{n}{2}}} \sum_{k=0}^{2^n-1} e^{\frac{2\pi ijk}{2^n}} |k\rangle,$$

$$= \frac{1}{2^{\frac{n}{2}}} \sum_{k_1=0}^{1} \ldots \sum_{k_n=0}^{1} e^{2\pi ij(\sum_{l=1}^{n} k_l 2^{-l})} |k_1 \ldots k_n\rangle$$

$$= \frac{1}{2^{\frac{n}{2}}} \sum_{k_1=0}^{1} \ldots \sum_{k_n=0}^{1} \prod_{l=1}^{n} e^{2\pi ijk_l 2^{-l}} |k_l\rangle,$$

$$= \frac{1}{2^{\frac{n}{2}}} \prod_{l=1}^{n} \left[\sum_{k_l=0}^{1} e^{2\pi ijk_l 2^{-l}} |k_l\rangle \right] = \frac{1}{2^{\frac{n}{2}}} \prod_{l=1}^{n} \left[|0\rangle + e^{2\pi ij 2^{-l}} |1\rangle \right].$$

where the following two-valued representation (notation) of the quantum state $|j\rangle$ has been used:

$$|j\rangle = j_1 j_2 \ldots j_n = j_1 2^{n-1} + j_2 2^{n-2} + \ldots + j_n 2^0,$$

and the two-valued fraction $\frac{j_l}{2} + \frac{j_{l+1}}{4} + \ldots + \frac{j_m}{2^{m-l+1}}$ is represented using the following quantum notation (representation):

$$j_l j_{l+1} \ldots j_m = j_l 2^{-1} + j_{l+1} 2^{-2} + \ldots + j_m 2^{-(m-l+1)}.$$

Efficient quantum circuit for the product representation of the quantum Fourier transform has been shown [167] using the Walsh-Hadamard gate $H = \frac{1}{\sqrt{2}} \begin{bmatrix} 1 & 1 \\ 1 & -1 \end{bmatrix}$, and the gate $R_k = \begin{bmatrix} 1 & 0 \\ 0 & e^{\frac{2\pi i}{2^k}} \end{bmatrix}$ (which is a generalization for the family that includes the phase gate and the π/8 gate). For example, one obtains the following unitary QFT matrix [ℑ] for a three qubit quantum Fourier transform (i.e., n = 3), where $v = e^{\frac{2\pi i}{2^3}} = \sqrt{i}$:

$$[\Im] = \frac{1}{\sqrt{2^3}} \begin{bmatrix} 1 & 1 & 1 & 1 & 1 & 1 & 1 & 1 \\ 1 & \nu^1 & \nu^2 & \nu^3 & \nu^4 & \nu^5 & \nu^6 & \nu^7 \\ 1 & \nu^2 & \nu^4 & \nu^6 & 1 & \nu^2 & \nu^4 & \nu^6 \\ 1 & \nu^3 & \nu^6 & \nu^1 & \nu^4 & \nu^7 & \nu^2 & \nu^5 \\ 1 & \nu^4 & 1 & \nu^4 & 1 & \nu^4 & 1 & \nu^4 \\ 1 & \nu^5 & \nu^2 & \nu^7 & \nu^4 & \nu^1 & \nu^6 & \nu^3 \\ 1 & \nu^6 & \nu^4 & \nu^2 & 1 & \nu^6 & \nu^4 & \nu^2 \\ 1 & \nu^7 & \nu^6 & \nu^5 & \nu^4 & \nu^3 & \nu^2 & \nu^1 \end{bmatrix}.$$

An example of the use of the QFT in quantum computing is as follows [249]. A two-valued (bipolar-valued) binary function f: $\{0,1\}^n \rightarrow \{-1,1\}$ can be represented as a Fourier expansion (two-valued Fourier expansion is called the Walsh-Hadamard expansion) as follows, where $\hat{f}(\vec{a})$ are the Fourier spectral coefficients and $\chi_{\vec{a}}(\vec{x})$ are the Fourier basis functions:

$$f(\vec{x}) = \sum_{\vec{a}\in\{0,1\}^n} \chi_{\vec{a}}(\vec{x}) \hat{f}(\vec{a}),$$

$$= \sum_{\vec{a}\in\{0,1\}^n} (-1)^{\vec{a}^T\vec{x}} \left(\frac{1}{2^n} \sum_{\vec{b}\in\{0,1\}^n} f(\vec{b}) \chi_{\vec{b}}(\vec{a}) \right),$$

$$= ([B]y)^* \left(\frac{1}{2^n} [B] f \right).$$

Where: $y_{\vec{b}} = \begin{cases} 1, & if \quad \vec{b} = \vec{x} \\ 0, & Otherwise \end{cases},$

$\chi_{\vec{a}}(\vec{x}) = (-1)^{\vec{a}^T\vec{x}}$, $\hat{f}(\vec{a}) = \frac{1}{2^n} \sum_{\vec{b}\in\{0,1\}^n} f(\vec{b}) \chi_{\vec{b}}(\vec{a})$, and * is the complex conjugate transpose (adjoint). The quantum formulation of the Walsh-Hadamard transform (QWHT) is as follows:

$$f(\vec{x}) = \left\langle \chi \middle| \hat{f} \right\rangle = \left\langle [\hat{B}]\vec{x} \middle| [\hat{B}] f \right\rangle,$$

where $\left| f \right\rangle = \sum_{\vec{x}\in\{0,1\}^n} c_{\vec{x}} \left| \vec{x} \right\rangle$, the amplitudes $c_{\vec{x}} = f(\vec{x}) = \{-1,1\}$ are properly normalized according to $\sum_i | c_i |^2 = 1$, $[\hat{B}]$ is the linear and unitary (quantum; normalized) Walsh-Hadamard transform matrix (operator; gate), and $\langle \mid \rangle$ is the inner (dot; scalar) product (projection) of the Bra component $\langle \mid$ and the Ket component $\mid \rangle$.

The comparative use of QWHT as a quantum computational learning algorithm for complete data versus quantum computational learning algorithm for incomplete (noisy) data was also shown [249].

11.2 New Two-Valued Quantum Evolution Processes

In the following theorems, the subscripts without parenthesis for evolution matrices resemble the order of the inputs and outputs that are used to generate the quantum evolution processes of the corresponding reversible gates from Chapt. 5.

Theorem 11.1. The following transformations represent the binary quantum processes for Fredkin_1 (Fig. 5.24b), Margolus_0 (Fig. 5.4h), Margolus_1 (Fig. 5.4i), and Margolus_2 (Fig. 5.4j), respectively.

$$F1_{abc/cf_{r0}f_{r1}}: \begin{bmatrix} 1 & 0 & 0 & 0 & 0 & 0 & 0 & 0 \\ 0 & 0 & 0 & 0 & 1 & 0 & 0 & 0 \\ 0 & 0 & 1 & 0 & 0 & 0 & 0 & 0 \\ 0 & 0 & 0 & 0 & 0 & 0 & 1 & 0 \\ 0 & 1 & 0 & 0 & 0 & 0 & 0 & 0 \\ 0 & 0 & 0 & 1 & 0 & 0 & 0 & 0 \\ 0 & 0 & 0 & 0 & 0 & 1 & 0 & 0 \\ 0 & 0 & 0 & 0 & 0 & 0 & 0 & 1 \end{bmatrix},$$

$$M0_{abc/f_0f_1f_2}: \begin{bmatrix} 1 & 0 & 0 & 0 & 0 & 0 & 0 & 0 \\ 0 & 0 & 0 & 0 & 1 & 0 & 0 & 0 \\ 0 & 1 & 0 & 0 & 0 & 0 & 0 & 0 \\ 0 & 0 & 0 & 0 & 0 & 0 & 1 & 0 \\ 0 & 0 & 1 & 0 & 0 & 0 & 0 & 0 \\ 0 & 0 & 0 & 1 & 0 & 0 & 0 & 0 \\ 0 & 0 & 0 & 0 & 0 & 1 & 0 & 0 \\ 0 & 0 & 0 & 0 & 0 & 0 & 0 & 1 \end{bmatrix},$$

$$M1_{abc/f_0f_1f_2}: \begin{bmatrix} 1 & 0 & 0 & 0 & 0 & 0 & 0 & 0 \\ 0 & 0 & 1 & 0 & 0 & 0 & 0 & 0 \\ 0 & 0 & 0 & 0 & 1 & 0 & 0 & 0 \\ 0 & 0 & 0 & 1 & 0 & 0 & 0 & 0 \\ 0 & 1 & 0 & 0 & 0 & 0 & 0 & 0 \\ 0 & 0 & 0 & 0 & 0 & 1 & 0 & 0 \\ 0 & 0 & 0 & 0 & 0 & 0 & 1 & 0 \\ 0 & 0 & 0 & 0 & 0 & 0 & 0 & 1 \end{bmatrix},$$

$$M2_{abc/f_0f_1f_2}: \begin{bmatrix} 1 & 0 & 0 & 0 & 0 & 0 & 0 & 0 \\ 0 & 1 & 0 & 0 & 0 & 0 & 0 & 0 \\ 0 & 0 & 1 & 0 & 0 & 0 & 0 & 0 \\ 0 & 0 & 0 & 0 & 0 & 1 & 0 & 0 \\ 0 & 0 & 0 & 0 & 1 & 0 & 0 & 0 \\ 0 & 0 & 0 & 0 & 0 & 0 & 1 & 0 \\ 0 & 0 & 0 & 1 & 0 & 0 & 0 & 0 \\ 0 & 0 & 0 & 0 & 0 & 0 & 0 & 1 \end{bmatrix}.$$

Proof: Utilizing Figs. 5.24b, 5.4h, 5.4i, and 5.4j for (3,3) qubit binary primitives, one obtains the following set of linearly independent Eqs. for each reversible gate:

$$\begin{bmatrix} \alpha1 & \alpha2 & \alpha3 & \alpha4 & \alpha5 & \alpha6 & \alpha7 & \alpha8 \\ \beta1 & \beta2 & \beta3 & \beta4 & \beta5 & \beta6 & \beta7 & \beta8 \\ \chi1 & \chi2 & \chi3 & \chi4 & \chi5 & \chi6 & \chi7 & \chi8 \\ \delta1 & \delta2 & \delta3 & \delta4 & \delta5 & \delta6 & \delta7 & \delta8 \\ \varepsilon1 & \varepsilon2 & \varepsilon3 & \varepsilon4 & \varepsilon5 & \varepsilon6 & \varepsilon7 & \varepsilon8 \\ \phi1 & \phi2 & \phi3 & \phi4 & \phi5 & \phi6 & \phi7 & \phi8 \\ \varphi1 & \varphi2 & \varphi3 & \varphi4 & \varphi5 & \varphi6 & \varphi7 & \varphi8 \\ \gamma1 & \gamma2 & \gamma3 & \gamma4 & \gamma5 & \gamma6 & \gamma7 & \gamma8 \end{bmatrix} \left|input\right\rangle_k = \left|output\right\rangle_k,$$

$$k = 1,\ldots,8, and \quad \left|input\right\rangle = \left|000\right\rangle,\ldots,\left|111\right\rangle.$$

Then by solving the set of linearly independent Eqs. over GF(2) for each reversible gate, one obtains the binary quantum evolution processes for the corresponding reversible gates: F_1 (F1 from Fig. 5.24b), $Margolus_0$ (M_0 from Fig. 5.4h), $Margolus_1$ (M_1 from Fig. 5.4i), and $Margolus_2$ (M_2 from Fig. 5.4j), respectively. **Q.E.D.**

The theorems that are introduced in this Sect. are necessary for automated analysis and verification of netlists of quantum gates.

They are also necessary for automated synthesis of a netlist described as an evolution matrix from quantum gates, especially for try-and-check (i.e., trial-and-error) methods such as evolutionary computations [252].

11.3 Novel Representations for Two-Valued Quantum Logic: Two-Valued Quantum Decision Trees and Diagrams

Since decision diagrams [217] allow for efficient representation of large sparse matrices, they have found applications in many computer aided design algorithms, and we believe that their quantum counterparts will be useful for quantum logic synthesis and analysis. Utilizing the binary Feynman and Swap evolution matrices that were presented previously, the following represents the binary initial state (Buffer) (which is equivalent to two wires), binary final state of a 2-qubit Feynman register, and binary final state of a 2-qubit Swap register using the binary computational basis states $[|00\rangle \ |01\rangle \ |10\rangle \ |11\rangle]$:

$$|\Psi\rangle_{Initial} = [|00\rangle \ |01\rangle \ |10\rangle \ |11\rangle]\begin{bmatrix}1&0&0&0\\0&1&0&0\\0&0&1&0\\0&0&0&1\end{bmatrix}\begin{bmatrix}\alpha_1\alpha_2\\\alpha_1\beta_2\\\beta_1\alpha_2\\\beta_1\beta_2\end{bmatrix}, \quad (11.13)$$

$$|\Psi\rangle_{Final-Feynman} = [|00\rangle \ |01\rangle \ |10\rangle \ |11\rangle]\begin{bmatrix}1&0&0&0\\0&1&0&0\\0&0&0&1\\0&0&1&0\end{bmatrix}\begin{bmatrix}\alpha_1\alpha_2\\\alpha_1\beta_2\\\beta_1\alpha_2\\\beta_1\beta_2\end{bmatrix}, \quad (11.14)$$

$$|\Psi\rangle_{Final-Swap} = [|00\rangle \ |01\rangle \ |10\rangle \ |11\rangle]\begin{bmatrix}1&0&0&0\\0&0&1&0\\0&1&0&0\\0&0&0&1\end{bmatrix}\begin{bmatrix}\alpha_3\alpha_4\\\alpha_3\beta_4\\\beta_3\alpha_4\\\beta_3\beta_4\end{bmatrix}. \quad (11.15)$$

Since various types of decision trees and diagrams are of fundamental importance in binary, multiple-valued (MV Reed-

Muller, Galois, arithmetic, etc) and fuzzy logic [215,217], it is obvious that they will be also useful in binary quantum logic.

Figures 11.16a and 11.16c illustrate examples of the computational quantum decision trees (CQDT) [15,26], and Fig. 11.16b illustrates an example of the computational decision diagram (CQDD) [15,26]. When traversing the tree in Fig. 11.16b two paths $|01\rangle$ and $|10\rangle$ lead to two leaves with same values, that is values $\alpha_1\beta_2$ and $\beta_1\beta_2$, respectively. Since the two paths lead to the same leaf then the two nodes are combined as a single leaf and thus a more compact representation of CQDD is created.

The new quantum representations can be useful in future algorithms for the synthesis of quantum circuits, analogous to the already existing algorithms that depend on such representations for the optimized synthesis of classical (non-quantum) circuits. This is because in order to perform complex operations one has to choose (1) a specific type of representation and (2) the corresponding set of basic operations that are associated with that particular representation. Since decision diagrams has proven in the classical domain their suitability as a representation that leads to efficient manipulation of large varieties of logic functions in terms of using minimal amount of space (memory) and minimal amount of time [45], the quantum decision diagrams would have the same effect for the representation and manipulation of functions in the quantum domain.

One notes that for a specific order of variables, the resulting CQDT (e.g., Figs. 11.16a and 11.16c) and CQDD (e.g., Fig. 11.16b) are canonical. Obviously, from the software implementation side, and similar to the tools for classical logics, quantum decision diagrams (Fig. 11.16b) can be realized on top of standard binary decision diagram (BDD) packages [231].

Figure 11.17 shows the binary quantum evolution decision tree for a serially-interconnected Swap gate followed by a Feynman gate [15,26]. One observes that the evolution matrices in Eqs. (11.13) through (11.15) force the quantum states $\{|00\rangle,|01\rangle,|10\rangle,|11\rangle\}$ and the probability amplitudes $\{\alpha_i, \beta_j\}$ to be in specific combinations (permutations) that are unique for the specific gates that are used.

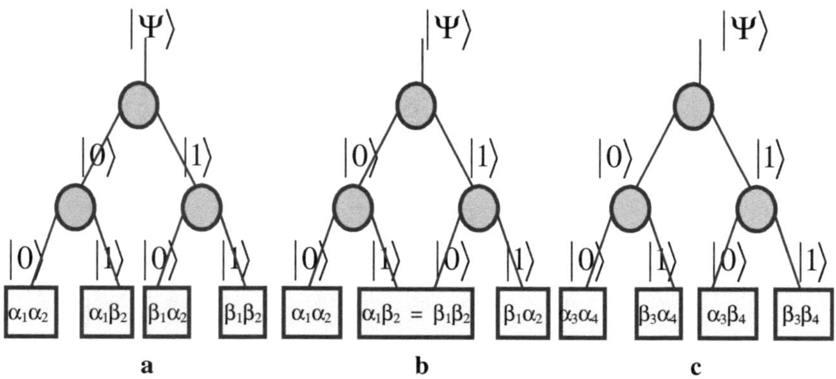

Fig. 11.16. Binary computational quantum decision tree (CQDT) and decision diagram (CQDD): **a** Buffer CQDT, **b** Feynman CQDD, and **c** Swap CQDT, for the binary quantum computational basis states $\{|00\rangle, |01\rangle, |10\rangle, |11\rangle\}$, where $\{\alpha_i, \beta_j\}$ are the probability amplitudes.

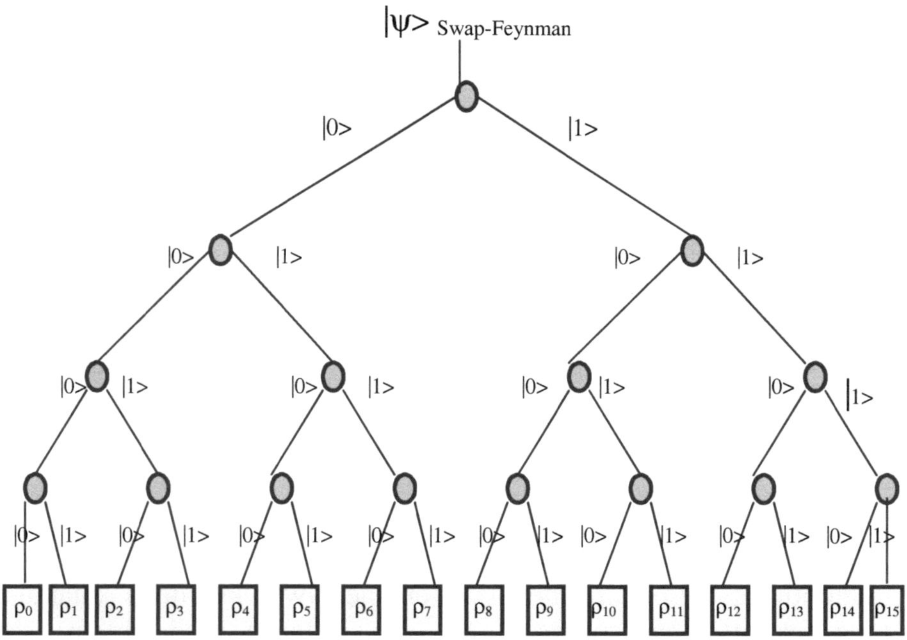

Fig. 11.17. Binary quantum evolution decision tree for Swap gate serially-interconnected with a Feynman gate, respectively, for the 2-qubit orthonormal computational basis states: $\{|00\rangle, |01\rangle, |10\rangle, |11\rangle\}$, where ρ_i are the probability amplitudes.

Note that the leaves in Fig. 11.17 represent the probability to obtain the state of the Swap-Feynman quantum register after measurement. So, for instance, by utilizing Fig. 11.17, one obtains the following states with the corresponding normalized probabilities:

$$|0000\rangle : P|0000\rangle = \rho_0 = \frac{|\alpha_3|^2|\alpha_4|^2|\alpha_1|^2|\alpha_2|^2}{(|\alpha_1|^2+|\beta_1|^2)(|\alpha_2|^2+|\beta_2|^2)(|\alpha_3|^2+|\beta_3|^2)(|\alpha_4|^2+|\beta_4|^2)},$$

$$|0001\rangle : P|0001\rangle = \rho_1 = \frac{|\alpha_3|^2|\alpha_4|^2|\alpha_1|^2|\beta_2|^2}{(|\alpha_1|^2+|\beta_1|^2)(|\alpha_2|^2+|\beta_2|^2)(|\alpha_3|^2+|\beta_3|^2)(|\alpha_4|^2+|\beta_4|^2)},$$

$$|0010\rangle : P|0010\rangle = \rho_2 = \frac{|\alpha_3|^2|\alpha_4|^2|\beta_1|^2|\beta_2|^2}{(|\alpha_1|^2+|\beta_1|^2)(|\alpha_2|^2+|\beta_2|^2)(|\alpha_3|^2+|\beta_3|^2)(|\alpha_4|^2+|\beta_4|^2)},$$

$$|0011\rangle : P|0011\rangle = \rho_3 = \frac{|\alpha_3|^2|\alpha_4|^2|\beta_1|^2|\alpha_2|^2}{(|\alpha_1|^2+|\beta_1|^2)(|\alpha_2|^2+|\beta_2|^2)(|\alpha_3|^2+|\beta_3|^2)(|\alpha_4|^2+|\beta_4|^2)},$$

$$|0100\rangle : P|0100\rangle = \rho_4 = \frac{|\beta_3|^2|\alpha_4|^2|\alpha_1|^2|\alpha_2|^2}{(|\alpha_1|^2+|\beta_1|^2)(|\alpha_2|^2+|\beta_2|^2)(|\alpha_3|^2+|\beta_3|^2)(|\alpha_4|^2+|\beta_4|^2)},$$

$$\cdots$$

$$|1110\rangle : P|1110\rangle = \rho_{14} = \frac{|\beta_3|^2|\beta_4|^2|\beta_1|^2|\beta_2|^2}{(|\alpha_1|^2+|\beta_1|^2)(|\alpha_2|^2+|\beta_2|^2)(|\alpha_3|^2+|\beta_3|^2)(|\alpha_4|^2+|\beta_4|^2)},$$

$$|1111\rangle : P|1111\rangle = \rho_{15} = \frac{|\beta_3|^2|\beta_4|^2|\beta_1|^2|\alpha_2|^2}{(|\alpha_1|^2+|\beta_1|^2)(|\alpha_2|^2+|\beta_2|^2)(|\alpha_3|^2+|\beta_3|^2)(|\alpha_4|^2+|\beta_4|^2)}.$$

One notes that for the condition $\sum_i |c_i|^2 = 1$, the denominators in the upper Eqs., for Fig. 11.17, will be equal to the value one. The quantum evolution decision tree in Fig. 11.17 can be computed for the composite basis states $\{\ |++\rangle, |+-\rangle, |-+\rangle, |--\rangle\ \}$ (that were produced in Example 11.2e), for which the states $\{\ |00\rangle, |01\rangle, |10\rangle, |11\rangle\ \}$ are replaced by the states $\{\ |++\rangle, |+-\rangle, |-+\rangle, |--\rangle\ \}$, respectively. For the composite basis states decision tree, the leaves represent the new probability amplitudes to obtain the state of the Swap-Feynman quantum register after measurement. So, for instance, one obtains the following state $|++++\rangle$ with the corresponding probability $P|++++\rangle$ for the composite basis states quantum evolution decision tree as follows (For more clarification, please refer to Example 11.2e):

$$|++++\rangle : P|++++\rangle = \Gamma_0 =$$

$$\frac{\frac{|\alpha_3+\beta_3|^2}{2}\frac{|\alpha_4+\beta_4|^2}{2}\frac{|\alpha_1+\beta_1|^2}{2}\frac{|\alpha_2+\beta_2|^2}{2}}{\left(\frac{|\alpha_3+\beta_3|^2}{2}+\frac{|\alpha_3-\beta_3|^2}{2}\right)\left(\frac{|\alpha_4+\beta_4|^2}{2}+\frac{|\alpha_4-\beta_4|^2}{2}\right)\left(\frac{|\alpha_1+\beta_1|^2}{2}+\frac{|\alpha_1-\beta_1|^2}{2}\right)\left(\frac{|\alpha_2+\beta_2|^2}{2}+\frac{|\alpha_2-\beta_2|^2}{2}\right)},$$

$$=$$

$$\frac{|\alpha_3+\beta_3|^2|\alpha_4+\beta_4|^2|\alpha_1+\beta_1|^2|\alpha_2+\beta_2|^2}{\left(|\alpha_3+\beta_3|^2+|\alpha_3-\beta_3|^2\right)\left(|\alpha_4+\beta_4|^2+|\alpha_4-\beta_4|^2\right)\left(|\alpha_1+\beta_1|^2+|\alpha_1-\beta_1|^2\right)\left(|\alpha_2+\beta_2|^2+|\alpha_2-\beta_2|^2\right)}.$$

One can note that, similar to the CQDD, although the values of the leaves in the CoQDT are not equal to each other in general, a binary composite quantum decision diagram (CoQDD) can be constructed for the corresponding CoQDT. The rules for such quantum decision diagrams are the same as in classical decision diagrams: (1) join isomorphic nodes, and (2) remove redundant nodes [45].

Also, one may note that having a variety of quantum decision diagrams can have an effect on the size of the representations of the corresponding quantum netlists and thus can lead to efficient quantum operations on such compact representations similar to the case of classical logic design [217].

In the following Sects., fundamentals of multiple-valued quantum computing are presented in Sect. 11.4, and new multiple-valued quantum evolution processes and orthonormal basis states are developed in Sect. 11.5. This includes the development of new multiple-valued 1-qubit and 2-qubit orthonormal quantum basis states called multiple-valued composite basis states, and multiple-valued Einstein-Podolsky-Rosen (EPR) basis states, respectively. This is achieved by using the new quantum Chrestenson transform (gate) (QChT) introduced in Sect. 11.5. (Further use of QChT in the generation of multiple-valued QFT will be discussed in Sect. 11.6.) Multiple-valued canonical quantum decision trees (QDTs) and quantum decision diagrams (QDDs) as efficient representations for multiple-valued quantum computations are also introduced.

Although the following Sects. are developed for GF(3), extensions to higher radices is very similar.

11.4 Fundamentals of Multiple-Valued Quantum Computing

In Sect. 11.1 of this Chapt., it was shown that in two-valued (binary) quantum logic, two qubits $|0\rangle$ and $|1\rangle$ are used. Similarly, in ternary quantum logic, the $|0\rangle$, $|1\rangle$, and $|2\rangle$ qubits are used. These qubits are represented by the vector that corresponds to the following [15,19,23,165]:

$$|0\rangle = \begin{bmatrix} 1 \\ 0 \\ 0 \end{bmatrix}, |1\rangle = \begin{bmatrix} 0 \\ 1 \\ 0 \end{bmatrix}, |2\rangle = \begin{bmatrix} 0 \\ 0 \\ 1 \end{bmatrix}.$$

As was shown previously, Figs. 11.3a, 11.3c, and 11.3d implements such multiple-valued quantum computations (MVQC). In ternary logic, an n-qubit ternary quantum register is an array of n ternary qubits. For a ternary quantum register composed of 2 ternary qubits, one obtains 9 possible states of the ternary quantum register $\{ |00\rangle, |01\rangle, |02\rangle, |10\rangle, |11\rangle, |12\rangle, |20\rangle, |21\rangle, |22\rangle \}$, where $\otimes$ is the tensor (Kronecker) product.

In general, a ternary quantum register that is composed of k ternary qubits can have up to 3^k distinct possible states. The ternary quantum register can be in any of the individual states at any instant of time or at all of the states at the same time. Due to the fact that the multiple-valued quantum register can be at all of the states at the same time is the major reason of the multi-valued parallelism that exists at the quantum level, and due to this parallelism, a ternary quantum processor can operate on all of the states of the quantum register at the same time (it can be modeled like having application-specific 3^k ternary parallel processors).

As was shown in Fig. 11.3, a physical system consisting of trapped ions under several laser excitations can be used to reliably apply MVQC. Also, a physical system in which a particle is exposed to a specific potential function can be used to implement MVQC. In such implementation, the resulting distinct energy levels are used as the set of orthonormal basis states (e.g., Fig. K.1 in Appendix K).

$$|00\rangle = |0\rangle \otimes |0\rangle = \begin{bmatrix} 1 \\ 0 \\ 0 \end{bmatrix} \otimes \begin{bmatrix} 1 \\ 0 \\ 0 \end{bmatrix} = [1 \quad 0 \quad 0 \quad 0 \quad 0 \quad 0 \quad 0 \quad 0 \quad 0]^T,$$

$$|01\rangle = |0\rangle \otimes |1\rangle = \begin{bmatrix} 1 \\ 0 \\ 0 \end{bmatrix} \otimes \begin{bmatrix} 0 \\ 1 \\ 0 \end{bmatrix} = [0 \quad 1 \quad 0 \quad 0 \quad 0 \quad 0 \quad 0 \quad 0 \quad 0]^T,$$

$$|02\rangle = |0\rangle \otimes |2\rangle = \begin{bmatrix} 1 \\ 0 \\ 0 \end{bmatrix} \otimes \begin{bmatrix} 0 \\ 0 \\ 1 \end{bmatrix} = [0 \quad 0 \quad 1 \quad 0 \quad 0 \quad 0 \quad 0 \quad 0 \quad 0]^T,$$

$$|10\rangle = |1\rangle \otimes |0\rangle = \begin{bmatrix} 0 \\ 1 \\ 0 \end{bmatrix} \otimes \begin{bmatrix} 1 \\ 0 \\ 0 \end{bmatrix} = [0 \quad 0 \quad 0 \quad 1 \quad 0 \quad 0 \quad 0 \quad 0 \quad 0]^T,$$

$$|11\rangle = |1\rangle \otimes |1\rangle = \begin{bmatrix} 0 \\ 1 \\ 0 \end{bmatrix} \otimes \begin{bmatrix} 0 \\ 1 \\ 0 \end{bmatrix} = [0 \quad 0 \quad 0 \quad 0 \quad 1 \quad 0 \quad 0 \quad 0 \quad 0]^T,$$

$$|12\rangle = |1\rangle \otimes |2\rangle = \begin{bmatrix} 0 \\ 1 \\ 0 \end{bmatrix} \otimes \begin{bmatrix} 0 \\ 0 \\ 1 \end{bmatrix} = [0 \quad 0 \quad 0 \quad 0 \quad 0 \quad 1 \quad 0 \quad 0 \quad 0]^T,$$

$$|20\rangle = |2\rangle \otimes |0\rangle = \begin{bmatrix} 0 \\ 0 \\ 1 \end{bmatrix} \otimes \begin{bmatrix} 1 \\ 0 \\ 0 \end{bmatrix} = [0 \quad 0 \quad 0 \quad 0 \quad 0 \quad 0 \quad 1 \quad 0 \quad 0]^T,$$

$$|21\rangle = |2\rangle \otimes |1\rangle = \begin{bmatrix} 0 \\ 0 \\ 1 \end{bmatrix} \otimes \begin{bmatrix} 0 \\ 1 \\ 0 \end{bmatrix} = [0 \quad 0 \quad 0 \quad 0 \quad 0 \quad 0 \quad 0 \quad 1 \quad 0]^T,$$

$$|22\rangle = |2\rangle \otimes |2\rangle = \begin{bmatrix} 0 \\ 0 \\ 1 \end{bmatrix} \otimes \begin{bmatrix} 0 \\ 0 \\ 1 \end{bmatrix} = [0 \quad 0 \quad 0 \quad 0 \quad 0 \quad 0 \quad 0 \quad 0 \quad 1]^T.$$

For a quantum register composed of 1-ternary qubit, and assuming, as in binary, the orthonormalization of the computational basis states, the evolution state ($|\Psi\rangle$) is represented as follows:

$$|\Psi\rangle_{ternary-qubit} = \alpha|0\rangle + \beta|1\rangle + \gamma|2\rangle, \tag{11.16}$$

where α, β, and γ are complex numbers called "probability amplitudes", and in general: $\alpha\alpha^* + \beta\beta^* + \gamma\gamma^* = |\alpha|^2 + |\beta|^2 + |\gamma|^2 = 1$.

$$|\alpha| = \sqrt{p_0}\sqrt{(|\alpha|^2 + |\beta|^2 + |\gamma|^2)},$$
$$|\beta| = \sqrt{p_1}\sqrt{(|\alpha|^2 + |\beta|^2 + |\gamma|^2)},$$
$$|\gamma| = \sqrt{p_2}\sqrt{(|\alpha|^2 + |\beta|^2 + |\gamma|^2)},$$
$$p(|0\rangle) = p_0 = \frac{|\alpha|^2}{|\alpha|^2 + |\beta|^2 + |\gamma|^2},$$
$$p(|1\rangle) = p_1 = \frac{|\beta|^2}{|\alpha|^2 + |\beta|^2 + |\gamma|^2},$$
$$p(|2\rangle) = p_2 = \frac{|\gamma|^2}{|\alpha|^2 + |\beta|^2 + |\gamma|^2}.$$

Where: $1 \geq p_i \geq 0$, $i \in \{0,1,2\}$, p_0 is the probability of the system being in state |0>, p_1 is the probability of the system being in state |1>, and p_2 is the probability of the system being in state |2>, and $p_0 + p_1 + p_2 = 1$. The quantum orthonormalization condition requires that $|\alpha|^2 + |\beta|^2 + |\gamma|^2 = 1$. Thus, if $\{|\alpha| = 0, |\gamma| = 0, |\beta| = 1\}$, $\{|\alpha| = 0, |\gamma| = 1, |\beta| = 0\}$, $\{|\alpha| = 1, |\gamma| = 0, |\beta| = 0\}$, $\{|\alpha| = 1/\sqrt{3}, |\gamma| = 1/\sqrt{3}, |\beta| = 1/\sqrt{3}\}$, or $\{|\alpha| = 1/\sqrt{9/4}, |\gamma| = 1/\sqrt{9/2}, |\beta| = 1/\sqrt{3}\}$, …etc, then $|\alpha|^2 + |\beta|^2 + |\gamma|^2 = p_0 + p_1 + p_2 = 1$.

$$\therefore |\Psi\rangle_{ternary-qubit} =$$
$$\sqrt{p_0}|0\rangle + \sqrt{p_1}|1\rangle + \sqrt{p_2}|2\rangle = \alpha|0\rangle + \beta|1\rangle + \gamma|2\rangle.$$

Equation (11.16) can be written as:

$$|\Psi\rangle_{ternary-qubit} = [|0\rangle \quad |1\rangle \quad |2\rangle][E]\begin{bmatrix}\alpha\\ \beta\\ \gamma\end{bmatrix}, \tag{11.17}$$

where [E] is the evolution matrix. For a 2-qubit ternary quantum register and giving the orthonormalization conditions $|\alpha_1|^2 + |\beta_1|^2 + |\gamma_1|^2 = 1$ and $|\alpha_2|^2 + |\beta_2|^2 + |\gamma_2|^2 = 1$, one obtains:

$$(|\alpha_1|^2 + |\beta_1|^2 + |\gamma_1|^2)(|\alpha_2|^2 + |\beta_2|^2 + |\gamma_2|^2) = 1,$$
$$\therefore |\alpha_1\alpha_2|^2 + |\alpha_1\beta_2|^2 + |\alpha_1\gamma_2|^2 + |\beta_1\alpha_2|^2 + |\beta_1\beta_2|^2 + |\beta_1\gamma_2|^2 + |\gamma_1\alpha_2|^2 + |\gamma_1\beta_2|^2 + |\gamma_1\gamma_2|^2 = 1.$$

Thus, for a ternary quantum register which is composed of two ternary qubits, the evolution quantum state $|\Psi\rangle$ is represented as follows:

$$|\Psi\rangle_{ternary-qubit1} = \alpha_1|0\rangle + \beta_1|1\rangle + \gamma_1|2\rangle,$$
$$|\Psi\rangle_{ternary-qubit2} = \alpha_2|0\rangle + \beta_2|1\rangle + \gamma_2|2\rangle.$$

For two ternary qubits, and by using the tensor (Kronecker) product , one obtains:

$$\begin{aligned}
|\Psi\rangle_{2-ternary-qubit} &= |\Psi\rangle_1 \otimes |\Psi\rangle_2 \\
&= (\alpha_1|0\rangle + \beta_1|1\rangle + \gamma_1|2\rangle) \otimes (\alpha_2|0\rangle + \beta_2|1\rangle + \gamma_2|2\rangle), \\
&= \alpha_1\alpha_2|0\rangle|0\rangle + \alpha_1\beta_2|0\rangle|1\rangle + \alpha_1\gamma_2|0\rangle|2\rangle + \beta_1\alpha_2|1\rangle|0\rangle + \beta_1\beta_2|1\rangle|1\rangle + \\
&\quad \beta_1\gamma_2|1\rangle|2\rangle + \gamma_1\alpha_2|2\rangle|0\rangle + \gamma_1\beta_2|2\rangle|1\rangle + \gamma_1\gamma_2|2\rangle|2\rangle, \\
&= \alpha_1\alpha_2|00\rangle + \alpha_1\beta_2|01\rangle + \alpha_1\gamma_2|02\rangle + \beta_1\alpha_2|10\rangle + \beta_1\beta_2|11\rangle + \\
&\quad \beta_1\gamma_2|12\rangle + \gamma_1\alpha_2|20\rangle + \gamma_1\beta_2|21\rangle + \gamma_1\gamma_2|22\rangle.
\end{aligned}$$

Therefore, for the case of two ternary qubits, and similarly to the matrix-based method that is used for two-valued representation in Eq. (11.10), $|\Psi\rangle$ can be represented using matrix-based form as follows:

$$|\Psi\rangle_{2-ternary-qubit} =$$

$$\begin{bmatrix} |00\rangle & |01\rangle & |02\rangle & |10\rangle & |11\rangle & |12\rangle & |20\rangle & |21\rangle & |22\rangle \end{bmatrix} [E] \begin{bmatrix} \alpha_1\alpha_2 \\ \alpha_1\beta_2 \\ \alpha_1\gamma_2 \\ \beta_1\alpha_2 \\ \beta_1\beta_2 \\ \beta_1\gamma_2 \\ \gamma_1\alpha_2 \\ \gamma_1\beta_2 \\ \gamma_1\gamma_2 \end{bmatrix}. \qquad (11.18)$$

where [E] is the evolution matrix, which is obtained through the solution of Schrodinger Eq.

For an N-ary quantum logic system the definition of quantum entanglement for multiple-valued quantum logic is a straightforward extension of Eq. (11.12). As the entanglement in the case of two-valued quantum systems seems to be the major factor behind the speedups of quantum computations by which many computations are performed in parallel, the same role of entanglement is expected to be observed in the case of multiple-valued quantum systems. In this aspect, entanglement will be a special new resource in multiple-valued quantum computing.

Example 11.5.

11.5a. Consider a ternary quantum system of two qubits, given as:

$$\frac{1}{3}\left(|00\rangle+|01\rangle+|02\rangle+|10\rangle+|11\rangle+|12\rangle+|20\rangle+|21\rangle+|22\rangle\right),$$

$$=\frac{1}{\sqrt{3}}\left(|0\rangle+|1\rangle+|2\rangle\right)\otimes\frac{1}{\sqrt{3}}\left(|0\rangle+|1\rangle+|2\rangle\right),$$

$$=\frac{1}{\sqrt{3}}\left(|0\rangle+|1\rangle+|2\rangle\right)\frac{1}{\sqrt{3}}\left(|0\rangle+|1\rangle+|2\rangle\right).$$

This system is decomposable, as the functions of the first and second qubits are disentangled according to Eq. (11.12).

11.5b. Consider now the ternary quantum system: $\frac{1}{\sqrt{3}}\left(|02\rangle+|10\rangle\right)$.

This system is entangled, as no decomposition according to Eq. (11.12) is possible.

11.5 New Multiple-Valued Quantum Chrestenson Evolution Process, Quantum Composite Basis States, and the Multiple-Valued Einstein-Podolsky-Rosen (EPR) Basis States

So far, not much has been published on multiple-valued quantum logic gates and especially their characterization and representation formalisms. It is the main goal of this Sect. and the following Sects. to start building a systematic theory of multiple-valued quantum gates, structures, and synthesis methods. Theorems 11.2 and 11.3

establish the ternary quantum composite basis states and the ternary quantum Einstein-Podolsky-Rosen (EPR) basis states, respectively. Although the results that are presented in the following Sects. are for the ternary case, generalization to higher radices is straightforward.

Theorem 11.2. The following represents the ternary composite basis states:

$$\{\ |+\rangle = \frac{|0\rangle + d_1|1\rangle + d_2|2\rangle}{\sqrt{3}}, |\|\rangle = \frac{|0\rangle + |1\rangle + |2\rangle}{\sqrt{3}}, |-\rangle = \frac{|0\rangle + d_2|1\rangle + d_1|2\rangle}{\sqrt{3}}\ \},$$

where:

$$|0\rangle = \frac{|+\rangle + |\|\rangle + |-\rangle}{\sqrt{3}}, |1\rangle = \frac{d_2|+\rangle + |\|\rangle + d_1|-\rangle}{\sqrt{3}}, |2\rangle = \frac{d_1|+\rangle + |\|\rangle + d_2|-\rangle}{\sqrt{3}},$$

$$d_2 = -(1+d_1) = -\frac{1}{2}(1+\sqrt{3}i) = e^{\frac{4\pi i}{3}}, d_1 = -(1+d_2) = -\frac{1}{2}(1-\sqrt{3}i) = e^{\frac{2\pi i}{3}}.$$

Proof. Utilizing the orthogonal ternary Chrestenson spectral transform [120,159,166] (ternary Walsh-Hadamard transform) for a single variable [15]:

$$C_{(1)}^{(3)} = \begin{bmatrix} 1 & 1 & 1 \\ 1 & d_1 & d_2 \\ 1 & d_2 & d_1 \end{bmatrix},$$

and due to the fact that the evolution process must be unitary, one obtains the following quantum (normalized) Chrestenson spectral transform (QChT) [15]:

$$C_{(1)\ normalized}^{(3)} = \frac{1}{\sqrt{3}} \begin{bmatrix} 1 & 1 & 1 \\ 1 & d_1 & d_2 \\ 1 & d_2 & d_1 \end{bmatrix},$$

where $C_{(1)\ normalized}^{(3)}$ means that QChT is for ternary radix (i.e., superscript is equal to 3) and a single variable (i.e., subscript is equal to 1).

By using the normalized Chrestenson transformation as the evolution matrix as follows in Fig. 11.18:

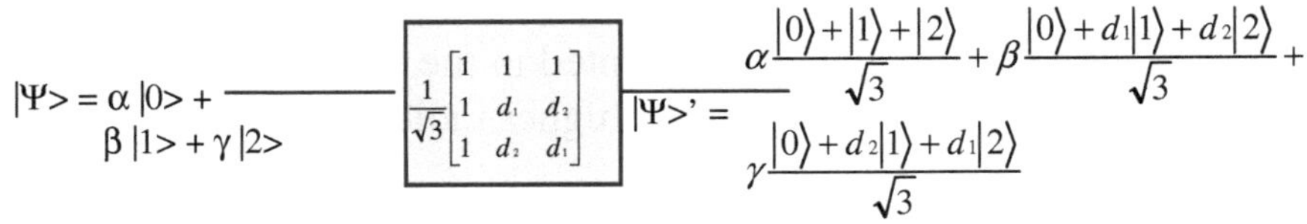

Fig. 11.18. Ternary quantum Chrestenson evolution matrix.

One obtains the corresponding output composite basis states in Theorem 11.2, for the corresponding ternary input: $|\Psi\rangle = \alpha|0\rangle + \beta|1\rangle + \gamma|2\rangle$. **Q.E.D.**

Theorem 11.3. For the following ternary inputs:

$$|00\rangle, |01\rangle, |02\rangle, |10\rangle, |11\rangle, |12\rangle, |20\rangle, |21\rangle, |22\rangle,$$

and by utilizing the QChT from Theorem 11.2, the following represents the set of ternary 2-qubit orthonormal Einstein-Podolsky-Rosen (EPR) basis states, respectively.

$$\frac{|00\rangle+|11\rangle+|22\rangle}{\sqrt{3}}, \frac{|01\rangle+|12\rangle+|20\rangle}{\sqrt{3}}, \frac{|02\rangle+|10\rangle+|21\rangle}{\sqrt{3}},$$
$$\frac{|00\rangle+d_1|11\rangle+d_2|22\rangle}{\sqrt{3}}, \frac{|01\rangle+d_1|12\rangle+d_2|20\rangle}{\sqrt{3}}, \frac{|02\rangle+d_1|10\rangle+d_2|21\rangle}{\sqrt{3}},$$
$$\frac{|00\rangle+d_2|11\rangle+d_1|22\rangle}{\sqrt{3}}, \frac{|01\rangle+d_2|12\rangle+d_1|20\rangle}{\sqrt{3}}, \frac{|02\rangle+d_2|10\rangle+d_1|21\rangle}{\sqrt{3}},$$

where:

$$d_2 = -(1+d_1) = -\frac{1}{2}(1+\sqrt{3}i) = e^{\frac{4\pi i}{3}}, d_1 = -(1+d_2) = -\frac{1}{2}(1-\sqrt{3}i) = e^{\frac{2\pi i}{3}}.$$

Proof. Analogously to the binary case (Fig. 11.7b) (where QWHT has been used), and by using the QChT in the following ternary quantum circuit:

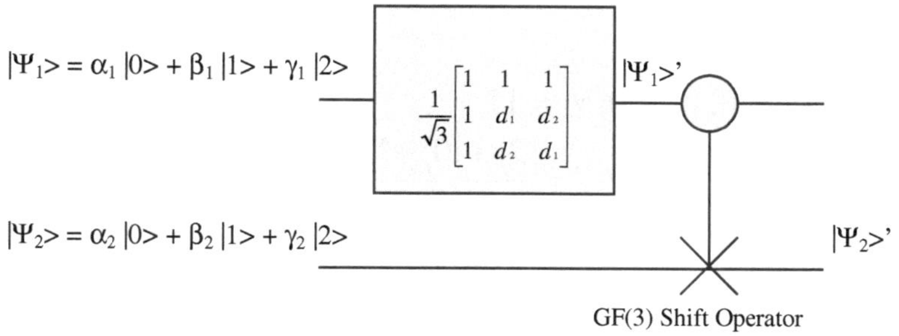

Fig. 11.19. Ternary quantum logic circuit for EPR production.

Where:

$$d_2 = -(1+d_1) = -\frac{1}{2}(1+\sqrt{3}i) = e^{\frac{4\pi i}{3}}, d_1 = -(1+d_2) = -\frac{1}{2}(1-\sqrt{3}i) = e^{\frac{2\pi i}{3}}.$$

By utilizing the shift operation over GF(3): {0→1, 1→2, 2→0}, one obtains the corresponding ternary EPR basis states as follows:

$$|\Psi_1\rangle = \alpha_1|0\rangle + \beta_1|1\rangle + \gamma_1|2\rangle, \quad |\Psi_2\rangle = \alpha_2|0\rangle + \beta_2|1\rangle + \gamma_2|2\rangle,$$

$$|\Psi 1\rangle' = \alpha_1 \frac{|0\rangle+|1\rangle+|2\rangle}{\sqrt{3}} + \beta_1 \frac{|0\rangle+d_1|1\rangle+d_2|2\rangle}{\sqrt{3}} + \gamma_1 \frac{|0\rangle+d_2|1\rangle+d_1|2\rangle}{\sqrt{3}},$$

$$|\Psi_2\rangle' = Shift_{|\Psi_2\rangle}(|\Psi_1\rangle').$$

Where $|\Psi_2\rangle' = Shift_{|\Psi_2\rangle}(|\Psi_1\rangle')$ means to shift the value of the basis states of $|\Psi_1\rangle'$ by the amount $|\Psi_2\rangle$ over GF(3).

For {$\alpha_1 = 1, \beta_1 = 0, \gamma_1 = 0$}: $|\Psi_1\rangle' = \frac{|0\rangle+|1\rangle+|2\rangle}{\sqrt{3}}$,

For {$\alpha_2 = 1, \beta_2 = 0, \gamma_2 = 0$}

$$\therefore |\Psi_2\rangle = |0\rangle \Rightarrow |\Psi_2\rangle' = \frac{|0\rangle+|1\rangle+|2\rangle}{\sqrt{3}} \Rightarrow |\Psi\rangle =$$

$$|\Psi_1{}'\Psi_2{}'\rangle = \frac{|00\rangle+|11\rangle+|22\rangle}{\sqrt{3}}.$$

For {$\alpha_2 = 0, \beta_2 = 1, \gamma_2 = 0$}

$$\therefore |\Psi_2\rangle = |1\rangle \Rightarrow |\Psi_2\rangle' = \frac{|1\rangle+|2\rangle+|0\rangle}{\sqrt{3}} \Rightarrow |\Psi\rangle =$$

$$|\Psi_1'\Psi_2'\rangle = \frac{|01\rangle + |12\rangle + |20\rangle}{\sqrt{3}}.$$

For $\{\alpha_2 = 0, \beta_2 = 0, \gamma_2 = 1\}$

$$\therefore |\Psi_2\rangle = |2\rangle \Rightarrow |\Psi_2\rangle' = \frac{|2\rangle + |0\rangle + |1\rangle}{\sqrt{3}} \Rightarrow |\Psi\rangle =$$

$$|\Psi_1'\Psi_2'\rangle = \frac{|02\rangle + |10\rangle + |21\rangle}{\sqrt{3}}.$$

For $\{\alpha_1 = 0, \beta_1 = 1, \gamma_1 = 0\}$: $|\Psi_1\rangle' = \dfrac{|0\rangle + d_1|1\rangle + d_2|2\rangle}{\sqrt{3}}$,

For $\{\alpha_2 = 1, \beta_2 = 0, \gamma_2 = 0\}$

$$\therefore |\Psi_2\rangle = |0\rangle \Rightarrow |\Psi_2\rangle' = \frac{|0\rangle + d_1|1\rangle + d_2|2\rangle}{\sqrt{3}} \Rightarrow |\Psi\rangle =$$

$$|\Psi_1'\Psi_2'\rangle = \frac{|00\rangle + d_1|11\rangle + d_2|22\rangle}{\sqrt{3}}.$$

For $\{\alpha_2 = 0, \beta_2 = 1, \gamma_2 = 0\}$

$$\therefore |\Psi_2\rangle = |1\rangle \Rightarrow |\Psi_2\rangle' = \frac{|1\rangle + d_1|2\rangle + d_2|0\rangle}{\sqrt{3}} \Rightarrow |\Psi\rangle =$$

$$|\Psi_1'\Psi_2'\rangle = \frac{|01\rangle + d_1|12\rangle + d_2|20\rangle}{\sqrt{3}}.$$

For $\{\alpha_2 = 0, \beta_2 = 0, \gamma_2 = 1\}$

$$\therefore |\Psi_2\rangle = |2\rangle \Rightarrow |\Psi_2\rangle' = \frac{|2\rangle + d_1|0\rangle + d_2|1\rangle}{\sqrt{3}} \Rightarrow |\Psi\rangle =$$

$$|\Psi_1'\Psi_2'\rangle = \frac{|02\rangle + d_1|10\rangle + d_2|21\rangle}{\sqrt{3}}.$$

For $\{\alpha_1 = 0, \beta_1 = 0, \gamma_1 = 1\}$: $|\Psi_1\rangle' = \dfrac{|0\rangle + d_2|1\rangle + d_1|2\rangle}{\sqrt{3}}$,

For $\{\alpha_2 = 1, \beta_2 = 0, \gamma_2 = 0\}$

$$\therefore |\Psi_2\rangle = |0\rangle \Rightarrow |\Psi_2\rangle' = \frac{|0\rangle + d_2|1\rangle + d_1|2\rangle}{\sqrt{3}} \Rightarrow |\Psi\rangle =$$

$$|\Psi_1{}'\Psi_2{}'\rangle = \frac{|00\rangle + d_2|11\rangle + d_1|22\rangle}{\sqrt{3}}.$$

For $\{\alpha_2 = 0, \beta_2 = 1, \gamma_2 = 0\}$

$$\therefore |\Psi_2\rangle = |1\rangle \Rightarrow |\Psi_2\rangle' = \frac{|1\rangle + d_2|2\rangle + d_1|0\rangle}{\sqrt{3}} \Rightarrow |\Psi\rangle =$$

$$|\Psi_1{}'\Psi_2{}'\rangle = \frac{|01\rangle + d_2|12\rangle + d_1|20\rangle}{\sqrt{3}}.$$

For $\{\alpha_2 = 0, \beta_2 = 0, \gamma_2 = 1\}$

$$\therefore |\Psi_2\rangle = |2\rangle \Rightarrow |\Psi_2\rangle' = \frac{|2\rangle + d_2|0\rangle + d_1|1\rangle}{\sqrt{3}} \Rightarrow |\Psi\rangle =$$

$$|\Psi_1{}'\Psi_2{}'\rangle = \frac{|02\rangle + d_2|10\rangle + d_1|21\rangle}{\sqrt{3}}.$$

Q.E.D.

One observes that, while in the two-valued case in Example 11. 2 and Fig. 11.7 the set of two-valued 2-qubit orthonormal Einstein-Podolsky-Rosen (EPR) basis states using QWHT consists of four basis states, the set of ternary 2-qubit orthonormal EPR basis states (from Theorem 11.3 and Fig. 11.19) consists of nine basis states. Therefore, in general, for N-valued 2-qubit EPR, one would have a set that contains N^2 of N-valued orthonormal basis states by performing the corresponding transformations on input qubits using $C^{(N)}_{(1)\ normalized}$ gate in a circuit topology similar to the one shown in Fig. 11.19 and by utilizing a shift operator over general Galois field of Nth radix.

Example 11.6. The following is a derivation of the probability amplitudes of the ternary composite basis states that were introduced in Theorem 11.2. Assuming the normalization of the probability amplitudes, and by using the ternary quantum signal $|\Psi\rangle$ = $\alpha|0\rangle + \beta|1\rangle + \gamma|2\rangle$ as an input to the ternary normalized quantum Chrestenson gate $C^{(3)}_{(1)\ normalized} = \frac{1}{\sqrt{3}}\begin{bmatrix} 1 & 1 & 1 \\ 1 & d_1 & d_2 \\ 1 & d_2 & d_1 \end{bmatrix}$, one obtains the following quantum signal at the output of the gate:

$$|\Psi\rangle' = [|0\rangle \quad |1\rangle \quad |2\rangle]\frac{1}{\sqrt{3}}\begin{bmatrix} 1 & 1 & 1 \\ 1 & d_1 & d_2 \\ 1 & d_2 & d_1 \end{bmatrix}\begin{bmatrix} \alpha \\ \beta \\ \gamma \end{bmatrix} = [|0\rangle \quad |1\rangle \quad |2\rangle]\begin{bmatrix} \frac{\alpha+\beta+\gamma}{\sqrt{3}} \\ \frac{\alpha+d_1\beta+d_2\gamma}{\sqrt{3}} \\ \frac{\alpha+d_2\beta+d_1\gamma}{\sqrt{3}} \end{bmatrix},$$

$$= \frac{\alpha+\beta+\gamma}{\sqrt{3}}|0\rangle + \frac{\alpha+d_1\beta+d_2\gamma}{\sqrt{3}}|1\rangle + \frac{\alpha+d_2\beta+d_1\gamma}{\sqrt{3}}|2\rangle.$$

$$|\Psi\rangle' = [|0\rangle \quad |1\rangle \quad |2\rangle]\frac{1}{\sqrt{3}}\begin{bmatrix} 1 & 1 & 1 \\ 1 & d_1 & d_2 \\ 1 & d_2 & d_1 \end{bmatrix}\begin{bmatrix} \alpha \\ \beta \\ \gamma \end{bmatrix},$$

$$= \begin{bmatrix} \frac{|0\rangle+|1\rangle+|2\rangle}{\sqrt{3}} & \frac{|0\rangle+d_1|1\rangle+d_2|2\rangle}{\sqrt{3}} & \frac{|0\rangle+d_2|1\rangle+d_1|2\rangle}{\sqrt{3}} \end{bmatrix}\begin{bmatrix} \alpha \\ \beta \\ \gamma \end{bmatrix},$$

$$= [|I\rangle \quad |+\rangle \quad |-\rangle]\begin{bmatrix} \alpha \\ \beta \\ \gamma \end{bmatrix} = \alpha|I\rangle + \beta|+\rangle + \gamma|-\rangle.$$

Where:

$$\{\ |+\rangle = \frac{|0\rangle+d_1|1\rangle+d_2|2\rangle}{\sqrt{3}}, |I\rangle = \frac{|0\rangle+|1\rangle+|2\rangle}{\sqrt{3}}, |-\rangle = \frac{|0\rangle+d_2|1\rangle+d_1|2\rangle}{\sqrt{3}}\ \}$$

, and:

$$\{|0\rangle = \frac{|+\rangle+|I\rangle+|-\rangle}{\sqrt{3}}, |1\rangle = \frac{d_2|+\rangle+|I\rangle+d_1|-\rangle}{\sqrt{3}}, |2\rangle = \frac{d_1|+\rangle+|I\rangle+d_2|-\rangle}{\sqrt{3}}\}.$$

Thus, one obtains at the input of the quantum Chrestenson gate the following quantum state:

$$\therefore |\Psi\rangle = \alpha|0\rangle + \beta|1\rangle + \gamma|2\rangle,$$

$$= \alpha\frac{|+\rangle + ||\rangle + |-\rangle}{\sqrt{3}} + \beta\frac{d_2|+\rangle + ||\rangle + d_1|-\rangle}{\sqrt{3}} + \gamma\frac{d_1|+\rangle + ||\rangle + d_2|-\rangle}{\sqrt{3}},$$

$$= \frac{\alpha + d_2\beta + d_1\gamma}{\sqrt{3}}|+\rangle + \frac{\alpha + \beta + \gamma}{\sqrt{3}}||\rangle + \frac{\alpha + d_1\beta + d_2\gamma}{\sqrt{3}}|-\rangle.$$

$$\therefore \{ \sqrt{p_{||\rangle}} = \frac{|\alpha + \beta + \gamma|}{\sqrt{3}}, \sqrt{p_{|-\rangle}} = \frac{|\alpha + d_1\beta + d_2\gamma|}{\sqrt{3}}, \sqrt{p_{|+\rangle}} = \frac{|\alpha + d_2\beta + d_1\gamma|}{\sqrt{3}} \}.$$

Consequently, measuring $|\Psi\rangle$ with respect to the new basis $||\rangle, \{|+\rangle, |-\rangle\}$ will result in the state (basis) $\{||\rangle\}$ with probability equals to $\frac{|\alpha + \beta + \gamma|^2}{3}$, will result in the state (basis) $\{|-\rangle\}$ with probability equals to $\frac{|\alpha + d_1\beta + d_2\gamma|^2}{3}$, and will result in the state (basis) $\{|+\rangle\}$ with probability equals to $\frac{|\alpha + d_2\beta + d_1\gamma|^2}{3}$, where:

$$d_2 = -(1 + d_1) = -\frac{1}{2}(1 + \sqrt{3}i) = e^{\frac{4\pi i}{3}},$$

$$d_1 = -(1 + d_2) = -\frac{1}{2}(1 - \sqrt{3}i) = e^{\frac{2\pi i}{3}}.$$

Figure 11.20 shows several factorizations of serially interconnected evolution processes that resemble equivalences between some ternary quantum logic circuits, using the ternary quantum Chrestenson operator $C^{(3)}_{(1)\ normalized} = \frac{1}{\sqrt{3}}\begin{bmatrix} 1 & 1 & 1 \\ 1 & d_1 & d_2 \\ 1 & d_2 & d_1 \end{bmatrix}$, which was presented in Theorem 11.2.

The equivalences of serially interconnected quantum Chrestenson primitives can be utilized in the synthesis of quantum circuits by replacing long serial gate interconnections with their equivalent circuits (i.e., technology mapping). For instance, such transformations can be applied to a quantum circuit that is created by a genetic algorithm (GA) or other evolutionary algorithms [207,252].

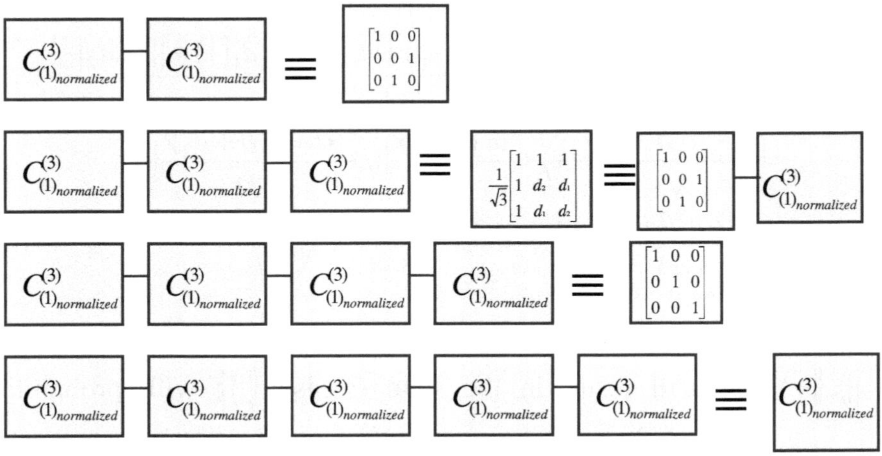

Fig. 11.20. Multiple-valued quantum logic circuit equivalences using quantum Chrestenson gate.

11.6 New Multiple-Valued Quantum Evolution Processes, Generalized Permuters, and their Circuit Analysis

The following presents theorems to obtain the ternary logic evolution processes for the ternary Feynman, Swap, Fredkin (Shannon), and Davio quantum gates (transformations), respectively.

The size of the following evolution matrices is governed by Eq. (11.9). For example, for a (2,2) ternary quantum gate (transformation) (as in Theorems 11.4 and 11.5), the size of the evolution matrix is equal to 3^2 rows · 3^2 columns = 9·9 = 81 elements, and for a (4,4) ternary quantum gate (transformation) (as in Theorems 11.6, 11.7, and 11.8), the size of the evolution matrix is equal to 3^4 rows · 3^4 columns = 81·81 = 6,561 elements.

Analogously to the binary case, the input qubit to the ternary quantum gate is the column index of the ternary evolution matrix, and the output qubit of the ternary quantum gate is the row index of the ternary evolution matrix. The column and row indices of the ternary evolution matrix take the following order for 1-input 1-output, 2-input 2-output, and 3-input 3-output gates, respectively.

$\Rightarrow 1-input \quad 1-output \quad gate: |0\rangle, |1\rangle, |2\rangle.$

$\Rightarrow 2-input \quad 2-output \quad gate: |00\rangle, |01\rangle, |02\rangle, |10\rangle, |11\rangle, |12\rangle, |20\rangle, |21\rangle, |22\rangle.$

$$\Rightarrow 3-input \quad 3-output \quad gate: |000\rangle, |001\rangle, |002\rangle, |010\rangle, |011\rangle, |012\rangle, |020\rangle, |021\rangle,$$
$$|022\rangle, |100\rangle, |101\rangle, |102\rangle, |110\rangle, |111\rangle, |112\rangle, |120\rangle,$$
$$|121\rangle, |122\rangle, |200\rangle, |201\rangle, |202\rangle, |210\rangle, |211\rangle, |212\rangle,$$
$$|220\rangle, |221\rangle, |222\rangle.$$

The following multiple-valued evolution matrices in Theorems 11.4 through 11.8 will be useful in the synthesis of multiple-valued quantum circuits. For instance, by using evolutionary algorithms for the synthesis of minimal size multiple-valued quantum circuits [252], one can consider the fitness function of the evolutionary algorithms (such as genetic programming or genetic algorithms from Appendix E) to contain two components: (1) first component is for the correctness of the resulting function (e.g., error is zero), and (2) the second component is for the cost of the resulting quantum circuit (e.g., number of gates). The synthesis of such multiple-valued quantum circuits, using evolutionary algorithms, is done through the calculation of the final evolution matrix of the whole circuit by using normal matrix multiplication (serial logic interconnects) and tensor multiplication (parallel logic interconnects) of the individual multiple-valued evolution matrices (i.e., gates). (Example 11.7 will illustrate such serial, and parallel algebraic manipulations for the analysis of the corresponding multiple-valued quantum logic circuits).

The representation of such evolutionary computations for the synthesis of multiple-valued quantum circuits can be done by using the evolutionary matrics as the chromosome of the evolutionary algorithm that was shown in Fig. 3.16.

Theorem 11.4. The following is the ternary Galois field Feynman evolution matrix:

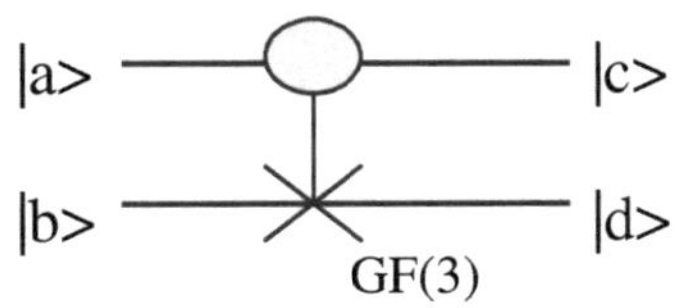

$$Feynman^{(3)}_{(2)\,ab/cd} = \begin{bmatrix} 1&0&0&0&0&0&0&0&0\\ 0&1&0&0&0&0&0&0&0\\ 0&0&1&0&0&0&0&0&0\\ 0&0&0&0&0&1&0&0&0\\ 0&0&0&1&0&0&0&0&0\\ 0&0&0&0&1&0&0&0&0\\ 0&0&0&0&0&0&0&1&0\\ 0&0&0&0&0&0&0&0&1\\ 0&0&0&0&0&0&1&0&0 \end{bmatrix}.$$

Proof. Utilizing the algebraic addition and multiplication operations over Galois field, one obtains the following quantum transformations of the ternary input qubits into the output qubits using GF(3) Feynman quantum register:

$$|00\rangle \to |00\rangle, |01\rangle \to |01\rangle, |02\rangle \to |02\rangle, |10\rangle \to |11\rangle, |11\rangle \to |12\rangle$$
$$|12\rangle \to |10\rangle, |20\rangle \to |22\rangle, |21\rangle \to |20\rangle, |22\rangle \to |21\rangle$$

Then by solving for the following set of linearly independent Eqs. over ternary Galois field:

$$\begin{bmatrix} \alpha1&\alpha2&\alpha3&\alpha4&\alpha5&\alpha6&\alpha7&\alpha8&\alpha9\\ \beta1&\beta2&\beta3&\beta4&\beta5&\beta6&\beta7&\beta8&\beta9\\ \chi1&\chi2&\chi3&\chi4&\chi5&\chi6&\chi7&\chi8&\chi9\\ \delta1&\delta2&\delta3&\delta4&\delta5&\delta6&\delta7&\delta8&\delta9\\ \varepsilon1&\varepsilon2&\varepsilon3&\varepsilon4&\varepsilon5&\varepsilon6&\varepsilon7&\varepsilon8&\varepsilon9\\ \phi1&\phi2&\phi3&\phi4&\phi5&\phi6&\phi7&\phi8&\phi9\\ \varphi1&\varphi2&\varphi3&\varphi4&\varphi5&\varphi6&\varphi7&\varphi8&\varphi9\\ \gamma1&\gamma2&\gamma3&\gamma4&\gamma5&\gamma6&\gamma7&\gamma8&\gamma9\\ \eta1&\eta2&\eta3&\eta4&\eta5&\eta6&\eta7&\eta8&\eta9 \end{bmatrix} |00\rangle = |00\rangle,$$

$$\begin{bmatrix} \alpha1 & \alpha2 & \alpha3 & \alpha4 & \alpha5 & \alpha6 & \alpha7 & \alpha8 & \alpha9 \\ \beta1 & \beta2 & \beta3 & \beta4 & \beta5 & \beta6 & \beta7 & \beta8 & \beta9 \\ \chi1 & \chi2 & \chi3 & \chi4 & \chi5 & \chi6 & \chi7 & \chi8 & \chi9 \\ \delta1 & \delta2 & \delta3 & \delta4 & \delta5 & \delta6 & \delta7 & \delta8 & \delta9 \\ \varepsilon1 & \varepsilon2 & \varepsilon3 & \varepsilon4 & \varepsilon5 & \varepsilon6 & \varepsilon7 & \varepsilon8 & \varepsilon9 \\ \phi1 & \phi2 & \phi3 & \phi4 & \phi5 & \phi6 & \phi7 & \phi8 & \phi9 \\ \varphi1 & \varphi2 & \varphi3 & \varphi4 & \varphi5 & \varphi6 & \varphi7 & \varphi8 & \varphi9 \\ \gamma1 & \gamma2 & \gamma3 & \gamma4 & \gamma5 & \gamma6 & \gamma7 & \gamma8 & \gamma9 \\ \eta1 & \eta2 & \eta3 & \eta4 & \eta5 & \eta6 & \eta7 & \eta8 & \eta9 \end{bmatrix} |01\rangle = |01\rangle,$$

$$\vdots$$

$$\begin{bmatrix} \alpha1 & \alpha2 & \alpha3 & \alpha4 & \alpha5 & \alpha6 & \alpha7 & \alpha8 & \alpha9 \\ \beta1 & \beta2 & \beta3 & \beta4 & \beta5 & \beta6 & \beta7 & \beta8 & \beta9 \\ \chi1 & \chi2 & \chi3 & \chi4 & \chi5 & \chi6 & \chi7 & \chi8 & \chi9 \\ \delta1 & \delta2 & \delta3 & \delta4 & \delta5 & \delta6 & \delta7 & \delta8 & \delta9 \\ \varepsilon1 & \varepsilon2 & \varepsilon3 & \varepsilon4 & \varepsilon5 & \varepsilon6 & \varepsilon7 & \varepsilon8 & \varepsilon9 \\ \phi1 & \phi2 & \phi3 & \phi4 & \phi5 & \phi6 & \phi7 & \phi8 & \phi9 \\ \varphi1 & \varphi2 & \varphi3 & \varphi4 & \varphi5 & \varphi6 & \varphi7 & \varphi8 & \varphi9 \\ \gamma1 & \gamma2 & \gamma3 & \gamma4 & \gamma5 & \gamma6 & \gamma7 & \gamma8 & \gamma9 \\ \eta1 & \eta2 & \eta3 & \eta4 & \eta5 & \eta6 & \eta7 & \eta8 & \eta9 \end{bmatrix} |22\rangle = |21\rangle.$$

One obtains the GF(3) Feynman evolution matrix:

$$\begin{bmatrix} 1 & 0 & 0 & 0 & 0 & 0 & 0 & 0 & 0 \\ 0 & 1 & 0 & 0 & 0 & 0 & 0 & 0 & 0 \\ 0 & 0 & 1 & 0 & 0 & 0 & 0 & 0 & 0 \\ 0 & 0 & 0 & 0 & 0 & 1 & 0 & 0 & 0 \\ 0 & 0 & 0 & 1 & 0 & 0 & 0 & 0 & 0 \\ 0 & 0 & 0 & 0 & 1 & 0 & 0 & 0 & 0 \\ 0 & 0 & 0 & 0 & 0 & 0 & 0 & 1 & 0 \\ 0 & 0 & 0 & 0 & 0 & 0 & 0 & 0 & 1 \\ 0 & 0 & 0 & 0 & 0 & 0 & 1 & 0 & 0 \end{bmatrix}.$$

Q.E.D.

Theorem 11.5. The following is the ternary Galois field Swap evolution matrix:

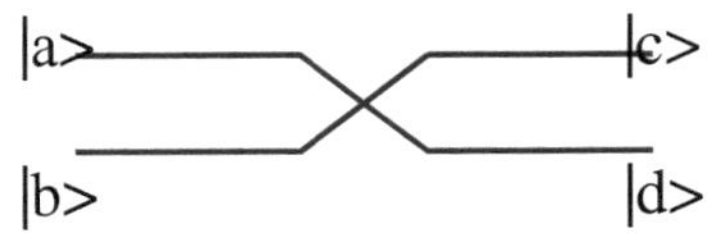

$$Swap^{(3)}_{(2)\,ab/cd} = \begin{bmatrix} 1 & 0 & 0 & 0 & 0 & 0 & 0 & 0 & 0 \\ 0 & 0 & 0 & 1 & 0 & 0 & 0 & 0 & 0 \\ 0 & 0 & 0 & 0 & 0 & 0 & 1 & 0 & 0 \\ 0 & 1 & 0 & 0 & 0 & 0 & 0 & 0 & 0 \\ 0 & 0 & 0 & 0 & 1 & 0 & 0 & 0 & 0 \\ 0 & 0 & 0 & 0 & 0 & 0 & 0 & 1 & 0 \\ 0 & 0 & 1 & 0 & 0 & 0 & 0 & 0 & 0 \\ 0 & 0 & 0 & 0 & 0 & 1 & 0 & 0 & 0 \\ 0 & 0 & 0 & 0 & 0 & 0 & 0 & 0 & 1 \end{bmatrix}.$$

Proof. Utilizing the algebraic addition and multiplication operations over Galois field, one obtains the following quantum transformations of the ternary input qubits into the output qubits using the ternary Swap quantum register:

$$|00\rangle \rightarrow |00\rangle, |01\rangle \rightarrow |10\rangle, |02\rangle \rightarrow |20\rangle, |10\rangle \rightarrow |01\rangle, |11\rangle \rightarrow |11\rangle$$
$$|12\rangle \rightarrow |21\rangle, |20\rangle \rightarrow |02\rangle, |21\rangle \rightarrow |12\rangle, |22\rangle \rightarrow |22\rangle$$

Similar to Theorem 11.4, by solving for the set of linearly independent Eqs. over a ternary Galois field, one obtains the ternary Swap Galois field evolution matrix. **Q.E.D.**

Using the previous approach in Theorems 11.4 and 11.5, one can construct the ternary evolution matrices for ternary reversible Fredkin gates, Toffoli gates, Davio gates, and Margolus gates that were presented in Chapt. 5. Since the ternary 4-qubit evolution matrices will be according to Eq. (11.9) of size: $3^4 \cdot 3^4 = 81 \cdot 81 =$ 6,561 elements, the documentation of the evolution process in a matrix form will be very difficult since the size of the matrix is very big. Alternatively, we write the evolution matrices in terms of the indices where the elements are of value “1”, where the remaining of the elements are understood to be of value “0”.

Theorem 11.6. The following is the ternary Galois field evolution matrix for a ternary Galois field reversible Shannon ($S_{1\,(1)}^{(3)}$) which is represented in Eq. (5.6) in the order of inputs/outputs $cf_0f_1f_2/cf_{r0}f_{r1}f_{r2}$, respectively.

Row index	Column index	Row index	Column index	Row index	Column index
0	0	27	27	54	54
1	1	28	30	55	63
2	2	29	33	56	72
3	3	30	36	57	55
4	4	31	39	58	64
5	5	32	42	59	73
6	6	33	45	60	56
7	7	34	48	61	65
8	8	35	51	62	74
9	9	36	28	63	57
10	10	37	31	64	66
11	11	38	34	65	75
12	12	39	37	66	58
13	13	40	40	67	67
14	14	41	43	68	76
15	15	42	46	69	59
16	16	43	49	70	68
17	17	44	52	71	77
18	18	45	29	72	60
19	19	46	32	73	69
20	20	47	35	74	78
21	21	48	38	75	61
22	22	49	41	76	70
23	23	50	44	77	79
24	24	51	47	78	62
25	25	52	50	79	71
26	26	53	53	80	80

Proof. Utilizing the algebraic addition and multiplication operations over a Galois field, one obtains the following quantum transformations of the ternary reversible Shannon decomposition (Eq. (5.6) and Fig. 5.26):

$|0000\rangle \to |0000\rangle, |0001\rangle \to |0001\rangle, |0002\rangle \to |0002\rangle, |0010\rangle \to |0010\rangle, |0011\rangle \to |0011\rangle, |0012\rangle \to |0012\rangle, |0020\rangle \to$
$|0020\rangle, |0021\rangle \to |0021\rangle, |0022\rangle \to |0022\rangle, |0100\rangle \to |0100\rangle, |0101\rangle \to |0101\rangle, |0102\rangle \to |0102\rangle, |0110\rangle \to |0110\rangle,$
$|0111\rangle \to |0111\rangle, |0112\rangle \to |0112\rangle, |0120\rangle \to |0120\rangle, |0121\rangle \to |0121\rangle, |0122\rangle \to |0122\rangle, |0200\rangle \to |0200\rangle, |0201\rangle \to$
$|0201\rangle, |0202\rangle \to |0202\rangle, |0210\rangle \to |0210\rangle, |0211\rangle \to |0211\rangle, |0212\rangle \to |0212\rangle, |0220\rangle \to |0220\rangle, |0221\rangle \to |0221\rangle,$
$|0222\rangle \to |0222\rangle, |1000\rangle \to |1000\rangle, |1001\rangle \to |1010\rangle, |1002\rangle \to |1020\rangle, |1010\rangle \to |1100\rangle, |1011\rangle \to |1110\rangle, |1012\rangle \to$
$|1120\rangle, |1020\rangle \to |1200\rangle, |1021\rangle \to |1210\rangle, |1022\rangle \to |1220\rangle, |1100\rangle \to |1001\rangle, |1101\rangle \to |1011\rangle, |1102\rangle \to |1021\rangle, |1110\rangle$
$\to |1101\rangle, |1111\rangle \to |1111\rangle, |1112\rangle \to |1121\rangle, |1120\rangle \to |1201\rangle, |1121\rangle \to |1211\rangle, |1122\rangle \to |1221\rangle, |1200\rangle \to |1002\rangle,$
$|1201\rangle \to |1012\rangle, |1202\rangle \to |1022\rangle, |1210\rangle \to |1102\rangle, |1211\rangle \to |1112\rangle, |1212\rangle \to |1122\rangle, |1220\rangle \to |1202\rangle, |1221\rangle \to$
$|1212\rangle, |1222\rangle \to |1222\rangle, |2000\rangle \to |2000\rangle, |2001\rangle \to |2100\rangle, |2002\rangle \to |2200\rangle, |2010\rangle \to |2001\rangle, |2011\rangle \to |2101\rangle,$
$|2012\rangle \to |2201\rangle, |2020\rangle \to |2002\rangle, |2021\rangle \to |2102\rangle, |2022\rangle \to |2202\rangle, |2100\rangle \to |2010\rangle, |2101\rangle \to |2110\rangle, |2102\rangle \to$
$|2210\rangle, |2110\rangle \to |2011\rangle, |2111\rangle \to |2111\rangle, |2112\rangle \to |2211\rangle, |2120\rangle \to |2012\rangle, |2121\rangle \to |2112\rangle, |2122\rangle \to |2212\rangle,$
$|2200\rangle \to |2020\rangle, |2201\rangle \to |2120\rangle, |2202\rangle \to |2220\rangle, |2210\rangle \to |2021\rangle, |2211\rangle \to |2121\rangle, |2212\rangle \to |2221\rangle, |2220\rangle \to$
$|2022\rangle, |2221\rangle \to |2122\rangle, |2222\rangle \to |2222\rangle.$

Then the proof follows the same method from Theorem 11.4.**Q.E.D.**

Theorem 11.7. The following is the ternary Galois field evolution matrix for a ternary Galois field reversible Shannon ($S_{4}{}^{(3)}_{(1)}$) which is represented in Eq. (5.9) in the order of inputs/outputs $cf_0f_1f_2/cf_{r0}f_{r1}f_{r2}$.

Row index	Column index	Row index	Column index	Row index	Column index
0	0	27	27	54	54
1	9	28	30	55	55
2	18	29	33	56	56
3	3	30	28	57	63
4	12	31	31	58	64
5	21	32	34	59	65
6	6	33	29	60	72
7	15	34	32	61	73
8	24	35	35	62	74
9	1	36	36	63	57
10	10	37	39	64	58
11	19	38	42	65	59
12	4	39	37	66	66
13	13	40	40	67	67
14	22	41	43	68	68
15	7	42	38	69	75
16	16	43	41	70	76
17	25	44	44	71	77
18	2	45	45	72	60
19	11	46	48	73	61
20	20	47	51	74	62
21	5	48	46	75	69
22	14	49	49	76	70
23	23	50	52	77	71
24	8	51	47	78	78
25	17	52	50	79	79
26	26	53	53	80	80

Proof. Using GF, one obtains the following quantum transformations:

$|0000\rangle \rightarrow |0000\rangle, |0001\rangle \rightarrow |0100\rangle, |0002\rangle \rightarrow |0200\rangle, |0010\rangle \rightarrow |0010\rangle, |0011\rangle \rightarrow |0110\rangle, |0012\rangle \rightarrow |0210\rangle, |0020\rangle \rightarrow$
$|0020\rangle, |0021\rangle \rightarrow |0120\rangle, |0022\rangle \rightarrow |0220\rangle, |0100\rangle \rightarrow |0001\rangle, |0101\rangle \rightarrow |0101\rangle, |0102\rangle \rightarrow |0201\rangle, |0110\rangle \rightarrow |0011\rangle,$
$|0111\rangle \rightarrow |0111\rangle, |0112\rangle \rightarrow |0211\rangle, |0120\rangle \rightarrow |0021\rangle, |0121\rangle \rightarrow |0121\rangle, |0122\rangle \rightarrow |0221\rangle, |0200\rangle \rightarrow |0002\rangle, |0201\rangle \rightarrow$
$|0102\rangle, |0202\rangle \rightarrow |0202\rangle, |0210\rangle \rightarrow |0012\rangle, |0211\rangle \rightarrow |0112\rangle, |0212\rangle \rightarrow |0212\rangle, |0220\rangle \rightarrow |0022\rangle, |0221\rangle \rightarrow |0122\rangle,$
$|0222\rangle \rightarrow |0222\rangle, |1000\rangle \rightarrow |1000\rangle, |1001\rangle \rightarrow |1010\rangle, |1002\rangle \rightarrow |1020\rangle, |1010\rangle \rightarrow |1001\rangle, |1011\rangle \rightarrow |1011\rangle, |1012\rangle \rightarrow$
$|1021\rangle, |1020\rangle \rightarrow |1002\rangle, |1021\rangle \rightarrow |1012\rangle, |1022\rangle \rightarrow |1022\rangle, |1100\rangle \rightarrow |1100\rangle, |1101\rangle \rightarrow |1110\rangle, |1102\rangle \rightarrow |1120\rangle, |1110\rangle$
$\rightarrow |1101\rangle, |1111\rangle \rightarrow |1111\rangle, |1112\rangle \rightarrow |1121\rangle, |1120\rangle \rightarrow |1102\rangle, |1121\rangle \rightarrow |1112\rangle, |1122\rangle \rightarrow |1122\rangle, |1200\rangle \rightarrow |1200\rangle,$
$|1201\rangle \rightarrow |1210\rangle, |1202\rangle \rightarrow |1220\rangle, |1210\rangle \rightarrow |1201\rangle, |1211\rangle \rightarrow |1211\rangle, |1212\rangle \rightarrow |1221\rangle, |1220\rangle \rightarrow |1202\rangle, |1221\rangle \rightarrow$
$|1212\rangle, |1222\rangle \rightarrow |1222\rangle, |2000\rangle \rightarrow |2000\rangle, |2001\rangle \rightarrow |2001\rangle, |2002\rangle \rightarrow |2002\rangle, |2010\rangle \rightarrow |2100\rangle, |2011\rangle \rightarrow |2101\rangle,$
$|2012\rangle \rightarrow |2102\rangle, |2020\rangle \rightarrow |2200\rangle, |2021\rangle \rightarrow |2201\rangle, |2022\rangle \rightarrow |2202\rangle, |2100\rangle \rightarrow |2010\rangle, |2101\rangle \rightarrow |2011\rangle, |2102\rangle \rightarrow$
$|2012\rangle, |2110\rangle \rightarrow |2110\rangle, |2111\rangle \rightarrow |2111\rangle, |2112\rangle \rightarrow |2112\rangle, |2120\rangle \rightarrow |2210\rangle, |2121\rangle \rightarrow |2211\rangle, |2122\rangle \rightarrow |2212\rangle,$
$|2200\rangle \rightarrow |2020\rangle, |2201\rangle \rightarrow |2021\rangle, |2202\rangle \rightarrow |2022\rangle, |2120\rangle \rightarrow |2021\rangle, |2211\rangle \rightarrow |2121\rangle, |2212\rangle \rightarrow |2122\rangle, |2220\rangle \rightarrow$
$|2220\rangle, |2221\rangle \rightarrow |2221\rangle, |2222\rangle \rightarrow |2222\rangle.$

Then the proof follows the same method from Theorem 11.4.**Q.E.D.**

Theorem 11.8. GF(3) evolution matrix for ternary reversible Davio$_0$ ($D_{0(1)}^{(3)}$) (Eq. (5.15)) in the order of $cf_0f_1f_2/cf_{r0}f_{r1}f_{r2}$ is as follows:

Row index	Column index	Row index	Column index	Row index	Column index
0	0	27	27	54	54
1	3	28	28	55	63
2	6	29	29	56	72
3	1	30	36	57	57
4	4	31	37	58	66
5	7	32	38	59	75
6	2	33	45	60	60
7	5	34	46	61	69
8	8	35	47	62	78
9	9	36	30	63	55
10	12	37	31	64	64
11	15	38	32	65	73
12	10	39	39	66	58
13	13	40	40	67	67
14	16	41	41	68	76
15	11	42	48	69	61
16	14	43	49	70	70
17	17	44	50	71	79
18	18	45	33	72	56
19	21	46	34	73	65
20	24	47	35	74	74
21	19	48	42	75	59
22	22	49	43	76	68
23	25	50	44	77	77
24	20	51	51	78	62
25	23	52	52	79	71
26	26	53	53	80	80

Proof. Using GF, one obtains the following quantum transformations:

$$|0000\rangle \rightarrow |0000\rangle, |0001\rangle \rightarrow |0010\rangle, |0002\rangle \rightarrow |0020\rangle, |0010\rangle \rightarrow |0001\rangle, |0011\rangle \rightarrow |0011\rangle, |0012\rangle \rightarrow |00212\rangle, |0020\rangle$$
$$\rightarrow |0002\rangle, |0021\rangle \rightarrow |0012\rangle, |0022\rangle \rightarrow |0022\rangle, |0100\rangle \rightarrow |0100\rangle, |0101\rangle \rightarrow |0110\rangle, |0102\rangle \rightarrow |0120\rangle, |0110\rangle \rightarrow |0101\rangle,$$
$$|0111\rangle \rightarrow |0111\rangle, |0112\rangle \rightarrow |0121\rangle, |0120\rangle \rightarrow |0102\rangle, |0121\rangle \rightarrow |0112\rangle, |0122\rangle \rightarrow |0122\rangle, |0200\rangle \rightarrow |0200\rangle, |0201\rangle \rightarrow$$
$$|0210\rangle, |0202\rangle \rightarrow |0220\rangle, |0210\rangle \rightarrow |0201\rangle, |0211\rangle \rightarrow |0211\rangle, |0212\rangle \rightarrow |0221\rangle, |0220\rangle \rightarrow |0202\rangle, |0221\rangle \rightarrow |0212\rangle,$$
$$|0222\rangle \rightarrow |0222\rangle, |1000\rangle \rightarrow |1000\rangle, |1001\rangle \rightarrow |1001\rangle, |1002\rangle \rightarrow |1002\rangle, |1010\rangle \rightarrow |1100\rangle, |1011\rangle \rightarrow |1101\rangle, |1012\rangle \rightarrow$$
$$|1102\rangle, |1020\rangle \rightarrow |1200\rangle, |1021\rangle \rightarrow |1201\rangle, |1022\rangle \rightarrow |1202\rangle, |1100\rangle \rightarrow |1010\rangle, |1101\rangle \rightarrow |1011\rangle, |1102\rangle \rightarrow |1012\rangle, |1110\rangle$$
$$\rightarrow |1110\rangle, |1111\rangle \rightarrow |1111\rangle, |1112\rangle \rightarrow |1112\rangle, |1120\rangle \rightarrow |1210\rangle, |1121\rangle \rightarrow |1211\rangle, |1122\rangle \rightarrow |1212\rangle, |1200\rangle \rightarrow |1020\rangle,$$
$$|1201\rangle \rightarrow |1021\rangle, |1202\rangle \rightarrow |1022\rangle, |1210\rangle \rightarrow |1120\rangle, |1211\rangle \rightarrow |1121\rangle, |1212\rangle \rightarrow |1122\rangle, |1220\rangle \rightarrow |1220\rangle, |1221\rangle \rightarrow$$
$$|1221\rangle, |1222\rangle \rightarrow |1222\rangle, |2000\rangle \rightarrow |2000\rangle, |2001\rangle \rightarrow |2100\rangle, |2002\rangle \rightarrow |2200\rangle, |2010\rangle \rightarrow |2010\rangle, |2011\rangle \rightarrow |2110\rangle,$$
$$|2012\rangle \rightarrow |2210\rangle, |2020\rangle \rightarrow |2020\rangle, |2021\rangle \rightarrow |2120\rangle, |2022\rangle \rightarrow |2220\rangle, |2100\rangle \rightarrow |2001\rangle, |2101\rangle \rightarrow |2101\rangle, |2102\rangle \rightarrow$$
$$|2201\rangle, |2110\rangle \rightarrow |2011\rangle, |2111\rangle \rightarrow |2111\rangle, |2112\rangle \rightarrow |2211\rangle, |2120\rangle \rightarrow |2021\rangle, |2121\rangle \rightarrow |2121\rangle, |2122\rangle \rightarrow |2221\rangle,$$
$$|2200\rangle \rightarrow |2002\rangle, |2201\rangle \rightarrow |2102\rangle, |2202\rangle \rightarrow |2202\rangle, |2210\rangle \rightarrow |2012\rangle, |2211\rangle \rightarrow |2112\rangle, |2212\rangle \rightarrow |2212\rangle, |2220\rangle \rightarrow$$
$$|2022\rangle, |2221\rangle \rightarrow |2122\rangle, |2222\rangle \rightarrow |2222\rangle.$$

Then the proof follows the same method from Theorem 11.4.**Q.E.D.**

Theorems 11.4, 11.5, 11.6, 11.7, and 11.8 provide the quantum computations of the reversible gates that were developed previously in Chapt. 5. Figure 11.21 shows the extensions for quantum computations of the gates that were provided in Fig. 5.28.

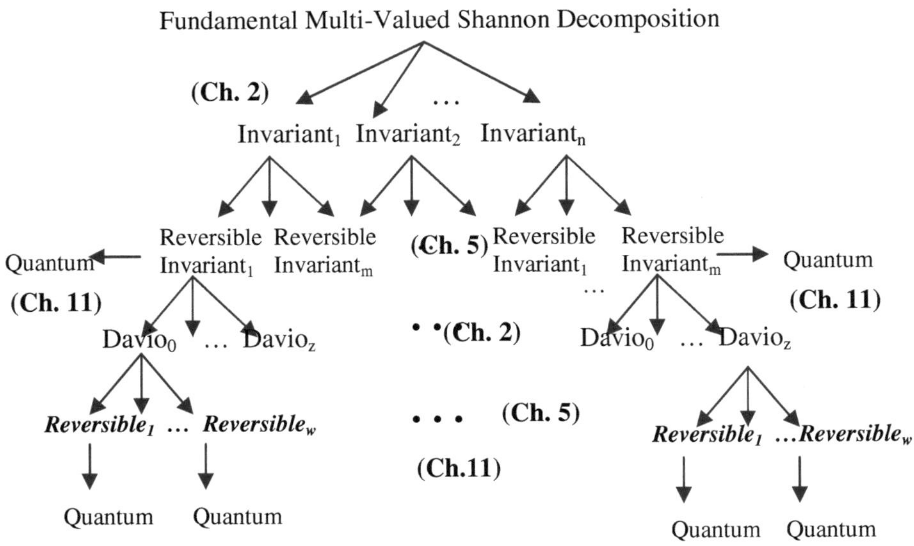

Fig. 11.21. A tree-based relationship between various decompositions.

Note in Fig. 11.21 that the quantum Shannon and Quantum Davio primitives, which are the extensions of the reversible Shannon and reversible Davio primitives from Sect. 5.4, are produced, and consequently quantum computations that use such quantum primitives can be implemented.

The quantum representations of reversible multiple-valued Shannon primitives, reversible multiple-valued Davio primitives, and other multiple-valued reversible primitives will be used to perform the multiple-valued quantum computing to analyze the multiple-valued quantum circuits and structures as will be shown in the following examples.

Example 11.7. The following circuits represent serial-like, parallel-like, and mixture of serial-like and parallel-like interconnects between multiple-valued quantum primitives.

11.7a. Let us evolve the input qubit |12> using the following serial-like interconnected multiple-valued quantum circuit:

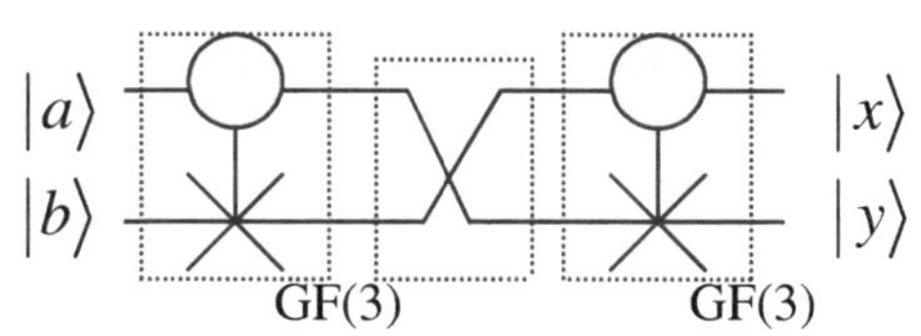

Fig. 11.22. Ternary Galois field quantum circuit composed of a Feynman gate interconnected serially with a Swap gate and then a Feynman gate, respectively.

Similar to two-valued quantum computing, the evolution of the input ternary qubit in Fig. 11.22 can be viewed in two equivalent perspectives, respectively. The first perspective is to evolve the input qubit step-by-step using the serially interconnected gates. The second perspective is to evolve the input qubit using the total quantum circuit at once. The second perspective is due to the fact that the total multiple-valued quantum evolution transformation $[M]$ of the total serially interconnected quantum circuit is equal to the normal matrix multiplication of the individual evolution matrices $[N_q]$ that correspond to the individual quantum primitives, i.e. $[M]_{serial} = \prod_q [N_q]$.

Perspective #1:

$$\begin{bmatrix} 1&0&0&0&0&0&0&0&0\\ 0&1&0&0&0&0&0&0&0\\ 0&0&1&0&0&0&0&0&0\\ 0&0&0&0&0&1&0&0&0\\ 0&0&0&1&0&0&0&0&0\\ 0&0&0&0&1&0&0&0&0\\ 0&0&0&0&0&0&0&1&0\\ 0&0&0&0&0&0&0&0&1\\ 0&0&0&0&0&0&1&0&0 \end{bmatrix} \begin{bmatrix}0\\0\\0\\0\\0\\1\\0\\0\\0\end{bmatrix} = \begin{bmatrix}0\\0\\0\\1\\0\\0\\0\\0\\0\end{bmatrix} \Rightarrow \begin{bmatrix} 1&0&0&0&0&0&0&0&0\\ 0&0&0&1&0&0&0&0&0\\ 0&0&0&0&0&0&1&0&0\\ 0&1&0&0&0&0&0&0&0\\ 0&0&0&0&1&0&0&0&0\\ 0&0&0&0&0&0&0&1&0\\ 0&0&1&0&0&0&0&0&0\\ 0&0&0&0&0&1&0&0&0\\ 0&0&0&0&0&0&0&0&1 \end{bmatrix} \begin{bmatrix}0\\0\\0\\1\\0\\0\\0\\0\\0\end{bmatrix} = \begin{bmatrix}0\\1\\0\\0\\0\\0\\0\\0\\0\end{bmatrix},$$

$$\Rightarrow \begin{bmatrix} 1&0&0&0&0&0&0&0&0\\ 0&1&0&0&0&0&0&0&0\\ 0&0&1&0&0&0&0&0&0\\ 0&0&0&0&0&1&0&0&0\\ 0&0&0&1&0&0&0&0&0\\ 0&0&0&0&1&0&0&0&0\\ 0&0&0&0&0&0&0&1&0\\ 0&0&0&0&0&0&0&0&1\\ 0&0&0&0&0&0&1&0&0 \end{bmatrix} \begin{bmatrix}0\\1\\0\\0\\0\\0\\0\\0\\0\end{bmatrix} = \begin{bmatrix}0\\1\\0\\0\\0\\0\\0\\0\\0\end{bmatrix} = |01\rangle.$$

So, the quantum circuit that is shown in Fig. 11.22 evolves the input qubit |12> into the output qubit |01>.

Perspective #2:

$$\begin{bmatrix} 1&0&0&0&0&0&0&0&0\\ 0&1&0&0&0&0&0&0&0\\ 0&0&1&0&0&0&0&0&0\\ 0&0&0&0&0&1&0&0&0\\ 0&0&0&1&0&0&0&0&0\\ 0&0&0&0&1&0&0&0&0\\ 0&0&0&0&0&0&0&1&0\\ 0&0&0&0&0&0&0&0&1\\ 0&0&0&0&0&0&1&0&0 \end{bmatrix} \bullet \begin{bmatrix} 1&0&0&0&0&0&0&0&0\\ 0&0&0&1&0&0&0&0&0\\ 0&0&0&0&0&0&1&0&0\\ 0&1&0&0&0&0&0&0&0\\ 0&0&0&0&1&0&0&0&0\\ 0&0&0&0&0&0&0&1&0\\ 0&0&1&0&0&0&0&0&0\\ 0&0&0&0&0&1&0&0&0\\ 0&0&0&0&0&0&0&0&1 \end{bmatrix} \bullet \begin{bmatrix} 1&0&0&0&0&0&0&0&0\\ 0&1&0&0&0&0&0&0&0\\ 0&0&1&0&0&0&0&0&0\\ 0&0&0&0&0&1&0&0&0\\ 0&0&0&1&0&0&0&0&0\\ 0&0&0&0&1&0&0&0&0\\ 0&0&0&0&0&0&0&1&0\\ 0&0&0&0&0&0&0&0&1\\ 0&0&0&0&0&0&1&0&0 \end{bmatrix} \bullet \begin{bmatrix}0\\0\\0\\0\\0\\1\\0\\0\\0\end{bmatrix} =$$

$$\begin{bmatrix} 1&0&0&0&0&0&0&0&0\\ 0&0&0&0&0&1&0&0&0\\ 0&0&0&0&0&0&0&1&0\\ 0&0&0&0&0&0&0&0&1\\ 0&1&0&0&0&0&0&0&0\\ 0&0&0&1&0&0&0&0&0\\ 0&0&0&0&1&0&0&0&0\\ 0&0&0&0&0&0&1&0&0\\ 0&0&1&0&0&0&0&0&0 \end{bmatrix} \begin{bmatrix}0\\0\\0\\0\\0\\1\\0\\0\\0\end{bmatrix} = \begin{bmatrix}0\\1\\0\\0\\0\\0\\0\\0\\0\end{bmatrix} = |01\rangle.$$

Identical to the result from perspective #1, the quantum circuit shown in Fig. 11.22 evolves the qubit |12> into the qubit |01>.

11.7b. Let us evolve the input qubit |122> using the multiple-valued quantum circuit in Fig. 11.23.

The total evolution transformation $[M]$ of the total parallel-interconnected quantum circuit is equal to the tensor (Kronecker) product of the individual evolution matrices $[N_q]$ that correspond to the individual quantum primitives, i.e., $[M]_{parallel} = \otimes[N]_q$. The evolution of the input ternary qubit, in Fig. 11.23, can be viewed in two equivalent perspectives, respectively. One perspective is to evolve the input qubit stage by stage. The second perspective is to evolve the input qubit using the total quantum circuit at once. The evolution matrices of the parallel-connected dashed boxes in (5) and (6), are as follows, respectively (Where the symbol || means parallel connection):

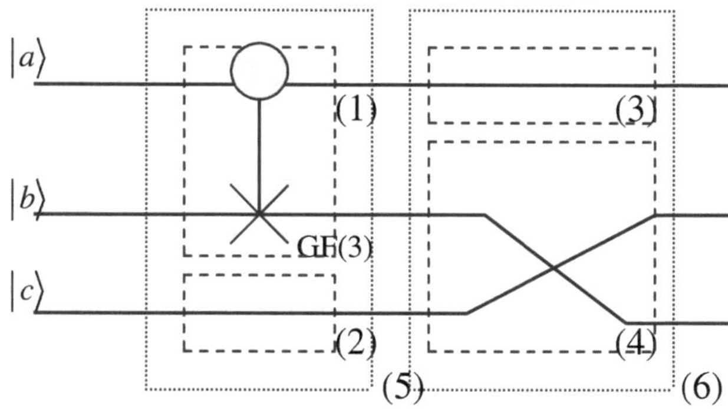

Fig. 11.23. Ternary Galois-field quantum circuit composed of serial interconnect of two parallel ternary Galois-field circuits: dashed boxes ((1),(2) and (3),(4)) in each sub-circuit are parallel connected, and dotted boxes (5) and (6) are serially interconnected.

$$\Rightarrow (5) = (1) \,||\, (2)\text{: Feynman} \otimes \text{Wire} = \begin{bmatrix} 1 & 0 & 0 & 0 & 0 & 0 & 0 & 0 & 0 \\ 0 & 1 & 0 & 0 & 0 & 0 & 0 & 0 & 0 \\ 0 & 0 & 1 & 0 & 0 & 0 & 0 & 0 & 0 \\ 0 & 0 & 0 & 0 & 0 & 1 & 0 & 0 & 0 \\ 0 & 0 & 0 & 1 & 0 & 0 & 0 & 0 & 0 \\ 0 & 0 & 0 & 0 & 1 & 0 & 0 & 0 & 0 \\ 0 & 0 & 0 & 0 & 0 & 0 & 0 & 1 & 0 \\ 0 & 0 & 0 & 0 & 0 & 0 & 0 & 0 & 1 \\ 0 & 0 & 0 & 0 & 0 & 0 & 1 & 0 & 0 \end{bmatrix} \otimes \begin{bmatrix} 1 & 0 & 0 \\ 0 & 1 & 0 \\ 0 & 0 & 1 \end{bmatrix},$$

$$= \begin{bmatrix} \begin{bmatrix}1&0&0\\0&1&0\\0&0&1\end{bmatrix} & & & & & & & & \\ & \begin{bmatrix}1&0&0\\0&1&0\\0&0&1\end{bmatrix} & & & & & & & \\ & & \begin{bmatrix}1&0&0\\0&1&0\\0&0&1\end{bmatrix} & & & & & & \\ & & & & & \begin{bmatrix}1&0&0\\0&1&0\\0&0&1\end{bmatrix} & & & \\ & & & \begin{bmatrix}1&0&0\\0&1&0\\0&0&1\end{bmatrix} & & & & & \\ & & & & \begin{bmatrix}1&0&0\\0&1&0\\0&0&1\end{bmatrix} & & & & \\ & & & & & & & \begin{bmatrix}1&0&0\\0&1&0\\0&0&1\end{bmatrix} & \\ & & & & & & & & \begin{bmatrix}1&0&0\\0&1&0\\0&0&1\end{bmatrix} \\ & & & & & & \begin{bmatrix}1&0&0\\0&1&0\\0&0&1\end{bmatrix} & & \end{bmatrix}.$$

$$\Rightarrow (6) = (3) \parallel (4): \text{Wire} \otimes \text{Swap} = \begin{bmatrix}1&0&0\\0&1&0\\0&0&1\end{bmatrix} \otimes \begin{bmatrix}1&0&0&0&0&0&0&0&0\\0&0&0&1&0&0&0&0&0\\0&0&0&0&0&0&1&0&0\\0&1&0&0&0&0&0&0&0\\0&0&0&0&1&0&0&0&0\\0&0&0&0&0&0&0&1&0\\0&0&1&0&0&0&0&0&0\\0&0&0&0&0&1&0&0&0\\0&0&0&0&0&0&0&0&1\end{bmatrix},$$

$$= \begin{bmatrix} \begin{bmatrix}1&0&0&0&0&0&0&0&0\\0&0&0&1&0&0&0&0&0\\0&0&0&0&0&0&1&0&0\\0&1&0&0&0&0&0&0&0\\0&0&0&0&1&0&0&0&0\\0&0&0&0&0&0&0&1&0\\0&0&1&0&0&0&0&0&0\\0&0&0&0&0&1&0&0&0\\0&0&0&0&0&0&0&0&1\end{bmatrix} & & \\ & \begin{bmatrix}1&0&0&0&0&0&0&0&0\\0&0&0&1&0&0&0&0&0\\0&0&0&0&0&0&1&0&0\\0&1&0&0&0&0&0&0&0\\0&0&0&0&1&0&0&0&0\\0&0&0&0&0&0&0&1&0\\0&0&1&0&0&0&0&0&0\\0&0&0&0&0&1&0&0&0\\0&0&0&0&0&0&0&0&1\end{bmatrix} & \\ & & \begin{bmatrix}1&0&0&0&0&0&0&0&0\\0&0&0&1&0&0&0&0&0\\0&0&0&0&0&0&1&0&0\\0&1&0&0&0&0&0&0&0\\0&0&0&0&1&0&0&0&0\\0&0&0&0&0&0&0&1&0\\0&0&1&0&0&0&0&0&0\\0&0&0&0&0&1&0&0&0\\0&0&0&0&0&0&0&0&1\end{bmatrix} \end{bmatrix}.$$

$\Rightarrow$ input = input$_1$ =

$$|1\rangle \otimes |2\rangle \otimes |2\rangle = [0\ 0\ 0\ 0\ 0\ 0\ 0\ 0\ 0\ 0\ 0\ 0\ 0\ 0\ 0\ 0\ 0\ 1\ 0\ 0\ 0\ 0\ 0\ 0\ 0\ 0\ 0]^T .$$

<u>Perspective #1</u>: $input_1 \Rightarrow (5) \Rightarrow output_1$, $output_1 = input_2 \Rightarrow (6) \Rightarrow output_2$.

Utilizing the same method in perspective #1 in Example 11.7a, the quantum circuit that is shown in Fig. 11.23 evolves input qubit |122> into the output qubit |120>.

<u>Perspective #2:</u> $input_1 \Rightarrow ((6)(5)) \Rightarrow output_2$.

Utilizing the same method in perspective #2 in Example 11.7a, the quantum circuit shown in Fig. 11.23 evolves the qubit |122> into the qubit |120> (which is the same result obtained in perspective #1).

The following are new multiple-valued quantum permuters that can be used in the future synthesis of multiple-valued quantum logic circuits, where ± 1 and $\pm i$ means that any combination of positive and negative 1 and any combination of positive and negative i can occur.

Theorem 11.9. The following are the ternary generalized inverters (permuters):

Class-A:

$$I_0:\begin{bmatrix}0&0&1\\0&1&0\\1&0&0\end{bmatrix}, I_1:\begin{bmatrix}0&1&0\\1&0&0\\0&0&1\end{bmatrix}, I_2:\begin{bmatrix}0&1&0\\0&0&1\\1&0&0\end{bmatrix}, I_3:\begin{bmatrix}0&0&1\\1&0&0\\0&1&0\end{bmatrix}, I_4:\begin{bmatrix}1&0&0\\0&0&1\\0&1&0\end{bmatrix}, I_5:\begin{bmatrix}1&0&0\\0&1&0\\0&0&1\end{bmatrix}.$$

Class-B:

$$I_6:\begin{bmatrix}0&0&\pm i\\0&\pm i&0\\\pm i&0&0\end{bmatrix}, I_7:\begin{bmatrix}0&\pm i&0\\\pm i&0&0\\0&0&\pm i\end{bmatrix}, I_8:\begin{bmatrix}0&\pm i&0\\0&0&\pm i\\\pm i&0&0\end{bmatrix}, I_9:\begin{bmatrix}0&0&\pm i\\\pm i&0&0\\0&\pm i&0\end{bmatrix}, I_{10}:\begin{bmatrix}\pm i&0&0\\0&0&\pm i\\0&\pm i&0\end{bmatrix}, I_{11}:\begin{bmatrix}\pm i&0&0\\0&\pm i&0\\0&0&\pm i\end{bmatrix}.$$

Class-C:

$$I_{12}:\begin{bmatrix}0&0&\pm 1\\0&\pm 1&0\\\pm 1&0&0\end{bmatrix}, I_{13}:\begin{bmatrix}0&\pm 1&0\\\pm 1&0&0\\0&0&\pm 1\end{bmatrix}, I_{14}:\begin{bmatrix}0&\pm 1&0\\0&0&\pm 1\\\pm 1&0&0\end{bmatrix}, I_{15}:\begin{bmatrix}0&0&\pm 1\\\pm 1&0&0\\0&\pm 1&0\end{bmatrix}, I_{16}:\begin{bmatrix}\pm 1&0&0\\0&0&\pm 1\\0&\pm 1&0\end{bmatrix}, I_{17}:\begin{bmatrix}\pm 1&0&0\\0&\pm 1&0\\0&0&\pm 1\end{bmatrix}.$$

Class-D:

$$I_{18}:\begin{bmatrix}0&0&\pm 1\\0&\pm i&0\\\pm i&0&0\end{bmatrix}, I_{19}:\begin{bmatrix}0&\pm 1&0\\\pm i&0&0\\0&0&\pm i\end{bmatrix}, I_{20}:\begin{bmatrix}0&\pm 1&0\\0&0&\pm i\\\pm i&0&0\end{bmatrix}, I_{21}:\begin{bmatrix}0&0&\pm 1\\\pm i&0&0\\0&\pm i&0\end{bmatrix}, I_{22}:\begin{bmatrix}\pm 1&0&0\\0&0&\pm i\\0&\pm i&0\end{bmatrix}, I_{23}:\begin{bmatrix}\pm 1&0&0\\0&\pm i&0\\0&0&\pm i\end{bmatrix},$$

$$I_{24}:\begin{bmatrix}0&0&\pm i\\0&\pm 1&0\\\pm i&0&0\end{bmatrix}, I_{25}:\begin{bmatrix}0&\pm i&0\\\pm 1&0&0\\0&0&\pm i\end{bmatrix}, I_{26}:\begin{bmatrix}0&\pm i&0\\0&0&\pm 1\\\pm i&0&0\end{bmatrix}, I_{27}:\begin{bmatrix}0&0&\pm i\\\pm 1&0&0\\0&\pm i&0\end{bmatrix}, I_{28}:\begin{bmatrix}\pm i&0&0\\0&0&\pm 1\\0&\pm i&0\end{bmatrix}, I_{29}:\begin{bmatrix}\pm i&0&0\\0&\pm 1&0\\0&0&\pm i\end{bmatrix},$$

$$I_{30}:\begin{bmatrix}0&0&\pm i\\0&\pm i&0\\\pm 1&0&0\end{bmatrix}, I_{31}:\begin{bmatrix}0&\pm i&0\\\pm i&0&0\\0&0&\pm 1\end{bmatrix}, I_{32}:\begin{bmatrix}0&\pm i&0\\0&0&\pm i\\\pm 1&0&0\end{bmatrix}, I_{33}:\begin{bmatrix}0&0&\pm i\\\pm i&0&0\\0&\pm 1&0\end{bmatrix}, I_{34}:\begin{bmatrix}\pm i&0&0\\0&0&\pm i\\0&\pm 1&0\end{bmatrix}, I_{35}:\begin{bmatrix}\pm i&0&0\\0&\pm i&0\\0&0&\pm 1\end{bmatrix},$$

Class-E:

$$I_{36}:\begin{bmatrix}0&0&\pm1\\0&\pm1&0\\\pm i&0&0\end{bmatrix}, I_{37}:\begin{bmatrix}0&\pm1&0\\\pm1&0&0\\0&0&\pm i\end{bmatrix}, I_{38}:\begin{bmatrix}0&\pm1&0\\0&0&\pm1\\\pm i&0&0\end{bmatrix}, I_{39}:\begin{bmatrix}0&0&\pm1\\\pm1&0&0\\0&\pm i&0\end{bmatrix}, I_{40}:\begin{bmatrix}\pm1&0&0\\0&0&\pm1\\0&\pm i&0\end{bmatrix}, I_{41}:\begin{bmatrix}\pm1&0&0\\0&\pm1&0\\0&0&\pm i\end{bmatrix},$$

$$I_{42}:\begin{bmatrix}0&0&\pm1\\0&\pm i&0\\\pm1&0&0\end{bmatrix}, I_{43}:\begin{bmatrix}0&\pm1&0\\\pm i&0&0\\0&0&\pm1\end{bmatrix}, I_{44}:\begin{bmatrix}0&\pm1&0\\0&0&\pm i\\\pm1&0&0\end{bmatrix}, I_{45}:\begin{bmatrix}0&0&\pm1\\\pm i&0&0\\0&\pm1&0\end{bmatrix}, I_{46}:\begin{bmatrix}\pm1&0&0\\0&0&\pm i\\0&\pm1&0\end{bmatrix}, I_{47}:\begin{bmatrix}\pm1&0&0\\0&\pm i&0\\0&0&\pm1\end{bmatrix},$$

$$I_{48}:\begin{bmatrix}0&0&\pm i\\0&\pm1&0\\\pm1&0&0\end{bmatrix}, I_{49}:\begin{bmatrix}0&\pm i&0\\\pm1&0&0\\0&0&\pm1\end{bmatrix}, I_{50}:\begin{bmatrix}0&\pm i&0\\0&0&\pm1\\\pm1&0&0\end{bmatrix}, I_{51}:\begin{bmatrix}0&0&\pm i\\\pm1&0&0\\0&\pm1&0\end{bmatrix}, I_{52}:\begin{bmatrix}\pm i&0&0\\0&0&\pm1\\0&\pm1&0\end{bmatrix}, I_{53}:\begin{bmatrix}\pm i&0&0\\0&\pm1&0\\0&0&\pm1\end{bmatrix}.$$

Proof. By performing all possible permutations of unitary 3-by-3 matrices that have each row and column composed of one element of value "1", one obtains the ternary generalized inverters (permuters) that are given in Theorem 11.9 **Q.E.D.**

While every quantum operator must be unitary, it is not claimed here that every unitary matrix is realizable as a quantum operator. Therefore, it might be possible that while some of the generalized unitary operators in Theorem 11.9 are realizable in quantum circuits, other unitary operators might be realizable only in other technologies.

Similar to the two-valued case, it is interesting to solve for the following multi-valued evolution process factorization (which is equivalent to multi-valued quantum circuit decomposition) problem: given the output (evolved) qubit, factorize the total composite multi-valued evolution process into a corresponding multi-valued evolution sub-processes. This type of quantum multi-valued decomposition can be very useful in the synthesis of multi-valued quantum logic circuits. Thus, similar to the binary case, decompositions that are commonly used in linear algebra can be utilized for the decomposition of multi-valued unitary evolution matrices, like: spectral theorem, Z-Y decomposition, LDU decomposition, fast Fourier-like decomposition, Jordan decomposition, Polar decomposition, Chinese Remainder Theorem (CRT) decomposition, and Singular Value Decomposition (SVD) [15].

Also, using the quantum Chrestenson operator [15,19,23], the theoretical development of the multiple-valued quantum Fourier transform (QFT) and the multiple-valued quantum circuit that generates the multiple-valued QFT have been shown [268].

The idea of performing super-fast multiple-valued operations (transformations) in quantum computing using the entanglement of several quantum Chrestenson operators, analogously to the role of the Walsh-Hadamard operator in super-fast two-valued quantum computing [167], has been also discussed [268].

11.7 Novel Representations for Multiple-Valued Quantum Logic: Multiple-Valued Quantum Decision Trees and Diagrams

Utilizing Theorems 11.4 and 11.5, the following is the GF(3) quantum Buffer (which is equivalent to two wires), Feynman (Theorem 11.4), and Swap (Theorem 11.5) evolution processes, for the ternary computational basis states $\{|00\rangle, |01\rangle, |02\rangle, |10\rangle, |11\rangle, |12\rangle, |20\rangle, |21\rangle, |22\rangle\}$, respectively:

$$|\Psi\rangle_{Buffer} = \begin{bmatrix} |00\rangle \\ |01\rangle \\ |02\rangle \\ |10\rangle \\ |11\rangle \\ |12\rangle \\ |20\rangle \\ |21\rangle \\ |22\rangle \end{bmatrix}^T \begin{bmatrix} 1&0&0&0&0&0&0&0&0 \\ 0&1&0&0&0&0&0&0&0 \\ 0&0&1&0&0&0&0&0&0 \\ 0&0&0&1&0&0&0&0&0 \\ 0&0&0&0&1&0&0&0&0 \\ 0&0&0&0&0&1&0&0&0 \\ 0&0&0&0&0&0&1&0&0 \\ 0&0&0&0&0&0&0&1&0 \\ 0&0&0&0&0&0&0&0&1 \end{bmatrix} \begin{bmatrix} \alpha 1\alpha 2 \\ \alpha 1\beta 2 \\ \alpha 1\gamma 2 \\ \beta 1\alpha 2 \\ \beta 1\beta 2 \\ \beta 1\gamma 2 \\ \gamma 1\alpha 2 \\ \gamma 1\beta 2 \\ \gamma 1\gamma 2 \end{bmatrix}, \tag{11.19}$$

$$|\Psi\rangle_{Final-Feynman} = \begin{bmatrix} |00\rangle \\ |01\rangle \\ |02\rangle \\ |10\rangle \\ |11\rangle \\ |12\rangle \\ |20\rangle \\ |21\rangle \\ |22\rangle \end{bmatrix}^T \begin{bmatrix} 1&0&0&0&0&0&0&0&0 \\ 0&1&0&0&0&0&0&0&0 \\ 0&0&1&0&0&0&0&0&0 \\ 0&0&0&0&0&1&0&0&0 \\ 0&0&0&1&0&0&0&0&0 \\ 0&0&0&0&1&0&0&0&0 \\ 0&0&0&0&0&0&0&1&0 \\ 0&0&0&0&0&0&0&0&1 \\ 0&0&0&0&0&0&1&0&0 \end{bmatrix} \begin{bmatrix} \alpha 1\alpha 2 \\ \alpha 1\beta 2 \\ \alpha 1\gamma 2 \\ \beta 1\alpha 2 \\ \beta 1\beta 2 \\ \beta 1\gamma 2 \\ \gamma 1\alpha 2 \\ \gamma 1\beta 2 \\ \gamma 1\gamma 2 \end{bmatrix}, \tag{11.20}$$

$$|\Psi\rangle_{Final-Swap} = \begin{bmatrix} |00\rangle \\ |01\rangle \\ |02\rangle \\ |10\rangle \\ |11\rangle \\ |12\rangle \\ |20\rangle \\ |21\rangle \\ |22\rangle \end{bmatrix}^T \begin{bmatrix} 1 & 0 & 0 & 0 & 0 & 0 & 0 & 0 & 0 \\ 0 & 0 & 0 & 1 & 0 & 0 & 0 & 0 & 0 \\ 0 & 0 & 0 & 0 & 0 & 0 & 1 & 0 & 0 \\ 0 & 1 & 0 & 0 & 0 & 0 & 0 & 0 & 0 \\ 0 & 0 & 0 & 0 & 1 & 0 & 0 & 0 & 0 \\ 0 & 0 & 0 & 0 & 0 & 0 & 0 & 1 & 0 \\ 0 & 0 & 1 & 0 & 0 & 0 & 0 & 0 & 0 \\ 0 & 0 & 0 & 0 & 0 & 1 & 0 & 0 & 0 \\ 0 & 0 & 0 & 0 & 0 & 0 & 0 & 0 & 1 \end{bmatrix} \begin{bmatrix} \alpha 1 \alpha 2 \\ \alpha 1 \beta 2 \\ \alpha 1 \gamma 2 \\ \beta 1 \alpha 2 \\ \beta 1 \beta 2 \\ \beta 1 \gamma 2 \\ \gamma 1 \alpha 2 \\ \gamma 1 \beta 2 \\ \gamma 1 \gamma 2 \end{bmatrix}. \quad (11.21)$$

Since various types of multiple-valued decision trees and diagrams are of fundamental importance in various algebraic systems [96,216], it is obvious that they will be also useful in multiple-valued quantum logic which is a generalization of multiple-valued logic, where the concepts of quantum decision trees and diagrams have not been introduced so far in the known literature.

Figures 11.24a and 11.24b represent the corresponding ternary Feynman gate (Eq. (11.20)) and ternary Swap gate (Eq. (11.21)) multiple-valued quantum decision trees (MvQDTs) for the ternary computational basis states, $\{|00\rangle,|01\rangle,|02\rangle,|10\rangle,|11\rangle,|12\rangle,|20\rangle,|21\rangle,|22\rangle\}$, respectively. One observes that the evolution matrices in Eqs. (11.19) through (11.21) force the quantum states $\{|00\rangle,|01\rangle,|02\rangle,|10\rangle,|11\rangle,|12\rangle,|20\rangle,|21\rangle,|22\rangle\}$ and the probability amplitudes $\{\alpha_i, \beta_j, \gamma_k\}$ to be in specific combinations (permutations) that are unique for the specific gates that are used.

The new quantum evolution decision tree representation can be useful in the future algorithms for the synthesis of quantum logic circuits, analogous to the already existing algorithms that depend on such multiple-valued representations for the optimized synthesis of classical multiple-valued (non-quantum) logic circuits.

Fig. 11.24c illustrates an example of the multiple-valued computational decision diagram (MvCQDD) [15,26]. When traversing the tree in Fig. 11.24c three paths: $|10\rangle$, $|11\rangle$, and $|12\rangle$ lead to three leaves with same values: $\alpha_1\beta_2$, $\beta_1\beta_2$, and $\gamma_1\beta_2$, respectively. Since the three paths lead to the same leaf value then the three nodes are combined as a single leaf and thus a more compact representation of MvCQDD is created.

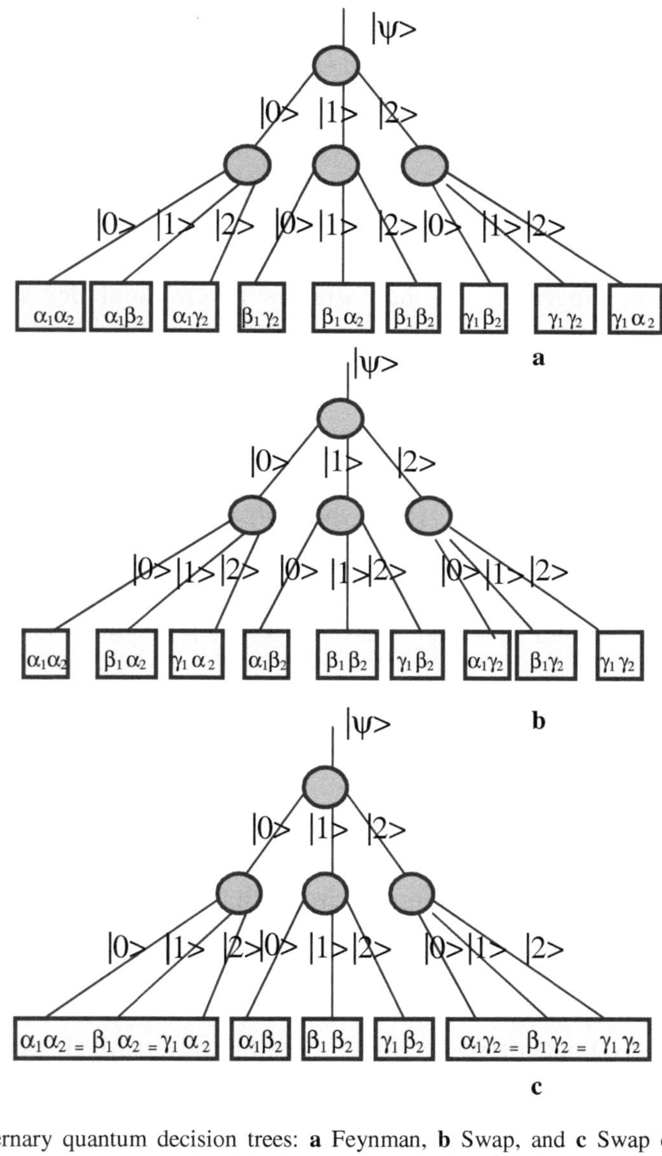

Fig. 11.24. Ternary quantum decision trees: **a** Feynman, **b** Swap, and **c** Swap decision diagram for the ternary computational basis states, where $\{\alpha_i, \beta_j, \gamma_k\}$ are the probability amplitudes.

The ternary quantum decision trees in Figs. 11.24a and 11.24b can be computed for the ternary composite basis states $\{ |\text{||}\rangle, |\text{|}+\rangle, |\text{|}-\rangle, |+\text{|}\rangle |++\rangle |+-\rangle |-\text{|}\rangle |-+\rangle |--\rangle \}$ (that were produced in Theorem 11.2), for which the states $|00\rangle, |01\rangle, |02\rangle, |10\rangle, |11\rangle, |12\rangle, |20\rangle, |21\rangle, |22\rangle$ are replaced by the states $\{ |\text{||}\rangle, |\text{|}+\rangle, |\text{|}-\rangle, |+\text{|}\rangle |++\rangle |+-\rangle |-\text{|}\rangle |-+\rangle |--\rangle \}$, respectively. Although the values in the leaves of the MvQDT in Figs. 11.24a and 11.24b are not equal to each other in general, multiple-valued quantum decision diagrams (MvQDDs) can be constructed for the corresponding multiple-valued quantum decision trees. The rules for such quantum decision diagrams are the same as in classical decision diagrams [45]: (1) join isomorphic nodes, and (2) remove redundant nodes. Figure 11.24c illustrates one case for the concept of ternary quantum evolution decision diagrams.

One notes that for specific orders of variables, the resulting MvQDTs (Figs. 11.24a and 11.24b) and MvQDDs (Fig. 11.24c) are canonical. Obviously, from the software implementation point of view, and similar to the tools for classical multiple-valued logic, quantum decision diagrams (Fig. 11.24c) can be realized on top of standard binary decision diagram (BDD) packages [231].

11.8 Automatic Synthesis of Two-Valued and Multiple-Valued Quantum Logic Circuits Using Evolutionary Algorithms

In order to design a quantum circuit that performs a desired quantum computation, it is necessary to find a decomposition of the unitary matrix that represents that computation in terms of a sequence of quantum gate operations. To date, such designs have either been found by hand or by exhaustive enumeration of all possible circuit topologies. It has been shown in [207,252] an automated approach to quantum circuit design using search heuristics based on principles abstracted from evolutionary genetics, which uses a genetic programming algorithm adapted specially for this problem. The method has been demonstrated on the task of discovering quantum circuit designs for quantum teleportation. It has been shown [207] that to find a given known circuit design (one which was hand-crafted by a human), the method considers roughly an order of magnitude fewer designs than naive enumeration. In addition, the method was shown to find novel circuit designs superior to those previously known.

The difficulties of applying GA for designing correct and minimal quantum circuits are the following: (1) a high time is needed for the of evaluation of the quantum circuit evolution matrix, especially when calculating the Kronecker product with matrices that possess sizes that grow exponentially for larger quantum circuits, (2) if the population is composed of high number of individuals then the synthesis result can be found in less number of generations but with longer times for the evaluation of fitness, and (3) using certain encodings for quantum gates leads to a big loss of time. To avoid extreme time consumption for calculations, one, for example, can limit the GA to a population that contains a relatively small number of individuals, and maintain the iterations of the algorithm to be limited to relatively small quantum circuits (i.e., with limited number of wires).

11.9 Quantum Computing for the New Two-Valued and Multiple-Valued Reversible Structures

As was demonstrated in previous Sects., the input qubits to any type of quantum circuit can be evolved from input to output by using the normal matrix product for serial interconnects and the Kronecker product for the parallel interconnects. These quantum evolutions are performed on the quantum matrix representations of the corresponding quantum primitives. Such matrix representations are a pure mathematical representation that can be realized physically using the corresponding quantum devices. The evolution operations can be implemented using the matrix representation or the corresponding QDTs or QDDs representations. This quantum evolution of the input qubits can be performed using the quantum counterparts of the reversible lattice structures from Chapt. 6, reversible Modified Reconstructability Analysis for Chapt. 7, and reversible Nets, Decision Diagrams, and Cascades from Chapt. 8, respectively. For this purpose, the quantum logic circuits from Chapt. 10 can be used.

The minimization of the quantum logic circuits from Chapt. 10 will lead to a minimum size of the quantum (scratchpad) register and thus to a minimum number of quantum computations. This reasoning is demonstrated in Fig. 11.25.

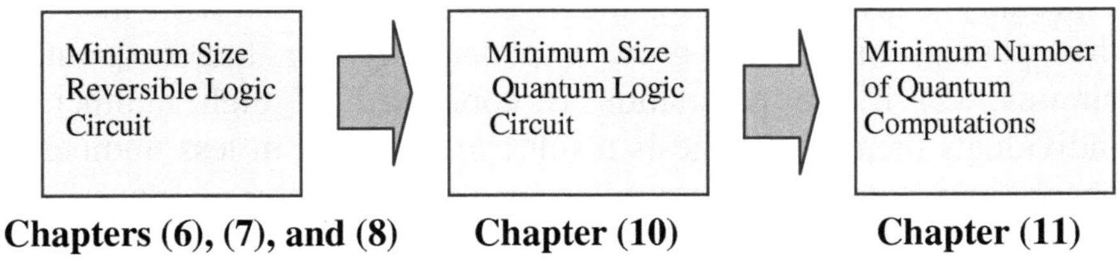

Fig. 11.25. The effect of minimization of the size of quantum logic circuits on the complexity of quantum computing.

The following example illustrates the point that was shown in Fig. 11.25.

Example 11.8. This example will demonstrate the effect of minimization of the size of two-valued quantum logic circuits on the total complexity of two-valued quantum computing (i.e., the total number of quantum arithmetic (addition and multiplication) operations that are needed). The method of functional minimization can be implemented using, for instance, the evolutionary algorithm for the minimization of general GFSOP forms using the IF polarity from the S/D trees that was introduced in Sect. 3.7. The effect of minimization is illustrated using the quantum logic circuits from Fig. 11.26. The quantum logic circuits in Fig. 11.26 are the same from Fig. 10.6, and thus one would observe the validity of the reasoning in Fig. 11.25.

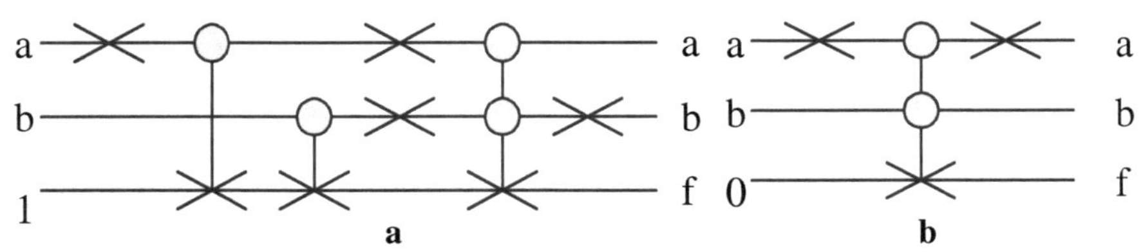

Fig. 11.26. Two equivalent two-valued quantum logic circuits.

The quantum evolution matrix for the circuit in Fig. 11.26a is obtained using 4 N (NOT) operations using the quantum primitive $\begin{bmatrix} 0 & 1 \\ 1 & 0 \end{bmatrix}$, 2 CN operations using the quantum primitive $\begin{bmatrix} 1 & 0 & 0 & 0 \\ 0 & 1 & 0 & 0 \\ 0 & 0 & 0 & 1 \\ 0 & 0 & 1 & 0 \end{bmatrix}$, and a single CCN operation using the quantum primitive $\begin{bmatrix} 1 & 0 & 0 & 0 & 0 & 0 & 0 & 0 \\ 0 & 1 & 0 & 0 & 0 & 0 & 0 & 0 \\ 0 & 0 & 1 & 0 & 0 & 0 & 0 & 0 \\ 0 & 0 & 0 & 1 & 0 & 0 & 0 & 0 \\ 0 & 0 & 0 & 0 & 1 & 0 & 0 & 0 \\ 0 & 0 & 0 & 0 & 0 & 1 & 0 & 0 \\ 0 & 0 & 0 & 0 & 0 & 0 & 0 & 1 \\ 0 & 0 & 0 & 0 & 0 & 0 & 1 & 0 \end{bmatrix}$. On the other hand, the quantum evolution matrix for the circuit in Fig. 11.26b is obtained using 2 N (NOT) operations using the quantum primitive $\begin{bmatrix} 0 & 1 \\ 1 & 0 \end{bmatrix}$, and a single CCN operation using the quantum primitive $\begin{bmatrix} 1 & 0 & 0 & 0 & 0 & 0 & 0 & 0 \\ 0 & 1 & 0 & 0 & 0 & 0 & 0 & 0 \\ 0 & 0 & 1 & 0 & 0 & 0 & 0 & 0 \\ 0 & 0 & 0 & 1 & 0 & 0 & 0 & 0 \\ 0 & 0 & 0 & 0 & 1 & 0 & 0 & 0 \\ 0 & 0 & 0 & 0 & 0 & 1 & 0 & 0 \\ 0 & 0 & 0 & 0 & 0 & 0 & 0 & 1 \\ 0 & 0 & 0 & 0 & 0 & 0 & 1 & 0 \end{bmatrix}$. Consequently, it is obvious that the total number of operations needed for evolving qubits using Fig. 11.26b is much less than the total number of operations that are needed for evolving qubits using Fig. 11.26a. This conclusion is also valid for the case of multiple-valued quantum computing as can be seen in Fig. 11.27 for example, where the notation follows from Chapt. 10.

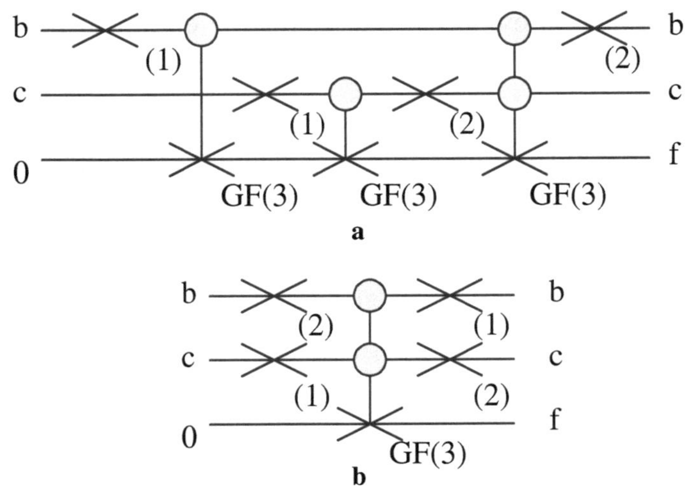

Fig. 11.27. Two equivalent ternary quantum logic circuits.

Analogous to the classical domain where (computational) complexity theory studies whether or not the total number of operations that are needed to execute certain computational task can be done in a polynomial time with respect to the increasing size of the problem (e.g., function), the domain of quantum complexity theory investigates whether or not the total number of quantum operations that are needed to execute certain quantum computational task can be done in a polynomial time with respect to the increasing size of the quantum problem (e.g., quantum function) [167].

Some of the quantum computing structures that have been introduced in Chapts. 10 and 11 possess certain advantages over the other quantum computing structures. Table 11.1 shows an initial evaluation that have been observed so far when performing quantum computing using such structures.

One observes that while in reversible circuits garbage do appear in the outputs, and thus one counts (1) number of garbage outputs, and (2) number of internal gates as the cost of the reversible circuit (Chapt. 9) and consequently as an efficiency measure of the method used for reversible logic synthesis, in quantum circuits garbage do not appear in the outputs, and thus one counts only number of internal gates (i.e., size of the quantum register) as the cost of the quantum circuit (i.e., as an efficiency measure of the method used for quantum logic synthesis).

Table 11.1. Initial evaluation for the use of the various two-valued and multiple-valued quantum structures to perform the corresponding two-valued and multiple-valued quantum computations.

Type	Advantages	Disadvantages
Quantum Shannon Lattices	(1) Utilizes exclusively the NOT gate and Controlled-Swap gate which are basic primitives for quantum circuits.	(1) Since quantum circuits do not allow for garbage, it needs the mirror image to cancel the garbage and thus requires more of the total number of basic quantum operations using the garbageless circuit, (2) requires big width of the scratchpad register.
Quantum Nets	(1) Good for quantum symmetric functions, and thus for highly symmetric functions can need less of total number of basic quantum operations to perform quantum computations.	(1) For highly non-symmetric quantum functions, it requires large number of the total basic quantum operations to perform quantum computations on the garbageless circuit, since the variable repetition is required and mirror image circuit is needed to eliminate the garbage, (2) requires big width of the scratchpad register.
Quantum Cascades	(1) Structurally fits quantum circuits since on avg. no garbage is created and the inputs propagate to the outputs thus requires less number of total number of basic quantum ops. to perform quantum computations, (2) the use of EXOR in cascades has strong relation to quantum structures, (3) requires relatively small width of the scratchpad register, (4) can be used for efficient realization of MIMO type of quantum functions (circuits).	(1) The quantum cascade circuit can be very long to realize quantum functions with many inputs, and thus need large number of basic quantum ops. to perform quantum computations for such functions.
Quantum DDs	(1) Very useful as good quantum data structure to perform fast simulations and ops. for quantum computing, (2) big variety of quantum DDs for the two-valued and multiple-valued computational basis states, composite basis states, and EPR basis states, besides the big types of DDs, to achieve both minimal size circuits and faster ops. on quantum data.	(1) create garbage, and thus it needs mirror image to cancel garbage and thus requires more of the total number of basic quantum ops. to perform quantum comps. using the garbageless circuit.

Table 11.1. (cont.)

Quantum MRA	(1) Due to its four circuit topologies, it can possess advantages to design quantum circuits by utilizing quantum processes that are low in cost and thus need less number of quantum ops., (2) in some cases it produces minimum size quantum circuits.	(1) Does not realize yet ESOP of quantum functions, (2) in general needs mirror image to cancel garbage and thus needs more number of basic quantum ops. using the garbageless circuit.

11.10 Summary

This Chapt. has introduced the following new results: (1) New two-valued and multiple-valued quantum primitives and evolution processes, (2) new multiple-valued composite basis states and Einstein-Podolsky-Rosen (EPR) basis states, (3) generalized multiple-valued quantum permuters, (4) various types of two-valued and multiple-valued canonical quantum decision trees (QDTs) and quantum decision diagrams (QDDs), and (5) the introduction of the mathematical operations for the analysis and synthesis of serial, parallel, and mixture of serial and parallel multiple-valued quantum circuits.

Results (1) and (3) are necessary for the automated analysis of netlists of quantum primitives. They are also necessary for automated synthesis of a netlist described as an evolution matrix from quantum gates, especially for try-and-check methods such as evolutionary algorithms [207,252]. Since decision diagrams allow for efficient representation of large sparse matrices, they found applications in many CAD algorithms, and we believe that their quantum counterparts in item (4) will be useful for quantum logic synthesis and analysis. Finally, result (2) is important because new forms of quantum decision trees and diagrams can be produced for the new multiple-valued EPR basis states, and thus allowing for further possible optimizations in the design of quantum circuits, analogous to the classical (non-quantum) case where different forms of decision trees and diagrams lead to different scales of optimizations in the design of logic circuits.

The new multiple-valued quantum EPR basis states have been achieved by utilizing the new quantum Chrestenson operator introduced in this Chapt. The new Galois-based quantum gates, evolution processes, and the corresponding canonical quantum decision trees and decision diagrams were introduced as a first attempt of developing a comprehensive set of (1) complete system of quantum logic elements (gates; primitives), (2) quantum representations, and (3) quantum synthesis methods. It has been also demonstrated that by minimizing the size of the two-valued and multiple-valued quantum circuits one would need minimum number of the corresponding arithmetic operations needed to perform the corresponding two-valued and multiple-valued quantum computing, respectively.

12 Conclusions

The biggest problems in system design today, and in the future, are the high rate of power consumption and the emergence of quantum effects for highly dense ICs. The real challenge is to design reliable systems that consume as little power as possible and in which the signals are processed and transmitted at very high speeds with very high signal integrity. The tools that are used to design ICs using the conventional design methodologies apply the area, delay, and power constraints. The synthesis approaches that are implemented using such computer-aided design tools are only for the classical type of synthesis.

This Book has proposed new methodologies for the synthesis of reversible binary and multiple-valued logic circuits and their implementations using quantum computing. The methodology presented for the proposed quantum computations evolved from the creation of necessary reversible primitives, then the structures to synthesize logic functions, binary and multiple-valued, using such primitives, next quantum logic circuits were constructed, and finally the quantum computations using such circuits were conducted.

First, Chapt. 4 provided the regular lattice structures for two-dimensional and three-dimensional synthesis to logic functions. Such structures exhibited high regularity that makes them fit for many applications, especially as such structures are synthesized on the nano-scale, which was shown in Chapt. 11. When the functions are deeply non-symmetric, one has to repeat variables so many times in order to realize the non-symmetric function using the lattice structure, and consequently lattices will grow larger very fast such that they do not fit the specific conventional layout boundaries any more. To solve this problem, a new algorithm called Iterative Symmetry Indices Decomposition (ISID) for two-valued and multiple-valued lattice structures was developed.

The development of new reversible primitives was accomplished in Chapt. 5. This development was necessary as such primitives were used later to synthesize more complex structures. The new reversible gates implement the Latin Square Property in their basis functions and consequently the permutation of cofactors.

Chapters 6 through 8 provided the foundation for synthesizing binary and multiple-valued logic functions using the reversible primitives that were introduced and developed in Chapt. 5. It was shown that among the proposed reversible structures, the reversible Cascade stands as one good method for synthesizing reversible logic functions without producing on average garbage in the outputs that are needed only for the purpose of reversibility. The disadvantage of such structures was shown to be the fact that they produce single-outputs where other reversible structures as the Nets can produce multiple-output functions, but this production of multiple-output functions is at the expense of having garbage in the outputs, and thus the mirror image reversible circuit has to be cascaded with the forward reversible circuit for the elimination of such garbage. This is important because in the quantum computations, that were provided in Chapt. 11, the garbage in outputs is not allowed.

The synthesis of logic functions using the reversible structures such as reversible Cascades requires the minimization of logic functions for optimal realization in such circuits, which is obtained for example in the case of reversible Cascades through the use of a minimal number of stages for the synthesis of binary and multiple-valued logic functions. The GFSOP minimizer that was proposed in Chapt. 3 uses the most general polarity of Inclusive Forms (IFs), where an evolutionary algorithm that implements the IF polarity chromosome was also introduced.

This Book started with a motivating research guidline that regularity in two-valued and multiple-valued reversible logic structures have an effect on the final complexity of the corresponding two-valued and multiple-valued quantum computing, respectively. Consequently, and to achieve this goal, the Book started with the invention of new methodologies for regular and semi-regular reversible structures and reversible logic synthesis methodologies, which was missing largely from the known previous literature, and then the exploration of the effect of such new methods on the total size of quantum logic circuits and consequently on the complexity of two-valued and multiple-valued quantum computing in terms of the count of the total number of operations required.

From the body of this Book, the conclusion was that, for small functions, highly regular reversible structures, such as the reversible lattice structures and reversible Nets, will require larger size of quantum logic circuits and thus more operations for the corresponding two-valued and multiple-valued quantum computing, as if compared to the size of the circuits in semi-regular reversible structures, such as reversible Cascades, and their corresponding quantum computations. Therefore, as the process of producing reversible structures will lead to specific internal symmetries in terms of certain levels of regularities, due to the implementation of the reverse as well as the forward lossless information retrieval process, regularity itself will lead by necessity in most cases to larger size structures that possess high internal symmetries, which can be good for many applications, such as testing, similar to the results in [221].

The main contributions of this Book can be summarized as follows:

- A generic methodology of generating new types of multi-valued Shannon and Davio expansions called the invariant Shannon and Davio spectral transforms, and the classification of the new types of multi-valued invariant Shannon and Davio spectral transforms into their corresponding families.
- The application of the new expansions into regular three-dimensional lattice structures, and the process of realizing non-symmetric ternary functions in three-dimensional lattice structures using a new 3-D joining operator. This was implemented for the regular and the new invariant Shannon and Davio 3-D lattice structures.
- The invention of a new methodology to generate and classify reversible multiple-valued Shannon decompositions that includes the binary case as a special case. This methodology implements the idea of the Latin Square Property of the Generalized Basis Function Matrix, which leads to the process of permutation of cofactors. The exhaustive classification of all possible reversible multi-valued Shannon gates into the corresponding classes was also provided.

- The generation of binary and multiple-valued reversible Davio decompositions, and the exhaustive classification of all possible reversible multi-valued Davio gates into the corresponding classes.
- The creation of new reversible structures including: (1) binary and multiple-valued reversible Lattice Structures, (2) reversible Modified Reconstructability Analysis (RMRA), (3) reversible Nets, (4) reversible Decision Diagrams (RDDs), and (5) multiple-Valued reversible Cascades.
- The invention of a new 2-valued decomposition: the Modified Reconstructability Analysis (MRA) decomposition. Two variants of binary MRA: 1-MRA and 0-MRA were also introduced. The extension to the multi-valued case is also achieved. The demonstration of the superiority of the Modified Reconstructability Analysis (MRA) decomposition to the Conventional Reconstructability Analysis (CRA) decomposition, in terms of complexity reduction for all 256 NPN-classified Boolean functions of three input variables has also been demonstrated.
- New multiple-valued 1-qubit and 2-qubit orthonormal quantum basis states: multi-valued composite basis states, and multi-valued Einstein-Podolsky-Rosen (EPR) basis states, respectively. The new quantum Chrestenson gate that produces such new multiple-valued basis states is also introduced.
- Two-valued and multiple-valued canonical quantum decision trees (QDTs) and quantum decision diagrams (QDDs).
- The serial interconnect and parallel interconnect operations for multiple-valued quantum computing, and the demonstration of these operations for the analysis of multiple-valued quantum logic circuits.
- New multiple-valued quantum gates and evolution processes including Feynman, Swap, Shannon (Fredkin), and Davio evolution processes, and generalized multi-valued inverters (permuters) were created. Synthesis of the new multi-valued quantum primitives in serial and parallel interconnects' topologies was also shown.
- The invention of several reversible combinational circuits like the reversible concurrent shift-left and shift-right Barrell shifter, reversible Sorter, reversible MIN/MAX tree, reversible pipelined

circuits, reversible systolic circuits, and reversible code converters.

- A novel Iterative Symmetry Indices Decomposition (ISID) needed for layout optimization.
- The creation of multiple-valued Shannon/Davio (S/D) trees, and their corresponding multiple-valued Inclusive Forms (IFs) and Generalized Inclusive Forms (GIFs). The very general count formula and the $IF_{n,2}$ triangles for the count of the total number of all binary and multiple-valued S/D forms that are generated have been also introduced.
- The synthesis of regular Boolean and multiple-valued optical classical and reversible circuits using: (1) total internal reflection, (2) optical polarizers, and (3) optical frequency shifters (See Appendix J).
- The implementation of Artificial Neural Networks using multiple-valued quantum computing (See Appendix K).

During the course of investigating new types of reversible logic methods and their corresponding two-valued and multiple-valued quantum computations, some secondary results were obtained (Appendices A through I provide such results). Although these contributions are not directly related to the theme of this Book, we provide their listing as follows:

- New theorems to count all possible families of the invariant multiple-valued Shannon and Davio expansions.
- Very general count formula that count the total number of multiple-valued Inclusive Forms that result from the corresponding S/D trees for any radix and an arbitrary number of variables.
- The invention of a new pattern for the count of IF forms for an arbitrary radix and two variables. We call it the $IF_{n,2}$ triangles. The relation of these triangles to the important Pascal triangle was also demonstrated.
- The creation of new types of Galois circuits.
- The creation of the multiple-valued Universal Logic Modules (ULMs) for ternary and quaternary S/D trees.
- An evaluation of the complexities of the new two-valued MRA decomposition versus the complexities obtained using Ashenhurst-Curtis (AC) like decompositions for all NPN classified Boolean functions of three vaiables. The evaluations are performed using

the log-functionality complexity measure which is suitable for machine learning, and using the complexity measure which is defined as the count of the total number of two-input primitives which is suitable for circuit design.

- The count of reversible Net structures.

In this Book, the goal was to explore the relationship between regularity and reversibility, which resulted in the development of new reversible logic synthesis techniques in order to synthesize functions using minimum size quantum logic circuits that will require minimum number of operations for quantum computing. However, one important factor for the evaluation of such reversible methods is the final total cost of the physical quantum circuits that will implement such reversible structures. While in conventional circuit design the cost of the design is measured by the total number of two-input gates that are used, in quantum circuits this is not the case, since in quantum circuits physical processes implement the quantum operations rather than simple hardware gates (e.g., CMOS) as in the case of the classical logic design. Quantum cost characterizes the physical process complexity that is needed to realize physically the corresponding reversible structures. Since little, if none, has been published on this quantum cost for the realization of the reversible structures, one very important question is still open on how much complex the quantum realization of the structures will be, and the answer to this question may very well lead to new cost-benefit conclusions.

This Book showed that the level of regularity in reversible logic structures has a direct effect on the size of the corresponding quantum circuits and their corresponding quantum computations. Thus this conclusion should be further explored with more quantitative analysis (i.e., numerical results) and qualitative analysis based on well-known two-valued and multiple-valued benchmarks. Since symmetries exist in reversible structures, due to the reversibility of lossless information retrieval, group-theoretic formulations of reversible primitives [243] from Chapt. 5, and reversible structures from Chapts. 6, 7, and 8 have to be further explored. Future work will include the construction of a comprehensive Computer-Aided Design (CAD) system [269] for reversible logic synthesis using various nano-based technologies as shown in Fig. 12.1.

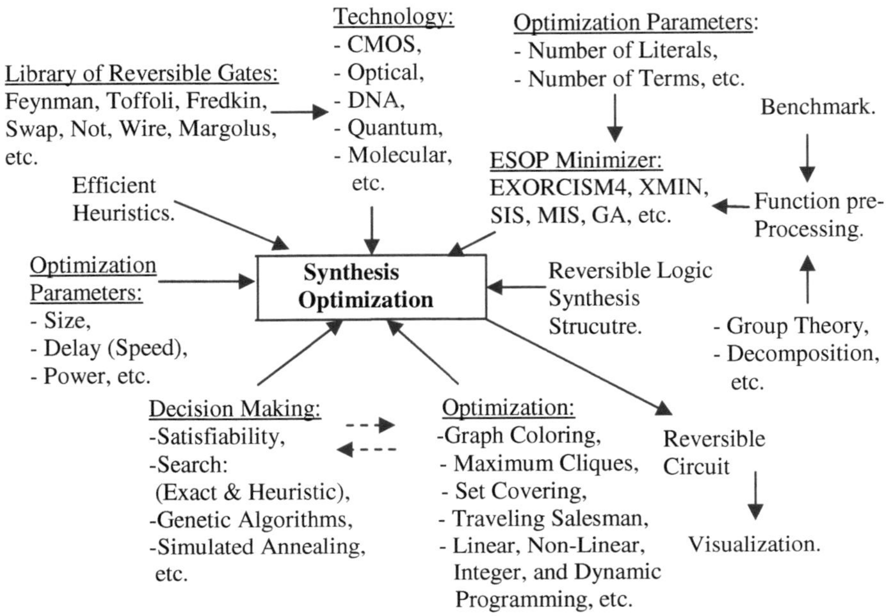

Fig. 12.1. Comprehensive CAD system for nano computing.

The design and fabrication of adiabatic (low-power) CMOS VLSI ICs using the new reversible logic synthesis methods will be conducted. Also future work will include the systematic methodology for the extension of current reversible logic synthesis methodologies to different types of quantum systems such as: multiple-input multiple-output (MIMO) quantum circuits, multiple-input single-output (MISO) quantum circuits, single-input multiple-output (SIMO) quantum circuits, and single-input single-output (SISO) quantum gates like Walsh, Phase, Pauli, etc, as shown in Fig. 12.2. The synthesis of a cascade of quantum gates to realize a quantum circuit for a specific length of a total quantum scratchpad register, as shown in Fig. 12.2, will be also investigated.

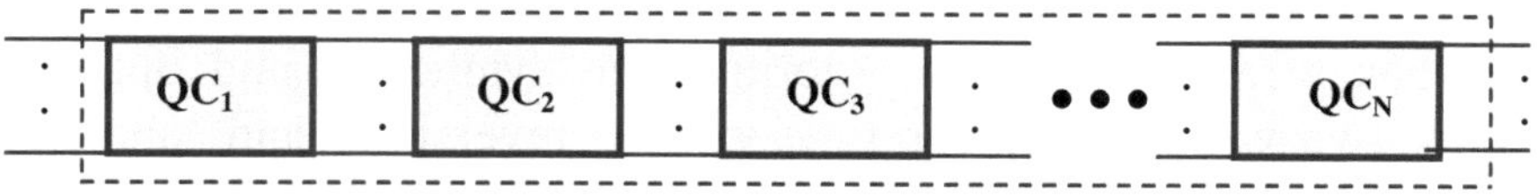

Fig. 12.2. Cascade of quantum circuits (QCts) for a specific length of a total scratchpad register.

Future work will include the implementation of the representations of binary and multiple-valued computational quantum decision trees (CQDTs) and diagrams (CQDDs), and binary and multiple-valued composite quantum decision trees (CoQDTs) and diagrams (CoQDDs), that are introduced in this Book, using the standard BDD packages [231]. Further research to find minimum and complete set of two-valued and multiple-valued quantum operators, from which complex two-valued and multiple-valued quantum circuits can be synthesized, will be conducted.

Making improvements of tasks to an existing system or synthesizing a new system to perform specific functions or tasks (e.g., recognition, prediction, diagnosis, robot control, planning, etc) is the main goal of Machine Learning (ML). Information retrieval from large data bases (sets) and finding data patterns in a data environment (i.e., extracting useful information from data, and fitting theories to data or enumerating patterns from data) is studied within specific ML area called Data Mining (DM). Knowledge Discovery in Database (KDD) is the field that focuses on the total process of information retrieval and data analysis (e.g., selection, preprocessing, and data transformation) [108]. Since the new two-valued and multiple-valued reversible and quantum computing methods, that are developed in this Book, were used for the purpose of circuit design, the use of these new methods for ML, DM, and KDD will be also investigated.

While mathematical formalisms can solve wide variety of synthesis problems (e.g., two-valued and multiple-valued reversible Shannon and Davio expansions from Chapt. 5), the solution and optimization of many two-valued and multiple-valued reversible and quantum synthesis problems (e.g., optimal variable ordering for the minimization of the total size of reversible lattice structure that reversibly realize large logic functions, in terms of minimization of the number of levels and minimization of the number of nodes in each level) are still very difficult to solve mathematically, and thus one has to rely on search algorithms to synthesize and optimize these two-valued and multiple-valued reversible and quantum synthesis problems. Since exhaustive search methods are suitable to solve algorithmically for a small dimension problem (that has none, little, or impractical mathematical formalisms), these exhaustive methods are not practical to synthesize and optimize large multi-input multi-output dimension problems. Consequently, future work will investigate the creation of suitable heuristics to perform heuristic search for the automatic synthesis and optimization of large multi-input multi-output dimension problems [269].

Future investigation should also involve the creation of systematic theories and formalisms for sequesntial reversible circuit synthesis; a problem that has not been yet totally resolved.

Further research on the realization of various abstract quantum primitives using other physically-realizable quantum primitives (such as the case in [230]) should also be conducted.

Appendix A

Count of the New Invariant Shannon and Davio Expansions

This Appendix provides the counts for the invariant Shannon and Davio Families of spectral transforms, that were presented in Chapt. 2, over an arbitrary Galois field. The following theorems provide counts for the new transform families, members per family, and total number of transforms, that belong to the sets of $\alpha\beta\ldots\gamma$ IS/D, $\alpha\beta\ldots\gamma$ ID/S, $\alpha_1\beta_1\ldots\gamma_1$IS/$\alpha_2\beta_2\ldots\gamma_2$ID families of spectral transforms, and $\alpha\beta\ldots\gamma$ IfS spectral transforms that are generated by the application of Theorems 2.2, 2.3, and 2.4, respectively. These counts can be used as a heuristic parameter and thus can play an important role in the search for optimal invariant expansions to create the corresponding minimal size three-dimensional lattice structures from Chapt. 4 [5,13] (For more clarifications, please refer to Definitions 2.1 through 2.7 in Chapt. 2, respectively).

Theorem A.1. For $\{\alpha, \beta, \ldots, \gamma\} \in GF(n)$, and for:
ϕ is the number of $\alpha\beta\ldots\gamma$ IS/D spectral transform families.
Θ is the number of members of Shannon transforms per family.
Γ is the number of members of Davio transforms per family.
Ω is the number of all members per $\alpha\beta\ldots\gamma$ IS/D transform family.
φ is the total number of spectral transforms for $\forall\alpha\beta\ldots\gamma$ IS/D families. Then we obtain:

$$\phi = (n - 1)^n, \quad (A.1)$$
$$\Theta = 1, \quad (A.2)$$
$$\Gamma = n, \quad (A.3)$$
$$\Omega = n + 1, \quad (A.4)$$
$$\varphi = \phi + \Omega - 1 = (n - 1)^n + n. \quad (A.5)$$

Proof. Utilizing the transform matrix for Generalized Shannon expansion (i.e. the identity matrix) over GF(n), the size of such a matrix is (n·n). In such matrix, there are n number of positions for nonzero-element and each position can take in general a number from (n-1) numbers (excluding the zero that leads to the trivial singular matrix). Consequently the total number of the resulting invariant Shannon transforms would be: $\phi = (n-1)\,(n-1)\,(n-1)\ldots(n-1) = (n-1)^n$. For GF(n), for each multi-valued fundamental Shannon expansion there are n corresponding multi-valued fundamental Davio expansions. By defining the invariant multi-valued Shannon transform and the corresponding multi-valued Davio transforms to be belonging to one Family, the number of transform members per family will be $\Omega = (n + 1)$. Consequently, the total number of transforms per Galois radix is $\varphi = \phi + \Omega - 1 = (n - 1)^n + n$. **Q.E.D.**

Theorem A.2. For $\{\alpha, \beta, \ldots, \gamma\} \in GF(n)$, and for:
X is the number of $\alpha\beta\ldots\gamma$ ID/S spectral transform families.
Δ is the number of members of Shannon transforms per family.
ϑ is the number of members of Davio transforms per family.
Ξ is the number of all members per $\alpha\beta\ldots\gamma$ ID/S transform family.
Ψ is the total number of transforms for $\forall\alpha\beta\ldots\gamma$ ID/S families.
Then we obtain:

$$X = (n-1)^{n^2}, \tag{A.6}$$
$$\Delta = 1, \tag{A.7}$$
$$\vartheta = n, \tag{A.8}$$
$$\Xi = n + 1, \tag{A.9}$$
$$\Psi = n(n - 1)^n + 1. \tag{A.10}$$

Proof. Using the transform matrix for any type of Generalized Davio expansion over GF(n), the size of such a matrix is n·n. In such matrix, there are (n^2-2n+3) number of positions for nonzero-element and (2n-3) of zero positions. Each position can take a number from (n-1) numbers (excluding the zero that leads to the trivial singular matrix). There are n basic Davio for GF(n). By multiplying the rows of each Davio type by $\{\alpha, \beta, \ldots, \gamma\} \in GF(n)$, we get consequently the total number of the resulting invariant Davio transforms per Davio type as: $\phi = (n-1)\ldots(n-1) = (n-1)^n$. So the total number of transforms for all Davio is $n(n - 1)^n$. For GF(n), for each MV Davio family there is one corresponding MV Shannon family. By defining the invariant Davio transforms and the corresponding Shannon transform to be belonging to one family, the number of transform members per family is $\Omega = (n + 1)$. Consequently, the total number of transforms is $\Psi = n(n - 1)^n + 1$. **Q.E.D.**

Theorem A.3. For $\{\alpha, \beta, \ldots, \gamma\} \in GF(n)$, and for:
χ is the number of $\alpha\beta\ldots\gamma$ ID/$\alpha\beta\ldots\gamma$IS spectral transform families.
δ is the number of members of Shannon transforms per family.
ε is the number of members of Davio transforms per family.
η is the number of all members per $\alpha\beta\ldots\gamma$ ID/$\alpha\beta\ldots\gamma$IS family.
ψ is the total number of transforms for $\forall$ $\alpha\beta\ldots\gamma$ ID/$\alpha\beta\ldots\gamma$IS families.
Then we obtain:

$$\chi = (n-1)^{n(n+1)}, \tag{A.11}$$
$$\delta = 1, \tag{A.12}$$
$$\varepsilon = n, \tag{A.13}$$
$$\eta = n + 1, \tag{A.14}$$
$$\psi = (n+1)(n-1)^n. \tag{A.15}$$

Proof. Since $\phi = (n - 1)^n$ and $X = (n-1)^{n^2}$, then the total number of $\alpha\beta\ldots\gamma$ ID/$\alpha\beta\ldots\gamma$IS spectral transform families is equal to $\chi = \phi \cdot X = (n-1)^{n(n+1)}$. Since there exists over GF(n), by definition, one Shannon transform and n Davio transforms per family, then the total number of transforms for all $\alpha\beta\ldots\gamma$ ID/$\alpha\beta\ldots\gamma$IS families is the union of all transforms. This will lead to the non-repetition of the existing transforms within the same set of total spectral transforms, and thus to the total number of $\psi = \varphi + \Psi - n - 1 = (n+1)(n-1)^n$. **Q.E.D.**

Theorem A.4. For $\{\alpha, \beta, \ldots, \gamma\} \in GF(n)$, and for ∂ is the number of $\alpha\beta\ldots\gamma$ IfS spectral transform families then we obtain:

$$\partial = (n - 1)^n. \tag{A.16}$$

Proof. Utilizing the transform matrix for Generalized flipped Shannon expansion over GF(n), the size of such a matrix is (n)(n). In such matrix, there are n number of positions for nonzero-element and each position can take in general a number from (n-1) numbers (excluding the zero that leads to the trivial singular matrix). Thus, the total number of the resulting invariant flipped Shannon transforms would be: $\partial = (n-1)(n-1)\ldots(n-1) = (n-1)^n$. **Q.E.D.**

The following example illustrates the various counts of the new set of spectral transforms.

Example A.1. Let us produce the counts of the new set of spectral transforms $\{\phi, \partial, \varphi, X, \Psi, \chi, \psi\}$ for Galois field of radices equal to 2, 3, 4, and 5, respectively.

For GF(2), one obtains:

$\phi = \partial = 1$, $\varphi = 3$, $X = 1$, $\Psi = 3$, $\chi = 1$, $\psi = 3$ (i.e., this is the familiar family of binary Shannon expansion, flipped Shannon expansion, and Davio expansions).

For GF(3), one obtains:

$\phi = \partial = 8$, $\varphi = 11$, $X = 512$, $\Psi = 25$, $\chi = 4{,}096$, $\psi = 32$.

For GF(4), one obtains:

$\phi = \partial = 81$, $\varphi = 85$, $X = 43{,}046{,}421$, $\Psi = 325$, $\chi = 3.4868 \cdot 10^{9}$, $\psi = 405$.

For GF(5), one obtains:

$\phi = \partial = 1{,}024$, $\varphi = 1{,}029$, $X = 1.1259 \cdot 10^{15}$, $\Psi = 5{,}121$, $\chi = 1.1529 \cdot 10^{18}$, $\psi = 6{,}144$.

Table A.1 provides a comparison of the spectral transforms: ϕ, ∂, φ, X, Ψ, χ, ψ and the total number of singular transforms ξ for various radices of Galois fields GF(p^k) (where p is a prime number and k is a natural number of value $k \geq 1$).

In general, the total number of the nonsingular spectral transforms will be much less than ξ. ξ is used in Table A.1 as an upper-extreme (Bound) reference to obtain an idea of how large the counts of the new families and their transforms are if compared to the whole space of spectral transforms.

Table A.1. Counts of the new invariant Shannon and Davio families of transformations for $GF(p^k)$.

GF	ϕ, ∂	φ	X	Ψ	χ	$\psi+\partial$	ξ
2	1	3	1	3	1	4	16
3	8	11	512	25	4,096	40	19,683
4	81	85	43,046,421	325	3.487×10^9	486	$\cong 4295 \times 10^6$
5	1024	1029	1.126×10^{15}	5,121	1.153×10^{18}	7,168	$\cong 2.98 \times 10^{17}$
7	279,936	279,943	1.347×10^{38}	1,959,553	3.77×10^{43}	2,519,424	$\cong 2.57 \times 10^{41}$
8	5,764,801	5,764,809	1.22×10^{54}	46,118,409	7.032×10^{60}	57,648,010	$\cong 6.28 \times 10^{57}$
9	134,217,728	134,217,737	1.4135×10^{73}	1.2080×10^9	1.897×10^{81}	$\cong 1.34 \times 10^9$	$\cong 1.97 \times 10^{77}$

Appendix B

Circuits for Quaternary Galois Field Sum-Of-Product (GFSOP) Canonical Forms

One interesting logic synthesis of quaternary GFSOPs are various circuit and filter implementations, similar to the work in [125], of which FIR filter is the simplest case. Higher radix extensions of this kind of filters can be useful for digital signal processing of 1-D multi-valued I/O signals or 2-D multi-valued I/O blocks of images (e.g., 4x4, and 8x8 blocks of a still image).

A general FIR filter for single variable GF(4) GFSOP-based expansion can be produced as in Fig. B.1 [125]. This realization is a direct implementation of the Shannon and Davio expansions that were obtained in Chapt. 2. Such realization can be useful in various applications such as applications that involve multi-valued I/O bio-medical signals. The addition and multiplication performed in Fig. B.1 are the Galois addition and multiplication operations defined in Chapt. 2. These additions and multiplications can be implemented using quaternary circuits, or using binary circuits. Yet, another implementation of GF(4) logical multiplication gate can be achieved by utilizing a third radix Galois addition operation. This can be achieved by utilizing the GF(3) addition which is presented in Chapt. 2, and noticing the relationship between the multiplication operation over GF(4) and the addition operation over GF(3). This relationship can be stated as follows: Excluding the first row of zeros and the first column of zeros in the GF(4) multiplication, then by subtracting one from the remaining rows and columns, the GF(3) addition operation is obtained, but next a value of "1" must be added to every entry of the GF(4) multiplication in order to obtain GF(3) addition.

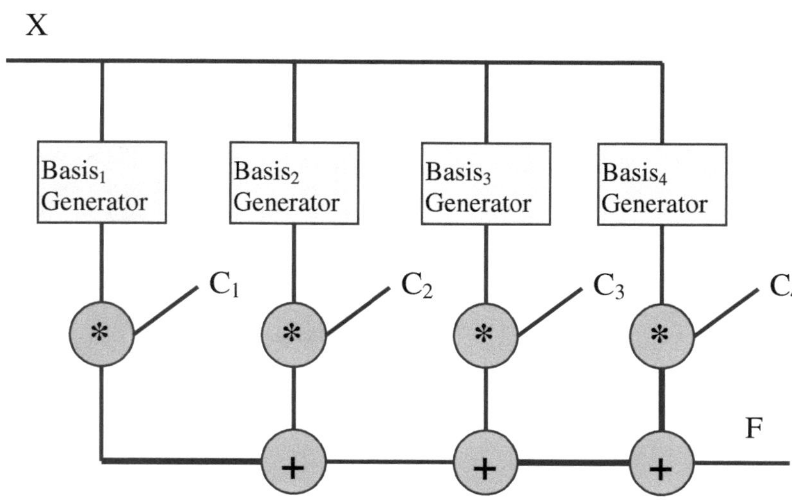

Fig. B.1. FIR Filter realization of four-valued GF(4) GFSOP canonical forms.

This can be formulated formally as follows:

$$a \cdot_{GF(4)} b = \begin{cases} 0, & \text{if } a=0 \text{ or } b=0, \\ [(a-1)\oplus(b-1)]+1 & ,\text{Otherwise.} \end{cases}$$

Where $\oplus$ is either GF(3) addition or mod-3 addition, + is either a shift up operation or mod-3 addition, and - is a shift down operation.

For simplicity of implementation, shift-up operation is used instead of mod-3 addition for + operation, and mod-3 addition is used instead of GF(3) addition for the $\oplus$ operation.

Figures B.2 and B.3 illustrate an implementation of the GF(4) addition and multiplication operations. Note that multi-input addition and multiplication circuits can be realized as trees of two-input addition and multiplication circuits as shown in Figs. B.2b and B.3b, respectively. Similar approaches can be used to realize addition and multiplication operators for higher radices of Galois fields.

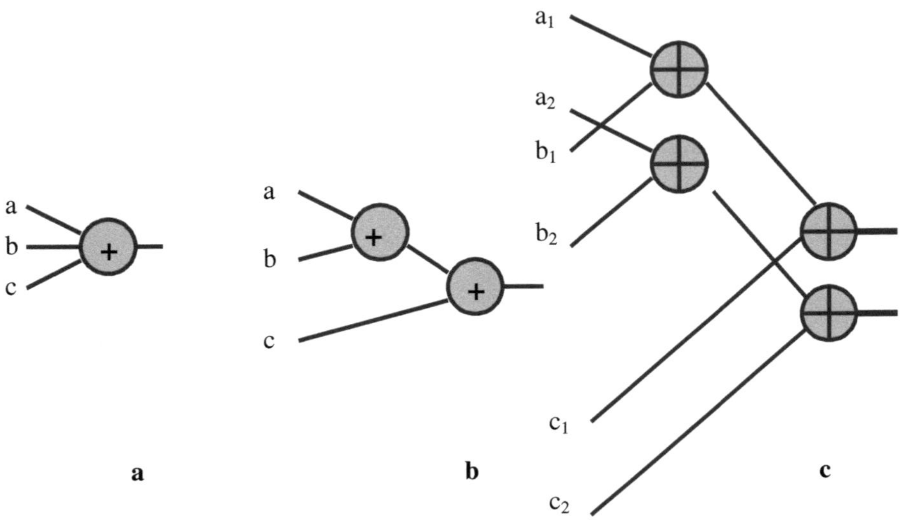

Fig. B.2. **a** Implementation of three input single GF(4) addition gate as two decomposed two-input tree-structured GF(4) addition gates as in **b**, and **c** consequently as quadruple two-input tree-structured vector of GF(2) EXORs.

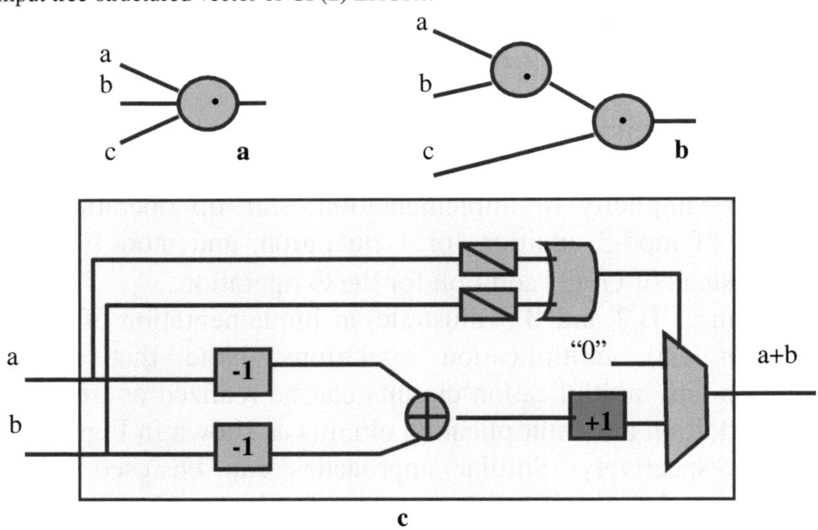

Fig. B.3. **a** Implementation of a 3-input single GF(4) logic multiplication gate as decomposed 2-input two tree-structured GF(4) multiplication gates as in **b**, and **c** the implementation of a single GF(4) logic multiplication using mod-3 addition and shift operations.

Where:

Is a –1 shifter

Is a +1 shifter

Is a modulo-3 addition

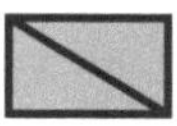
Is a comparator to value “0”

Is a 1/0 binary multiplexer

Logical addition and multiplication gates are used in many applications such as the realization of 2^r-bit arithmetic addition and 2^r-bit arithmetic multiplication operations, which are the fundamental arithmetic operations by which the complex arithmetic and logic unit (ALU) is designed.

Complex arithmetic operations such as the convolution operation are the base of linear transformations used in many digital signal processing and digital image processing applications. Instead of standard arithmetics, Galois logic can be used for fast and efficient realizations which use linear algebra.

The physical implementation of the circuit in Fig. B.3 can be implemented utilizing various choices of technologies: micro technologies, deep sub-micron technologies, or nano technologies. One of the options is the pass transistor logic (PTL) that have recently found popularity within the CMOS technologies due to circuit optimization issues. Also, other future technologies such as: (1) single electron transistors (SET), (2) Josephson junction, and (3) low-power VLSI design [206,262] technologies require efficient realization of multiple-valued Shannon and Davio canonical expressions and Galois gates.

Appendix C

Count of the Number of S/D Inclusive Forms and the Novel $IF_{n,2}$ Triangles

This Appendix provides the count for the numbers of Inclusive Forms (IF) in Chapt. 3, and provide a generic way of counting such forms when the number of variables is very large in a way such that an ordinary computer routine will not perform such counts in a polynomial time, and thus the need for the $IF_{n,2}$ triangles as a pattern-based way for performing a count. These counts can be used as numerical parameters (e.g., upper-bounds) in search heuristics that search for minimum GFSOP expressions from Chapt. 3.

Theorem C.1. For GF(3) and N variables, the total number of TIFs per variable order is:

$$\# TIFs =$$
$$\sum_{k_1=0}^{(3)^{N-1}} \sum_{k_2=0}^{(3)^{N-2}} \sum_{k_3=0}^{(3)^{N-3}} \cdots \sum_{K_N=0}^{(3)^0} \{[\frac{3^{(N-1)}!}{(3^{(N-1)}-k_1)!k_1!} \frac{3^{(N-2)}!}{(3^{(N-2)}-k_2)!k_2!} \frac{3^{(N-3)}!}{(3^{(N-3)}-k_3)!k_3!} \cdots$$
$$\frac{3^0!}{(3^0-k_N)!k_N!}][(3^{2.(3)^0})^{k_1}(3^{2.(3)^1})^{k_2}(3^{2.(3)^2})^{k_3} \ldots (3^{2.(3)^{(N-1)}})^{k_N}]\}$$

(C.1)

Proof. The following is the derivation of the general formula (C.1) to calculate #TIFs per variable order: the total number of nodes for any GF(3) tree with N levels (i.e., N variables) equals to:

$$\sum_{k=0}^{N-1} (3)^k . \qquad (C.2)$$

For any *S-type* node there is only one type of nodes (as the branches have the possibility of single value each). Yet, for *D-type* node there are N possible types of nodes (where N is the number of variables, which is equal to the number of levels). The highest possible number of forms for the D-type node is when the D-type node exists in the first (highest) level, and the lowest possible number of forms for the D-type node is when the D-type node exists in the Nth-level (lowest level).

Therefore, for certain number (M) of S-type nodes the following formula describes the number of the D-type nodes for N variables:

$$\# S = M \quad \Rightarrow \quad \# D = [\sum_{k=0}^{N-1} (3)^k \ - \ M]. \tag{C.3}$$

It can be shown that for GF(3) (i.e., ternary decision tree (TDT)) and N-levels (i.e., N-variables), the general formulas that count the number of D-type nodes, and the number of all possible forms for the D-type node in the kth level of the Nth-level TDT are, respectively:

$$\# D_k = (3)^{(K-1)}, \tag{C.4}$$

$$|\, D_k \,|\, per\ node = (3)^{2.(3)^{(N-K)}}, \tag{C.5}$$

where:

$\# D_k$: is the number of D-type nodes in the kth level,

$|D_k|$: is the number of all possible forms for the D-type node in the kth level.

Let us define S/D tree category to be: the S/D trees that have in common the same number of S-type nodes and the same number of D-type nodes within the same variable order. Also, let us define the following:

$$\Psi = \textit{number of variable orders.} \tag{C.6}$$

$$\Omega = \textit{number of S/D tree categories per variable order.} \tag{C.7}$$

$$\phi = \textit{number of S/D trees per category.} \tag{C.8}$$

$$\Phi = \textit{number of TIFs per variable order.} \tag{C.9}$$

From Eqs. (C.2) through (C.5), and using some elementary count rules, we can derive by mathematical induction the following general formulas for N being the number of variables:

$$\psi = N!, \tag{C.10}$$

$$\Omega = \sum_{k=0}^{N-1}(3)^k + 1, \tag{C.11}$$

$$\phi = \frac{[\sum_{k=0}^{N-1}(3)^k]!}{[\sum_{k=0}^{N-1}(3)^k - k]!k!},$$

$$where\ k = 0, 1, 2, 3, \ldots, \sum_{k=0}^{N-1}(3)^k, \tag{C.12}$$

$$\Phi =$$

$$\sum_{k_1=0}^{(3)^{N-1}} \sum_{k_2=0}^{(3)^{N-2}} \sum_{k_3=0}^{(3)^{N-3}} \cdots \sum_{K_N=0}^{(3)^0} \{[\frac{3^{(N-1)}!}{(3^{(N-1)}-k_1)!k_1!} \frac{3^{(N-2)}!}{(3^{(N-2)}-k_2)!k_2!}$$

$$\frac{3^{(N-3)}!}{(3^{(N-3)}-k_3)!k_3!} \cdots \frac{3^0!}{(3^0-k_N)!k_N!}][(3^{2.(3)^0})^{k_1}(3^{2.(3)^1})^{k_2}(3^{2.(3)^2})^{k_3} \ldots$$

$$(3^{2.(3)^{(N-1)}})^{k_N}]\}$$

(C.13)

Q.E.D.

From Eqs. (C.10) through (C.13), it can be noticed that the total number of TIFs for all variable orders is equal to {[N!][#TIFs per order]}, and the number of TGIFs is bounded according to the inequality {#TIFs per variable order < #TGIFs < #TIFs for all variable orders}.

Example C.1. For number of variables equal to two (N = 2), Eq. (C.12) reduces to:

$$\Phi =$$

$$\sum_{k_1=0}^{3^1}\sum_{k_2=0}^{1}\{\frac{3^1!}{(3^1-k_1)!k_1!}\;\frac{3^0!}{(3^0-k_2)!k_2!}\;(3^{2.(3)^0})^{k_1}\;(3^{2.(3)^1})^{k_2}\}$$

$$\Phi = \Phi\Big|_{k_1=0,k_2=0} + \Phi\Big|_{k_1=1,k_2=0} + \Phi\Big|_{k_1=2,k_2=0} +$$
$$\Phi\Big|_{k_1=3,k_2=0} + \Phi\Big|_{k_1=0,k_2=1} + \Phi\Big|_{k_1=1,k_2=1} + \Phi\Big|_{k_1=2,k_2=1} +$$
$$\Phi\Big|_{k_1=3,k_2=1},$$

$$= \Phi_{00} + \Phi_{10} + \Phi_{20} + \Phi_{30} + \Phi_{01} + \Phi_{11} + \Phi_{21} + \Phi_{31},$$

$$= 1 + 27 + 243 + 729 + 729 + 19683 + 177147 + 531441,$$

$$= 730{,}000\,.$$

Utilizing multi-valued map representation, there are, in general, for N^{th}-valued input-output logic: $(N)^{\#Minterms}$ different functions. Therefore, for ternary logic, there are $3^9 = 19{,}683$ different ternary functions of two variables, and 730,000 ternary Inclusive Forms generated by the S/D trees. Thus, on the average every function of two variables can be realized in approximately 37 ways.

Theorem C.2. For GF(4) and N variables, the total number of QIFs per variable order is equal to:

$$\#QIFs = \Phi =$$

$$\sum_{k_1=0}^{(4)^{N-1}}\sum_{k_2=0}^{(4)^{N-2}}\cdots\sum_{k_N=0}^{(4)^{0}}\{\frac{4^{(N-1)}!}{(4^{(N-1)}-k_1)!k_1!}\;\frac{4^{(N-2)}!}{(4^{(N-2)}-k_2)!k_2!}\cdots\frac{4^{0}!}{(4^{0}-k_N)!k_N!}$$
$$(4^{3.(4)^0})^{k_1}(4^{3.(4)^1})^{k_2}\cdots(4^{3.(4)^{(N-1)}})^{k_N}\}. \qquad (C.14)$$

Proof. A general proof that will include the quaternary Galois field as a special case will be provided later in this Appendix. **Q.E.D.**

Properties and extended Green/Sasao hierarchy for quaternary S/D trees and their corresponding forms can be developed similar to the work in [4] and the properties of ternary S/D trees shown in Chapt. 3. The extension of the concept of S/D trees to higher radices of Galois fields (i.e., higher than four) is a systematic and direct process that follows the same methodology developed for the ternary case [3,4] and the quaternary case.

The following example demonstrates the counts of QIFs using Theorem C.2.

Example C.2. For number of variables equal to two (N=2), Eq. (C.14) reduces to:

$$\Phi = \sum_{k_1=0}^{(4)^1} \sum_{k_N=0}^{(4)^0} \{ \frac{4^{(1)}!}{(4^{(1)}-k_1)!k_1!} \frac{4^{(0)}!}{(4^{(0)}-k_2)!k_2!} \cdot (4^{3.(4)^0})^{k_1} (4^{3.(4)^1})^{k_2} \}$$

$$= \Phi\Big|_{k1=0,k2=0} + \Phi\Big|_{k1=1,k2=0} + \Phi\Big|_{k1=2,k2=0} + \Phi\Big|_{k1=3,k2=0} + \Phi\Big|_{k1=4,k2=0} + \Phi\Big|_{k1=0,k2=1} + \Phi\Big|_{k1=1,k2=1} + \Phi\Big|_{k1=2,k2=1} + \Phi\Big|_{k1=3,k2=1} + \Phi\Big|_{k1=4,k2=1},$$

$$= \Phi_{00} + \Phi_{10} + \Phi_{20} + \Phi_{30} + \Phi_{40} + \Phi_{01} + \Phi_{11} + \Phi_{21} + \Phi_{31} + \Phi_{41},$$

$$= 1 + 256 + 24{,}576 + 1{,}048{,}576 + 16{,}777{,}216 + 16{,}777{,}216 + 4{,}294{,}967{,}296 + 412{,}316{,}860{,}416 + 1.75921860444 * 10^{13} + 2.81477976711 * 10^{14},$$

$$= 2.99483809211 * 10^{14}.$$

Utilizing MVL map representation, we can easily prove that there are 4^{16} = 4,294,967,296 quaternary functions of two variables, and $2.99483809211 * 10^{14}$ quaternary Inclusive Forms generated by the S/D trees. Thus, on the average every function of two variables can be synthesized (realized) in approximately 69,729 ways. This high number of realizations means that most functions of two variables are realized with less than five expansions, and all functions with at most five expansions.

C.1 General Formula to Compute the Number of IFs for an Arbitrary Number of Variables and Arbitrary Galois Field Radix

Although the S/D trees and Inclusive Forms that were developed in Chapt. 3 are for GF(4), the same concept can be directly and systematically extended to the case of nth radix of Galois fields and N variables. Theorem C.3 provides the total number of IFs per variable order for N variables (i.e., N decision tree levels) and n^{th} radix of any arbitrary algebraic field, including $GF(p^k)$ where p is a prime number and k is a natural number ≥ 1. The generality of Theorem C.3 comes from the fact that algebraic structures specify the type of operations (e.g., addition and multiplication operations) in the functional expansions but do not specify the counts which are an intrinsic property of the tree structure and are independent of the algebraic operations performed. Thus, Theorem C.3 is valid, among others, for Galois fields of arbitrary radix (p^k) where p is a prime number and k is a natural number ≥ 1 (e.g., 3, 4, 5, 7, 8, 9, 11, 13, etc).

Theorem C.3. The total number of Inclusive Forms for N variables and n^{th} radix Galois field logic is equal to:

$$\# QIFs = \Phi_{n,N} =$$

$$\sum_{k_1=0}^{(n)^{N-1}} \sum_{k_2=0}^{(n)^{N-2}} \cdots \sum_{k_N=0}^{(n)^{0}} \{ \frac{n^{(N-1)}!}{\left(n^{(N-1)}-k_1\right)!k_1!} \frac{n^{(N-2)}!}{\left(n^{(N-2)}-k_2\right)!k_2!} \cdots \frac{n^{0}!}{\left(n^{0}-k_N\right)!k_N!}$$

$$\left(n^{(n-1).(n)^{0}}\right)^{k_1} \left(n^{(n-1).(n)^{1}}\right)^{k_2} \cdots \left(n^{(n-1).(n)^{(N-1)}}\right)^{k_N} \}. \tag{C.15}$$

Proof. The following is the derivation of the general Eq. to calculate the number of IFs per variable order: The total number of nodes for any n^{th} radix Galois field (GF(n)) tree with N levels (i.e., N variables) equals to:

$$\sum_{k=0}^{N-1} (n)^{k}. \tag{C.16}$$

For any S-type (i.e., Shannon type) node there is only one type of nodes as the branches of the Shannon node have the possibility of single value each. Yet, for D-type (i.e., Davio type) node there are N possible types of nodes (where N is the number of variables, which is equal to the number of levels). The highest possible number of forms for the D-type node exists when the Davio node exists in the first (highest) level, and the lowest possible number of forms for the D-type node is when the Davio node exists in the Nth-level (lowest level). Therefore, for certain number (M) of S-type nodes the following formula describes the number of the D-type nodes for N variables:

$$\# S = M \Rightarrow \# D = \sum_{k=0}^{N-1} (n)^k \text{ - } M. \tag{C.17}$$

It can be shown that for GF(n) (n-ary decision tree with N-levels (i.e., N variables), the general formulas that count the number of D-type nodes, and the number of all possible forms for the D-type node in the kth level (where k is less than or equal the total number of levels N) are, respectively:

$$\# D_k = (n)^{(k-1)}, \tag{C.18}$$

$$|D_k| = (n)^{(n-1).(n)^{(N-K)}}, \tag{C.19}$$

where $\# D_k$ is the number of D-type nodes in the kth level, $|D_k|$ is the number of all possible forms (per node) for the D-type node in the kth level, Let us define the S/D tree category to be the S/D trees that have in common the same number of S-type nodes and the same number of D-type nodes within the same variable order. Let us define the following entities for n^{th} radix Galois field and N variables (i.e., N decision tree levels):

$\psi_{n,N}$ = *number of variable orders.* (C.20)

$\Omega_{n,N}$ = *number of S/D tree categories per variable order.* (C.21)

$\phi_{n,N}$ = *number of S/D trees per category.* (C.22)

$\Phi_{n,N}$ = *number of IFs per variable order.* (C.23)

From the previous Eqs., and using elementary count rules, one can derive by mathematical induction the following general formulas for N being the number of variables, and n being the field radix:

$$\Psi_{n,N} = N!, \tag{C.24}$$

$$\Omega_{n,N} = \sum_{k=0}^{N-1} (n)^k + 1, \tag{C.25}$$

$$\phi_{n,N} = \frac{[\sum_{k=0}^{N-1} n^k]!}{[\sum_{k=0}^{N-1} n^k - k]!k!}, \text{ where } k = 0, 1, 2, 3, \ldots, N-1, \tag{C.26}$$

$$\Phi_{n,N} =$$

$$\sum_{k_1=0}^{(n)^{N-1}} \sum_{k_2=0}^{(n)^{N-2}} \cdots \sum_{k_N=0}^{(n)^0} \{ \frac{n^{(N-1)}!}{\left(n^{(N-1)} - k_1\right)!k_1!} \frac{n^{(N-2)}!}{\left(n^{(N-2)} - k_2\right)!k_2!} \cdots \frac{n^0!}{\left(n^0 - k_N\right)!k_N!}$$

$$\left(n^{(n-1).(n)^0}\right)^{k_1} \left(n^{(n-1).(n)^1}\right)^{k_2} \cdots \left(n^{(n-1).(n)^{(N-1)}}\right)^{k_N} \}. \tag{C.27}$$

Q.E.D.

One can note that the formula in Eq. (C.27) used to obtain the total number of Inclusive Forms for N variables and n^{th} radix of Galois field is a very general fomula that includes the ternary case (Eq. (C.1)) and the quaternary case (Eq. (C.14)) as special cases.

Numerical counting results that are obtained from Eq. (C.27) can be used in search heuristics as numerical bounds that could be incorporated into efficient search of S/D trees (from Chapt. 3) in order to obtain minimal GFSOP forms for specific multiple-valued logic functions. Since such search for minimal forms is already a difficult problem in two-valued logic (for example using binary S/D trees) especially when the number of variables is large, the search for minimal GFSOP forms in multiple-valued Galois logic will be very difficult. Thus, further numerical evaluations have to be conducted in order to estimate the usefulness of numerical bounds obtained from Eq. (C.27) in such multiple-valued search heuristics.

Example C.3. Let us produce the number of QIFs over GF(4) for two variables (i.e., N=2 and n = 4), then one obtains:

$$\Phi_{4,2} = \sum_{k_1=0}^{(4)^{2-1}} \sum_{k_2=0}^{(4)^{2-2}} \left\{ \frac{4^{(2-1)}!}{\left(4^{(2-1)}-k_1\right)!k_1!} \frac{4^{(2-2)}!}{\left(4^{(2-2)}-k_2\right)!k_2!} \cdot \left(4^{(4-1).(4)^0}\right)^{k_1} \left(4^{(4-1).(4)^1}\right)^{k_2} \right\}$$

$= \Phi_{00|4,2} + \Phi_{10|4,2} + \Phi_{20|4,2} + \Phi_{30|4,2} + \Phi_{40|4,2} + \Phi_{01|4,2} + \Phi_{11|4,2} + \Phi_{21|4,2} + \Phi_{31|4,2} + \Phi_{41|4,2}$,
$= 1 + 256 + 24{,}576 + 1{,}048{,}576 + 16{,}777{,}216 + 16{,}777{,}216 + 4{,}294{,}967{,}296 + 412{,}316{,}860{,}416 + 1.75921860444 * 10^{13} + 2.81477976711 * 10^{14}$,
$= 2.99483809211 * 10^{14}$.

Corollary C.1. From [52] and Eq. (C.15), the following mathematical corollary can be obtained to count the number of Inclusive Forms for N variables and second radix:

$$\prod_{k=0}^{n-1} \left(1+2^{2^{n-k-1}}\right)^{2^k} = \sum_{k_1=0}^{(2)^{N-1}} \sum_{k_2=0}^{(2)^{N-2}} \cdots \sum_{k_N=0}^{(2)^0} \left\{ \frac{2^{(N-1)}!}{\left(2^{(N-1)}-k_1\right)!k_1!} \frac{2^{(N-2)}!}{\left(2^{(N-2)}-k_2\right)!k_2!} \cdots \frac{2^0!}{\left(2^0-k_N\right)!k_N!} \left(2^{(2-1).(2)^0}\right)^{k_1} \left(2^{(2-1).(2)^1}\right)^{k_2} \cdots \left(2^{(2-1).(2)^{(N-1)}}\right)^{k_N} \right\}. \quad \text{(C.28)}$$

In general, expansions of functions can be produced over basis functions of a single variable, two variables, or any number of variables. The interesting case of expansions of functions utilizing pairs of variables can be produced using the general procedure for expansions over Linearly Independent (LI) logic. This can be achieved by the recursive expansions of a multi-variable function over bases of two variables. The advantage of such expansions is the regular usage of universal blocks with two control variables that generalize multiplexers (data selectors) with two control variables (i.e., variables of control functions) and four data inputs (data functions). The following Sect. introduces a fast method to calculate the number of IFs (that are special cases of Linearly Independent (LI) logic) for an arbitrary Galois field logic for functions with two variables.

C.2 A Fast Method to Calculate the Number of IFs for an Arbitrary Radix of Galois Field GF(p^k) for Functions of Two Variables Using $IF_{n,2}$ Triangles

The count of the number of IFs is important in many applications, especially in providing upper numerical boundaries for efficient search of a minimum GFSOP. Calculating the numbers of Inclusive Forms (IFs) can be very time consuming due to the time required to perform the mathematical operations in the general Eq. (C.15). This is why a fast method to generate the number of IFs is needed. Because functions with two variables find an important application in the ULMs for pairs of variables, the following subsection provides a fast method to calculate the number of IFs over an arbitrary radix of Galois field GF(p^k) for two variable functions (i.e., N = 2).

C.2.1 $IF_{n,2}$ Triangles

Functions with two variables are attractive in logic synthesis since many functional decomposition methods exist that produce two control inputs for primitive cells in a standard library of standard cells (such as in a multiplexer with two address lines). It is shown in [3,4] how to produce ULMs for pairs of control variables that generalize Shannon and Davio expansion modules. Theorem C.4 will introduce a fast method to calculate the number of Inclusive Forms for functions with two input variables over an arbitrary radix of Galois field [3].

These triangles are important because the complexity of count using Eq. (C.15) for high dimensions is very high, and thus the ability of a personal computer to compute the counts for number of variables greater than five in a reasonable amount of time becomes questionable. Consequently, the $IF_{n,2}$ triangles provide an alternative numerical and geometrical pattern of computing [3].

Theorem C.4. The following $IF_{n,2}$ Triangles provide a fast method to calculate the number of IFs over an arbitrary n^{th} radix of Galois field ($GF(p^k)$) for two variable functions (N=2).

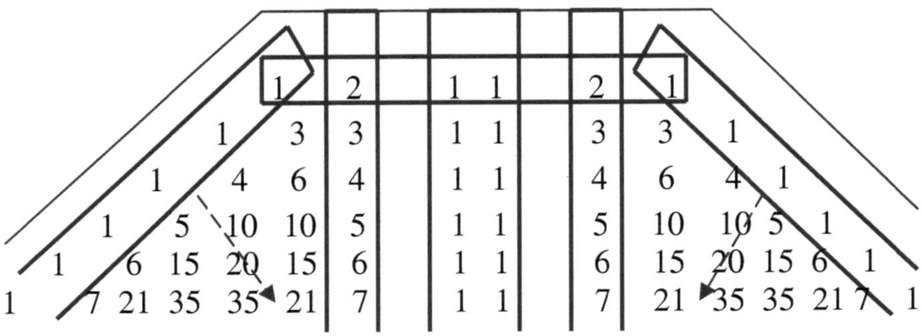

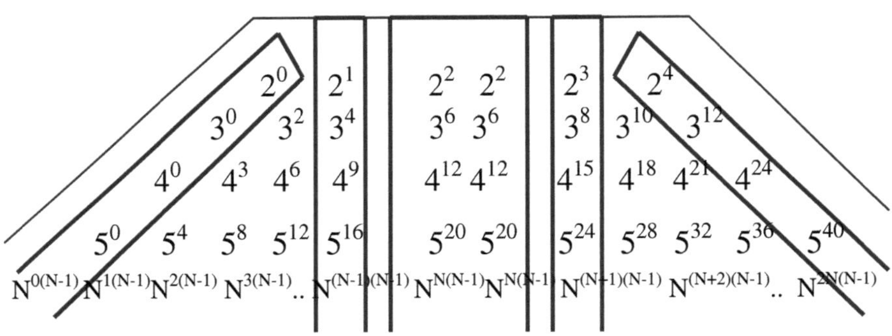

Fig. C.1. IFn,2 Triangles: **a** the Triangle of Coefficients, and **b** the Triangle of Values for a fast calculation of the number of Inclusive Forms for an arbitrary radix Galois field and functions of two input variables (N=2).

Proof. The proof of Theorem C.4 follows directly from the mathematical induction of the number of IFs over an arbitrary radix of Galois field $GF(p^k)$ for two variable functions. This can be deduced directly from Eq. (C.15); if the $IF_{n,2}$ Triangles are valid for $n = q$ then they will be also valid for $n = q + 1$, where $n = p^k$ (p is a prime number and k is a natural number ≥ 1). **Q.E.D.**

It can be observed that the $IF_{n,2}$ Triangle of Coefficients possesses a close similarity to the well known Pascal Triangle. This occurs as follows: if one omits the first two rows of the Pascal Triangle and duplicates each row into another horizontally adjacent row, the $IF_{n,2}$ Triangle of Coefficients will be obtained. This fact helps in creating computer algorithms that generates the $IF_{n,2}$ Triangle of Coefficients since many efficient and optimized algorithms already exist to generate the Pascal Triangle. The following example illustrates the concept of count of the number of IFs over an arbitrary radix of Galois field $GF(p^k)$ for two variable functions through the $IF_{n,2}$ Triangles that were demonstrated in Fig. C.1.

Example C.4. Utilizing the $IF_{n,2}$ Triangles from Fig. C.1, we can calculate the number of Inclusive Forms for GF(2), GF(3), and GF(4) for two variables:

$$\Phi_{2,2} = 1\cdot 2^0+2\cdot 2^1+1\cdot 2^2+1\cdot 2^2+2\cdot 2^3+1\cdot 2^4 = 1 + 4 + 4 + 4 + 16 + 16 = 45.$$

$$\Phi_{3,2} = 1\cdot 3^0+3\cdot 3^2+3\cdot 3^4+1\cdot 3^6+1\cdot 3^6+3\cdot 3^8+3\cdot 3^{10}+1\cdot 3^{12} = 730{,}000.$$

$$\Phi_{4,2} = 1\cdot 4^0+4\cdot 4^3+6\cdot 4^6+4\cdot 4^9+1\cdot 4^{12}+1\cdot 4^{12}+4\cdot 4^{15}+6\cdot 4^{18}+4\cdot 4^{21}+1\cdot 4^{24},$$
$$= 2.99483809211*10^{14}.$$

One can observe that the results from Eqs. (C.29), (C.30), and (C.31) are the same results that were obtained previously.

C.2.2 Properties of $IF_{n,2}$ Triangles

(1) The number of positions (elements) in each row of the triangles in Figs. C.1a and C.1b are even starting from six.

(2) The sum of elements in each row in Fig. C.1a equals to the number of S/D trees per variable order.
(3) The triangle in Fig. C.1a possesses even symmetry around an imaginary vertical axis in the middle of the triangle.
(4) The minimum number of columns required to generate the whole triangle in Fig. C.1a is equal to three (due to even symmetry): one wing, one column neighbor to the middle column, and one middle column.
(5) The triangle in Fig. C.1a can be generated by the process of "Shift Diagonally and Add Diagonally" (SDAAD); shift the left wing diagonally from west to southeast direction and add two numbers diagonally from east to southwest direction, and shift the right wing diagonally from east to southwest direction and add two numbers diagonally from west to southeast direction.
(6) The difference in powers in the triangle in Fig. C.1b per row element is (N-1).
(7) The first number in each row of the triangle in Fig. C.1b is N^0 and the last number per row is $N^{2N(N-1)}$, where N is the number of variables.
(8) The middle two numbers in each row of the triangle in Fig. C.1b are always equal to $N^{N(N-1)}$, where N is the number of variables.

Appendix D

Universal Logic Modules (ULMs) for Circuit Realization of Shannon/Davio (S/D) Trees

This Appendix provides logic circuit realizations of Universal Logic Modules (ULMs) for the S/D expansions of multiple-valued Shannon and Davio spectral transforms from Chapt. 3.

D.1 S/D Universal Logic Modules for Ternary Radix

The nonsingular expansions of Ternary Shannon (S) and Ternary Davio (D_0, D_1, and D_2), can be realized using a "Universal Logic Module" (ULM) with control variables corresponding to the variables of the basis functions (i.e., the variables we are expanding upon). We call it a universal logic module, because similarly to a multiplexer, all functions of two variables can be realized with two-level trees of such modules using constants on the second-level data inputs. ULMs are complete systems, because they can implement all possible functions with certain number of variables. The concept of the universal logic module was used for binary RM logic (over GF(2)), as well as the very general case of Linearly Independent (LI) logic [172,174], that includes R-M logic as a special case. Binary LI logic extended the universal logic module from just being a multiplexer (Shannon Expansion), AND/EXOR gate (positive Davio expansion), and AND/EXOR/NOT gate with inverted control variable (negative Davio expansion), to the universal logic modules for any expansion over any linearly independent basis functions. Analogously to the binary case, Figs. D.1 and D.2 present the universal logic modules for ternary Shannon (S), and ternary Davio (D_0, D_1, and D_2), respectively.

We can note, as seen from Chapt. 3, that any function f can be produced by the application of the independent variable {x} and the cofactors {f_i, f_j, and f_k } as inputs to a ULM. The form of the resulting function depends on our choice of the shift and power operations that we choose inside the ULM for the input independent variable, and on our choice of the weighted combinations of the input cofactors. Utilizing this note, we can combine all Davio ULMs to create the single all-Davio ULM. Figure D.3 illustrates this ULM. An even more general Universal Logic Module can be generated to implement all Ternary Shannon and Davio expansions over GF(3). Figure D.4 illustrates such a ULM.

In general, the gates in the ULMs can be implemented, among other circuit technologies, by using binary logic over GF(2) [3,4], or using multi-valued circuit gates. Each ternary ULM corresponds to a single node in the nodes of TDTs that were illustrated in Chapt. 3. The main advantage of such powerful ULMs is in high layout regularity that is required by future nano technologies. The trees can be realized in layout because they do not grow exponentially for practical functions. For instance, assuming a ULM from Fig. D.4, although every function of two variables can be realized with four such modules, it is highly probable that most of the functions of two variables will require less than four modules. Because of these properties, this approach should give very good results when applied to incompletely specified functions and multi-valued relations.

Multiplexers and Davio gates are used to design new reconfigurable structures, such as FPGAs with their well-known applications in memory-based Ping-Pong architectures and parallel processing systems (such as DEC-PERLE system), and regular data path blocks, besides many of the multi-level structures that are based on them. Similarly, the new ULMs can find various implementations in different sorts of regular structures, such as: iterative circuits, Cellular Automata (CA), Lattices (or Pseudo-Symmetric Decision Diagrams (PSDDs) from Chapt. 4), Pipelining, and Systolic Architectures. One important implementation is one-to-one mapping of regular layout of functions into Lattices. Regular structures are most favorable in circuit design as they allow the ease of many tasks, such as: (1) fault detection (circuit diagnosis; circuit testing), (2) fault localization, (3) circuit self-repairing, (4) evolvable hardware, and (5) circuit manufacturing. Such properties are essential in the future advanced technologies for sensitive applications such as space-oriented applications (e.g., satellite circuits that are immune to galactic/cosmic radiation) and bio-medical applications (e.g., human-IC interface). ULMs can be also created for pairs of variables and their larger sets.

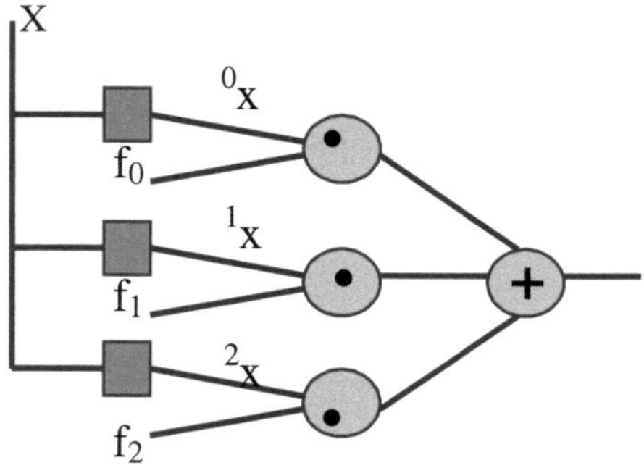

Fig. D.1. ULM of ternary Shannon over GF(3).

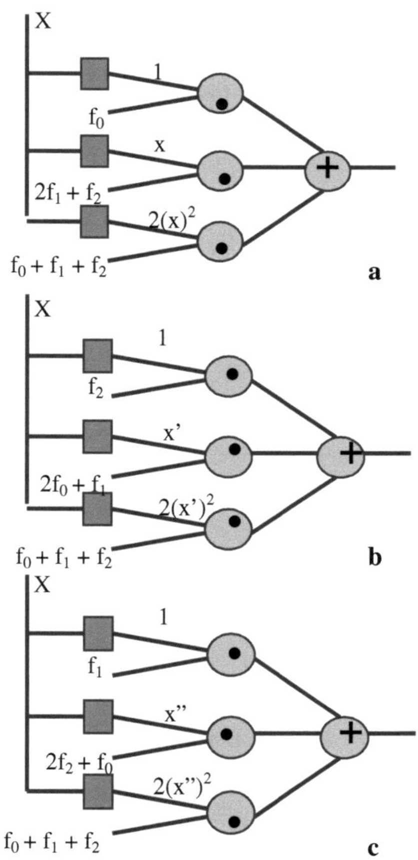

Fig. D.2. ULMs of Davio over GF(3): **a** D_0, **b** D_1, and **c** D_2.

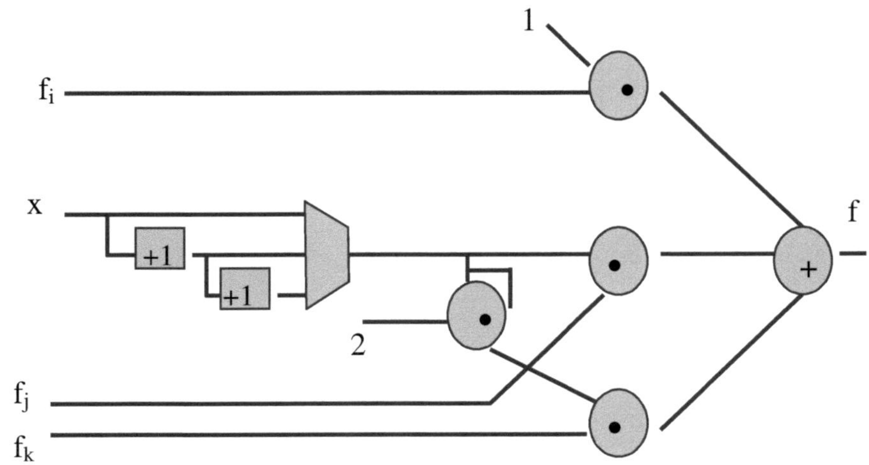

Fig. D.3. Universal Logic Module for all ternary Davio expansions over GF(3).

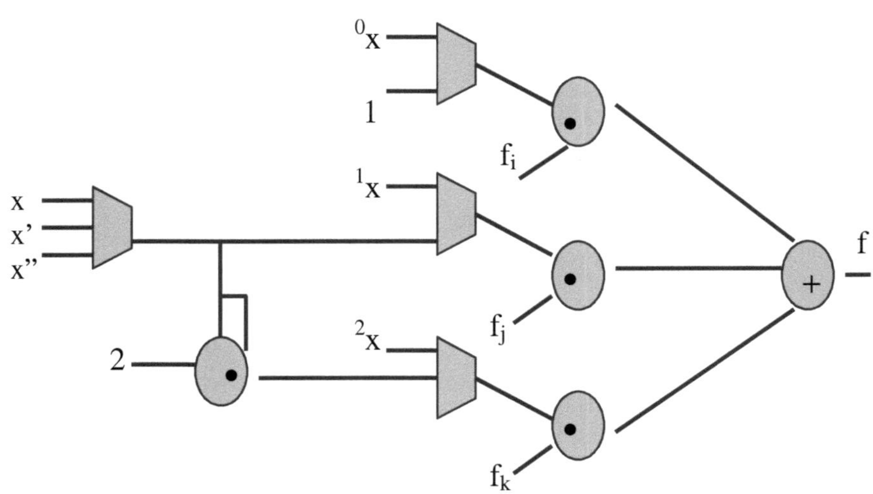

Fig. D.4. Ternary S/D ULM over GF(3).

D.2 Logic Synthesis of Quaternary GFSOPs

The nonsingular expansions of quaternary Shannon and Davio expansions, can be realized using a quaternary Universal Logic Module (ULM) with control variables corresponding to the variables of the basis functions (i.e., the variables that are expanded upon).

Similarly to the ternary case, quaternary ULMs (i.e, ULMs with quaternary gates) for quaternary Shannon and quaternary Davio expansions can be designed. Yet, because of the "maturity" and the intensive usage of the binary-based (i.e., Boolean) technologies, we will illustrate the implementation of the quaternary ULMs as binary-based logic circuits. This can be done through the encoding of a single 4-valued variable into two 2-valued variables.

Analogously to the ternary case, the general ULM that covers the quaternary Shannon and all Davio expansions can be created. Note that by utilizing Figs. D.5 and D.6, all the quaternary logical addition and multiplication gates in Fig. D.7 can be converted into the corresponding binary logical addition and multiplication gates, respectively, similar to the encoding of a single 4-valued variable into two 2-valued variables presented in (for simplicity of illustration, the internal 4/2 encoders and 2/4 decoders in Fig. D.7 are cancelled and the internal structure is simplified). It also can be observed in Fig. D.7 that Shannon ULM is a two level circuit, Davio ULM (i.e., per Davio) is a two level circuit, all Davio ULM (i.e., ULM for all Davio types) is a four level ULM circuit, and all Shannon and Davio ULM is a five level circuit (excluding the levels of literals' generators as seen in Fig. D.7).

In general, each quaternary ULM corresponds to a single node of QuDTs that were illustrated in Chapt. 3. The main advantage of such powerful ULMs is in high layout regularity that will be required by future nano-technologies. The trees can be realized in layout because they do not grow exponentially for practical functions. For instance, although every function of two variables can be realized with five such modules, it is highly probable that most of the functions of two variables will require less than five modules. Because of these properties, this approach should give very good results when applied to incompletely specified functions and multi-valued relations.

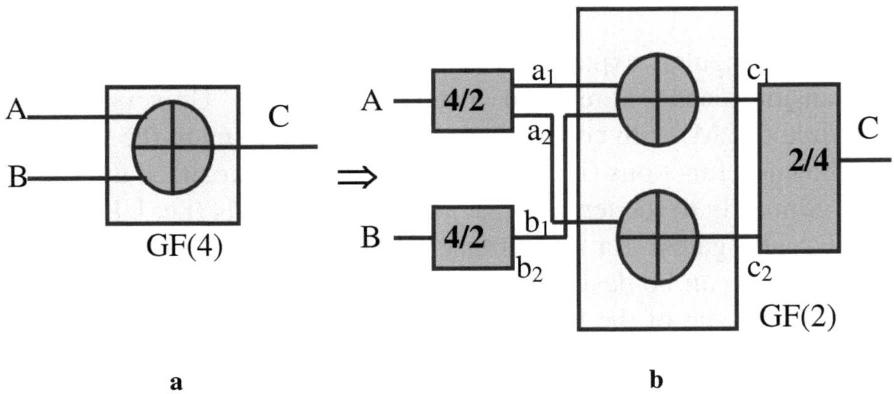

Fig. D.5. Realization of GF(4) addition in **a** as GF(2) addition in **b** (i.e., vector of EXORs).

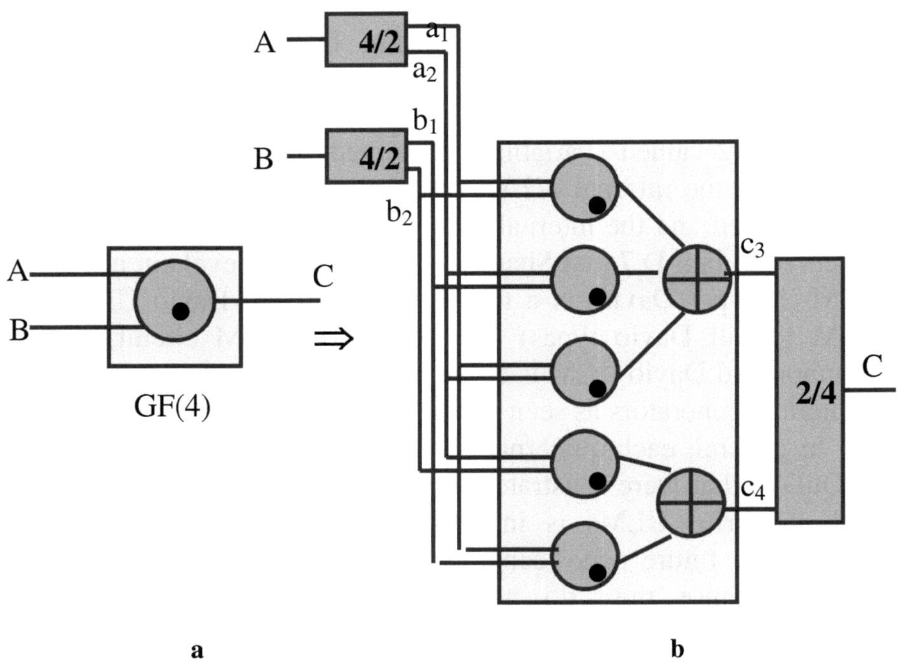

Fig. D.6. Realization of GF(4) multiplication in **a** using GF(2) operations from **b**.

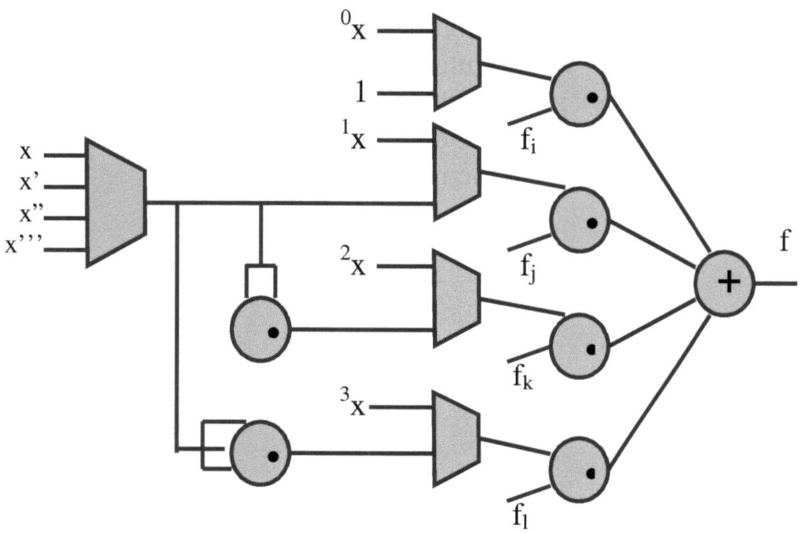

Fig. D.7. Quaternary ULM that produces quaternary Shannon expansion (Eq. (2.27)), and all quaternary Davio expansions (Eqs. (2.44), (2.45), (2.46), and (2.47)).

Where:

Is a quaternary-to-binary encoder.

Is a binary-to-quaternary decoder.

Is a binary Mux.

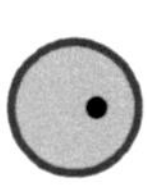

Is a GF(4) logical multiplier.

Is a GF(4) logical adder.

The ULM shown in Fig. D.7 is used repeatedly (systematically) as the processing node in the corresponding canonical multi-level QuDTs. Also, nodes of the corresponding optimized (minimized), non-canonical, and non-regular Reduced Quaternary Decision Diagrams (RQuDDs) can be implemented using the ULM in Fig. D.7. RQuDD circuits are important in the case when they are one-to-one mapped into an isomorphic hardware.

Appendix E

Evolutionary Computing: Genetic Algorithm (GA) and Genetic Programming (GP)

This Appendix provides a basic background for the evolutionary-based algorithms in Chapts. 3, 8, and 11, respectively. Evolutionary Computing (EC) is one type of "black box" global optimization methods that has been successfully implemented to solve for many nonlinear difficult problems [102].

EC implements the idea which was proposed by Darwin as an explanation of the biological world surrounding us: Evolution by Natural Selection [66]. By evolution we mean the change of the genes that produce a structure. The result of this evolution is the Survival of the Fittest and the Elimination of the Unfit. Darwin's theory of evolutionary selection holds that variation within species occurs randomly and that the survival or extinction of each organism is determined by that organism's ability to adapt to its environment.

This very simple but very powerful idea has been implemented in algorithms which are called Genetic Algorithms (GA) and Genetic Programming (GP). The only difference between GA and GP is the representation of the problem and consequently the set of genetic operators used to obtain the solution. This is because GA uses string representation and the consequent genetic operators [102], and GP uses tree representation and the consequent genetic operators [137].

Figure E.1 represents the general optimization or synthesis EC method, where iterations on this flow diagram are made until the actual output matches exactly the desired output (i.e., no error) or the actual output mismatches the desired output within an acceptable range of error.

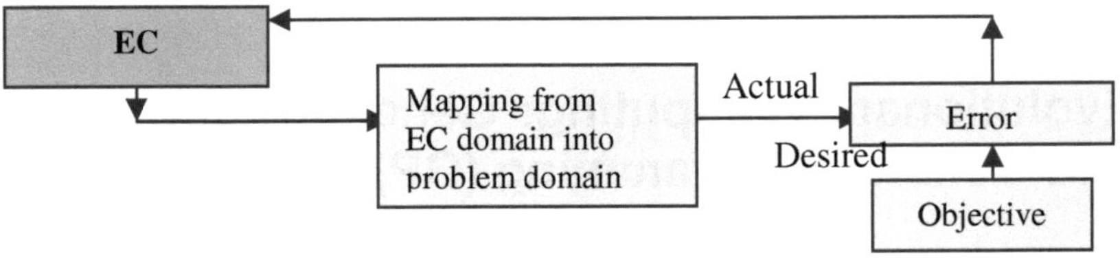

Fig. E.1. Block diagram that illustrates the mechanism of solving a problem using Evolutionary Computing (EC).

The main operations that are used in EC are copying, mutation (or copying error), and crossover. By copying we mean a reproduction of an exact copy of the individual. Mutation means a reproduction of an erroneous copy of the individual. Crossover means the combination of genes from two parents to produce offsprings. Figure E.2 demonstrates a general flow diagram of an EC, where Run is the current run number, N is the maximum number of runs, Gen. is the current generation number, M is the population size, i is the current individual in the population, P_r is the probability of reproduction, P_c is the probability of crossover, P_m is the probability of mutation, and $P_r + P_c + P_m = 1.0$.

In Fig. E.2, the result of looping over Gen. is best-of-run individual, the result of looping over Run is best-of-all individual, and the result of looping over i is the best-of-generation individual. Iterations in Fig. E.2 continues until optimal solution is obtained. EC algorithms are try-and-check (try-and-error) probabilistic search algorithms (i.e., depends on the reduction of error in the search process to produce a solution), and the EC program may have to perform so many iterations (as in Fig. E.2) to produce the desired solution to a problem. Thus, and although EC methods produce in many occasions new solutions that humans never made before, it is in general highly advisable to consider EC as one final option for problem solving (i.e., when other methods fail to solve the problem), since EC acts like a "black box" that produces solutions without showing methodology (i.e., EC does not provide a detailed step-by-step algorithm (analytical or procedural) to solve a problem and it only shows the final solution).

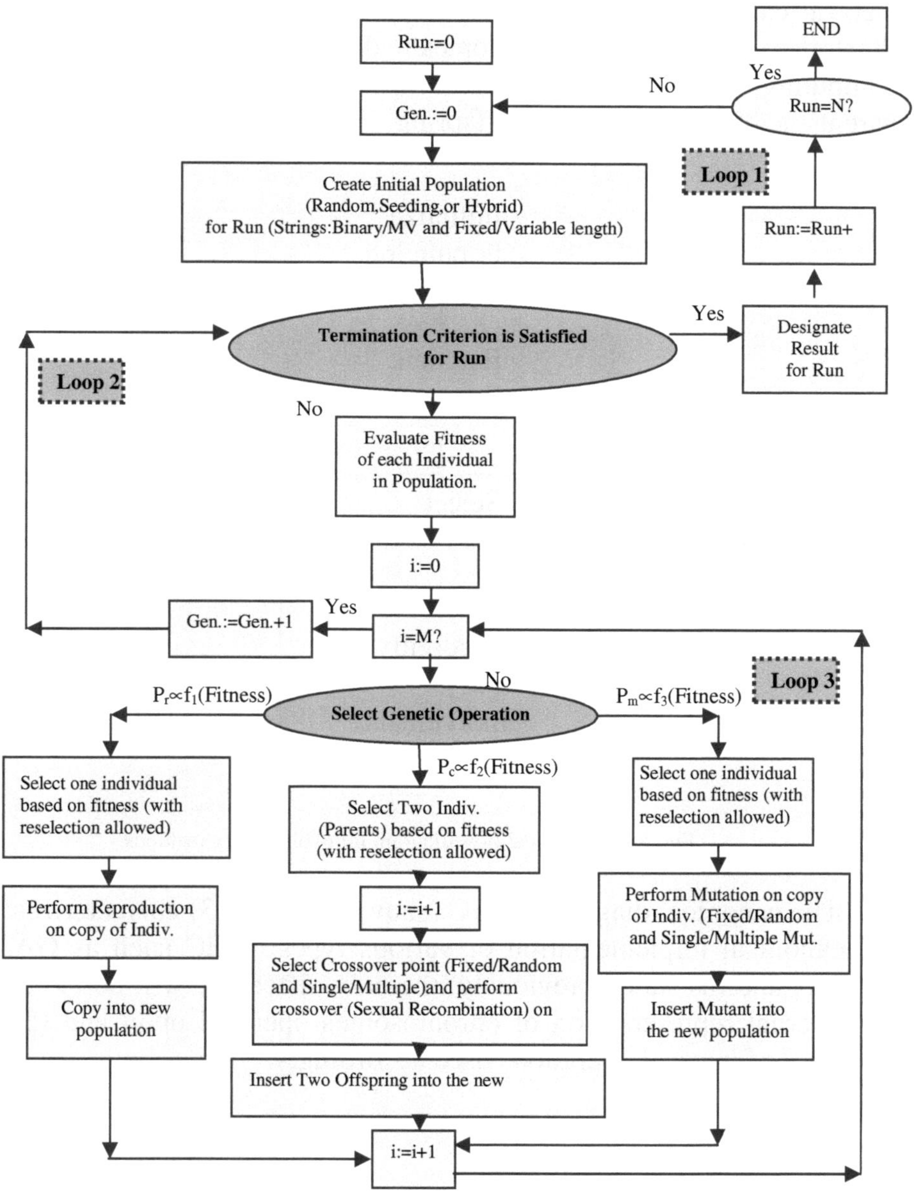

Fig. E.2. Flowgraph of a general GA and GP.

The evolutionary algorithm from Fig. E.2 has many variants. Yet, a canonical form in all these variants exist. Figure E.3 illustrates one possible canonical diagram for evolutionary computing, where select survivors means the selection of (1) parents, and (2) generation of offspring.

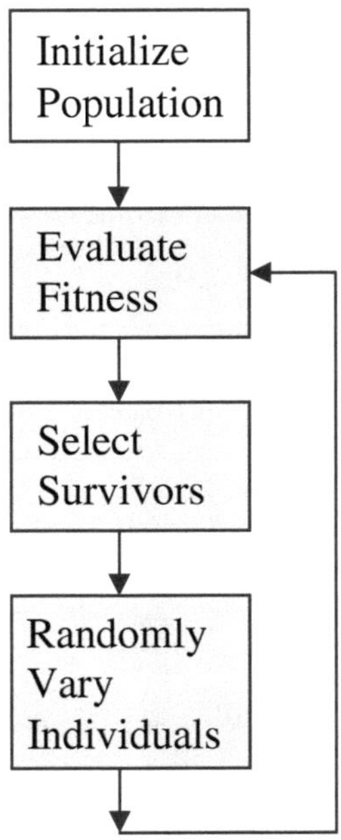

Fig. E.3. Canonical flow diagram for evolutionary methods.

The canonical diagram for EC (shown in Fig. E.3) characterizes the canonical implementation of various types of EC such as GA and GP, and (as stated previously) the only difference will be in (1) the internal representation of chromosomes operated upon and (2) the types of internal operations used accordingly.

Appendix F

Count for the New Multiple-Valued Reversible Shannon and Davio Decompositions

This Appendix provides the count for the new families of reversible Shannon and Davio spectral transforms from Chapt. 5. These counts can be used as a heuristic for efficient search for specific families of multiple-valued reversible Shannon and Davio expansions.

F.1 Counts for the new families of reversible spectral transforms

The following theorems count the number of the new reversible Shannon and Davio spectral transforms over GF(n) where $n = p^k$, where p is a prime number and k is a natural number of value $k \geq 1$.

Theorem F.1. There exists n! reversible fundamental Shannon expansions over GF(n).

Proof. To obtain reversible Shannon expansions we should have reversible basis functions matrices (as was shown in Chapt. 5). This implies that the total count of correct permutations of the rows in the basis functions matrix that satisfy the Cyclic Group Property for GF(n) is equal to:

$$\sum_{k=0}^{n-1} (n-1)! \;=\; n(n-1)! \;=\; n!. \qquad \textbf{Q.E.D.}$$

Theorem F.2. For each type of reversible invariant multi-valued Davio expansion D_n there exists n reversible invariant multi-valued Davio expansions of that type (D_n). There exists n^2 total reversible fundamental Davio expansions of all types per reversible fundamental Shannon expansion, and total of $n^2n!$ for all possible reversible fundamental Shannon expansions.

Proof. Since each row of the reversible fundamental Shannon expansion over GF(n) corresponds to n possible fundamental Davio expansions, then the total reversible fundamental Davio expansions for the whole rows in the basis functions matrix per reversible Shannon expansion is equal to $n \cdot n = n^2$. Since there exists n! total reversible fundamental Shannon expansions, so the total number of reversible fundamental Davio expansions is equal to $n^2n!$. **Q.E.D.**

For the generation of the total number of counts for all reversible invariant multi-valued Shannon and Davio expansions one just needs to refer to the results presented previously.

Example F.1. The following are counts of the corresponding reversible fundamental Shannon and Davio expansions:

GF(2): there exists 2 possible reversible Shannon expansions and 8 possible reversible Davio expansions.

GF(3): there exists 6 possible reversible Shannon expansions and 54 possible reversible Davio expansions.

GF(4): there exists 24 possible reversible Shannon expansions and 384 possible reversible Davio expansions.

Table F.1 provides counts for the new reversible families of fundamental multi-valued Shannon and Davio spectral transforms. The large numbers of the new reversible multi-valued spectral transforms that are shown in Table F.1 imply that one can have a very large space of total reversible spectral transforms to choose from for any applications that involve spectral methods (like constructing 3-D regular structures as was shown in Chapt. 4).

Table F.1. Counts of the new reversible multi-valued GF-based classes of decompositions.

GF(n)	Reversible Fundamental Multi-Valued Shannon Expansions	Total Reversible Fundamental Multi-Valued Davio Expansions
2	2	8
3	6	54
4	24	384
5	120	3,000
7	5,040	251,460

This can be good on one hand in the sense that one can have wide variety of reversible spectral transforms to choose from to meet certain optimization criteria such as reduction of area, delay, and power, and improving testability, but can be challenging on the other hand in that one needs “smart” search heuristics and strategies to search such large space for the optimal spectral transforms that meet certain optimization criteria.

Appendix G

NPN Classification of Boolean Functions and Complexity Measures

This Appendix provides the NPN-classifification of Boolean functions and the complexity measures that were used in Chapt. 7 and Appendix H.

G.1 NPN-Classification of Logic Functions

There exist many classification methods to cluster logic functions into families of functions [118,164]. Two important operations that produce equivalence classes of logic functions are *negation* and *permutation*. Accordingly, the following classification types result:
(1) P-Equivalence class: a family of identical functions obtained by the operation of permutation of variables.
(2) NP-Equivalence class: a family of identical functions obtained by the operations of negation or permutation of one or more variables.
(3) NPN-Equivalence class: a family of identical functions obtained by the operations of negation or permutation of one or more variables, and also negation of function.

NPN-Equivalence classification is used in this Book. Table G.1 lists 3-variable Boolean functions, for the non-degenerate classes (i.e., the classes depending on all three variables).

Example G.1. The following steps produce the sets of all possible Boolean functions that are included in class #1 in Table G.1 for the representative function: $F = x_1x_2 + x_2x_3 + x_1x_3$.

Table G.1. NPN equivalence classes for non-degenerate Boolean functions of three binary variables.

NPN Class	Representative Function	Number of Functios
1	$F = x_1x_2 + x_2x_3 + x_1x_3$	8
2	$F = x_1 \oplus x_2 \oplus x_3$	2
3	$F = x_1 + x_2 + x_3$	16
4	$F = x_1(x_2 + x_3)$	48
5	$F = x_1x_2x_3 + x_1'x_2'x_3'$	8
6	$F = x_1'x_2x_3 + x_1x_2' + x_1x_3'$	24
7	$F = x_1(x_2x_3 + x_2'x_3')$	24
8	$F = x_1x_2 + x_2x_3 + x_1'x_3$	24
9	$F = x_1'x_2x_3 + x_1x_2'x_3 + x_1x_2x_3'$	16
10	$F = x_1x_2'x_3' + x_2x_3$	48

(1) Negation of variables (N): $\{F_1 = x_1'x_2 + x_2x_3 + x_1'x_3,\ F_2 = x_1x_2' + x_2'x_3 + x_1x_3,\ F_3 = x_1x_2 + x_2x_3' + x_1x_3',\ F_4 = x_1'x_2' + x_2'x_3 + x_1'x_3,\ F_5 = x_1'x_2' + x_2x_3' + x_1'x_3',\ F_6 = x_1x_2' + x_2'x_3' + x_1x_3',\ F_7 = x_1'x_2' + x_2'x_3' + x_1'x_3'\}$.
(2) Permutation of variables (P): $\{F_8 = x_1x_2 + x_2x_3 + x_1x_3\}$.
(3) Negation of functions (N): $\{F_9 = x_1'x_2' + x_2'x_3' + x_1'x_3'\}$.
Thus the union of the three types of sets from steps 1-3 (which produces a set with irredundant functions as its elements) is $\{F_1 = x_1'x_2 + x_2x_3 + x_1'x_3,\ F_2 = x_1x_2' + x_2'x_3 + x_1x_3,\ F_3 = x_1x_2 + x_2x_3' + x_1x_3',\ F_4 = x_1'x_2' + x_2'x_3 + x_1'x_3,\ F_5 = x_1'x_2' + x_2x_3' + x_1'x_3',\ F_6 = x_1x_2' + x_2'x_3' + x_1x_3',\ F_7 = x_1'x_2' + x_2'x_3' + x_1'x_3',\ F_8 = x_1x_2 + x_2x_3 + x_1x_3\}$, which encompasses a total of eight functions.

G.2 Complexity Measures

Decomposability means complexity reduction. Many complexity measures exist for the purpose of evaluating the efficiency of the decomposition of complex systems into simpler sub-systems. Such complexity measures include: (1) the Cardinality complexity measure (DFC), (2) the Log-Functionality (LF) complexity measure [108], and (3) the Sigma complexity measure [271]. In the first two measures, complexity is a *count* of the total number of possible functions realizable by all of the sub-blocks; the third just indicates the level of decomposition in the lattice of possible structures. The complexity of the decomposed structures is always less or equal to the complexity of the original Look-Up-Table (LUT) that represents the mapping of the non-decomposed structure. That is, if a decomposed structure has higher complexity than the original structure, then the original structure is said to be non-decomposable. Although the DFC measure is easier and more familiar, LF is a better measure because it more properly deals with non-disjoint systems [108]. Consequently, the LF measure is used in Chapt. 7 and Appendix H. The DFC and LF complexity measures are illustrated using Fig. G.1, which exemplifies an example of Ashenhurst-Curtis (AC) decomposition, as follows:

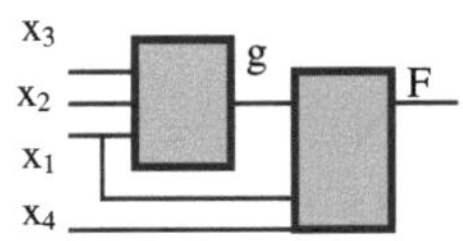

Fig. G.1. Generic non-disjoint decomposition.

In Fig. G.1, for the first block, the total number of possible functions for three 2-valued input variables is $2^{2^3} = 256$. Also, for the second block, the total number of possible functions is similarly 256. The total possible number of functions for the whole structure is equal to 256·256 = 65,536. The DFC measure is defined as:

$$DFC = O \cdot 2^I, \tag{G.1}$$

$$C_{DFC} = \sum_n DFC_n, \tag{G.2}$$

where O is the number of outputs to a block, I is the number of inputs to the same block, Eq. (G.1) is the complexity for every block, and Eq. (G.2) is the complexity for the total decomposed structure. For instance, the DFC for Fig. G.1 is: $C_{DFC} = 1 \cdot 2^3 + 1 \cdot 2^3 = \log_2(65{,}536) = 16$. It was shown in [108] that, for Fig. G.1, the Log-Functionality complexity measure (C_{LF}) for Boolean functions can be expressed as follows:

$$C_{LF} = \log_2(C_F), \tag{G.3}$$

where:

$$C_F = (C'_F)^{p_{X_3}}, C'_F = \sum_{i=0}^{p_{Y1}-1} P(p_{Y2}^{\frac{p_{X2}}{p_{X3}}}, p_{Y1}-i)S(\frac{p_{X1}}{p_{X3}}, p_{Y1}-i),$$

$$P(n,k) = \frac{n!}{(n-k)!}, S(n,k) = \frac{1}{k!}\sum_{i=0}^{k}(-1)^i \binom{k}{i}(k-i)^n, \binom{k}{i} = \frac{k!}{i!(k-i)!},$$

$$X_1 = \{x_1, x_2, x_3\}, X_2 = \{x_1, x_4\}, X_3 = X_1 \cap X_2 = \{x_1\},$$

$$p_{X1} = \prod_{x_i \in X_1} |x_i|, p_{X2} = \prod_{x_i \in X_2} |x_i|, p_{X3} = \prod_{x_i \in X_3} |x_i|,$$

$$p_{Y1} = \prod_{y_i \in Y1} |y_i|, \; p_{Y2} = \prod_{y_i \in Y2} |y_i|,$$

where X_1 is the set of input variables to the first block, X_2 is the set of input variables to the second block, X_3 is the set of overlapping variables between sets X_1 and X_2, P_{Xi} is the product of cardinalities of the input variables in set X_i, and P_{Yi} is the product of cardinalities of output variables in set Y_i. For example, the LF for Fig. G.1 is:

$$X_1 = \{x_1, x_2, x_3\}, X_2 = \{x_1, x_4\}, X_3 = X_1 \cap X_2 = \{x_1\}.$$

$$\therefore p_{X1} = 2 \cdot 2 \cdot 2 = 8, p_{X2} = 2 \cdot 2 = 4, p_{X3} = 2,$$

$$p_{Y1} = 2, p_{Y2} = 2,$$

$$C'_F = \sum_{i=0}^{1} P(2^2, 2-i)S(4, 2-i) = 88,$$

$$\therefore C_F = 7{,}744 \Rightarrow C_{LF} = \log_2(7{,}744) = 12.92.$$

Figure G.1 shows a four input function, where the variable sets for the first and second blocks are not disjoint. Note that the variable sets for the two blocks with outputs g and F are necessarily disjoint, because if the two blocks shared one input variable, F would have three inputs and the decomposed structure would be more complex than the original non-decomposed 3-input function.

Example G.2.

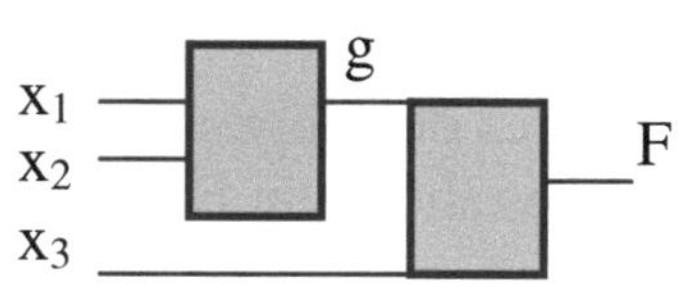

Fig. G.2. A decomposed structure.

The Log-Functionality complexity measure of the structure in Fig. G.2 is obtained as follows: Each sub-block in Fig. G.2 has a total of $2^{2^2} = 16$ possible Boolean functions. Figure G.3 illustrates all of the possible 16 two-variable Boolean functions per sub-block in Fig. G.2.

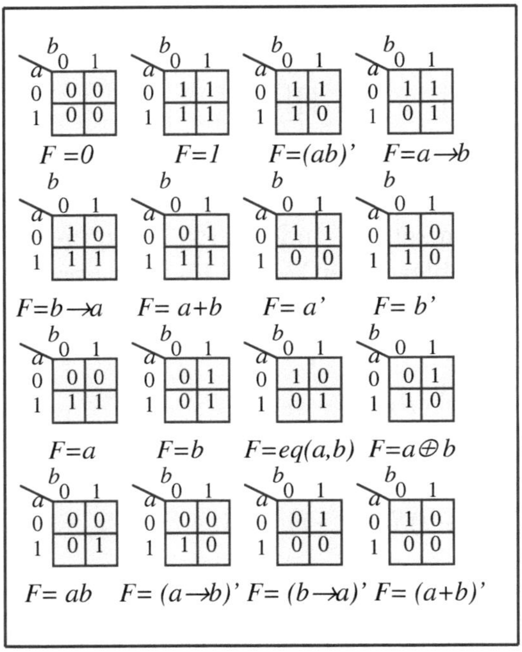

Fig. G.3. Maps of all 16 possible Boolean functions of two variables.

By allowing g and F in Fig. G.2 to take on all possible maps from Fig. G.3, one obtains the following count of total non-repeated (irredundant) 3-variable functions, as follows: $C_F = 88 \Rightarrow C_{LF} = 6.5$. This answer agrees with the result which is obtained previously [108].

Example G.3. RA produces decompositions for 3-variable functions that resemble the structures shown in Fig. G.4.

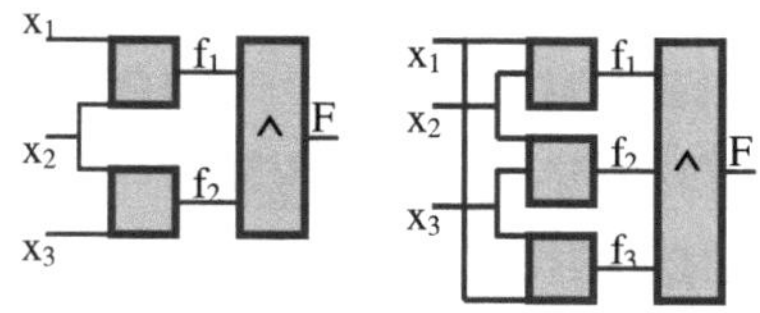

Fig. G.4. Some RA decomposed structures.

The Log-Functionality complexity measure for the structures in Fig. G.4, is obtained as shown in Fig. G.5. Figure G.5 represents a tree that generates all possible functions for the structures in Figs. G.4a and G.4b, respectively. (Superscripts of functions denote the specific edge between two nodes in the tree).

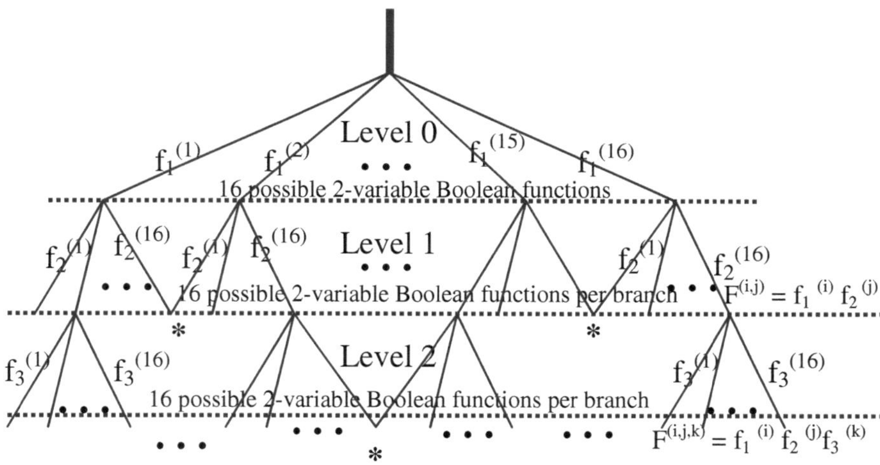

Fig. G.5. All possible combinations of sub-functions $f_1^{(i)}$, $f_2^{(j)}$, and $f_3^{(k)}$ in Figs. G.4a and G.4b, respectively. Log-functionality complexity measure represents the count of all possible irredundant functions, that is all different sub-functions and $F^{(i,j)}$ within Fig. G.4a, and all different sub-functions and $F^{(i,j,k)}$ within Fig. G.4b, where two nodes of the tree are superposed (*), they are counted only once. At level 2, 100 of the $(16)^2$ possible nodes are irredundant, and at level 3, 152 out of $(16)^3$ are irredundant.

Utilizing this methodology of removing redundant functions, one obtains the following results for Log-Functionality: for Fig. G.4a, the total number of irredundant sub-functions is $C_F = 100 \Rightarrow \therefore C_{LF} = \log_2(100) = 6.6$, and for Fig. G.4b, the total number of irredundant sub-functions is $C_F = 152 \Rightarrow \therefore C_{LF} = \log_2(152) = 7.2$.

Appendix H

Initial Evaluation of the New Modified Reconstructability Analysis and Ashenhurst-based Decompositions: Ashenhurst, Curtis, and Bi-Decomposition

This Appendix provides the necessary background for Ashenhurst-Curtis (AC) decomposition and Bi-decomposition (BD). This Appendix also introduces the comparisons between the Modified Reconstructability Analysis (MRA) (Chapt. 7), Ashenhurst-Curtis (AC) decomposition, and Bi-decomposition for the decomposition of 3-variable NPN-classified Boolean functions from Table G.1. Although the evaluation results provided in this Appendix are only for all NPN-classified Boolean functions of three variables, and thus the results are a big simplification if compared to real life problems which have hundreds of inputs and hundreds of outputs, the results serve as an important initial insight into the comparative various complexities of MRA and AC-like decompositions.

H.1 Binary Ashenhurst-Curtis Decomposition

Ashenhurst-Curtis (AC) decomposition [32,33,34,59,60,61] is one of the major techniques for the decomposition of functions that is commonly used in the field of logic synthesis. Other decompositions exist, such as the Bi-decomposition, composition (or Bottom-Up decomposition), and others [175]. The main idea of AC decomposition is to decompose logic functions into simpler logic blocks using the compression of the number of cofactors in the corresponding representation. This compression is achieved through exploiting the logical compatibility (i.e., redundancy) of cofactors (i.e., column multiplicity). As a result of AC decomposition (i.e., as a result of column compression), intermediate constructs (latent variables) are created, and learning is achieved as a result of these variables [96,108]. A general algorithm of the AC decomposition utilizing map representation, for instance, is as follows:

(1) Select the type of the variant of AC decomposition required: Ashenhurst, Curtis, generalized Curtis, etc.

(2) Specify the optimization criterion (i.e., termination condition) to be optimized by the process of decomposition. Such optimization criterion can be: area, speed, power, or testability.

(3) Remove the vacuous (redundant) variables.

(4) Decompose (partition) the input set of variables into free set and bound set.

(5) Label columns (cofactors) of the map.

(6) Decompose the bound set and create a new map for the decomposed bound set. Utilizing minimum graph coloring, maximum clique, or any other algorithm, combine similar cofactors into single cofactor. Each cell in the new map represents a labeled column in the original map.

(7) Encode the columns (cofactors) in the cells of the new map. This stage will produce the input decomposed sub-block. These intermediate variables are shown as g and h in Example H.1 (Fig. H.1).

(8) Using the intersection of the cell values of the compatible columns produce the output decomposed sub-block (Express the intermediate variables as functions of the bound set variables).

(9) Produce the total decomposed structure.

(10) Iterate until decomposed blocks, within the optimization criterion, are obtained.

In general, steps (4) and (7) determine the optimality of the AC decomposition (i.e., whether the resulting decomposed blocks are of minimal complexity or not).

Example H.1. For the following logic function: $F = x_2x_3 + x_1x_3 + x_1x_2$, let the sub-set of variables $\{x_2, x_3\}$ be the Bound Set, and the sub-set of variables $\{x_1\}$ be the Free Set. The following is the disjoint AC decomposition of F (where $\{-\}$ means don't care).

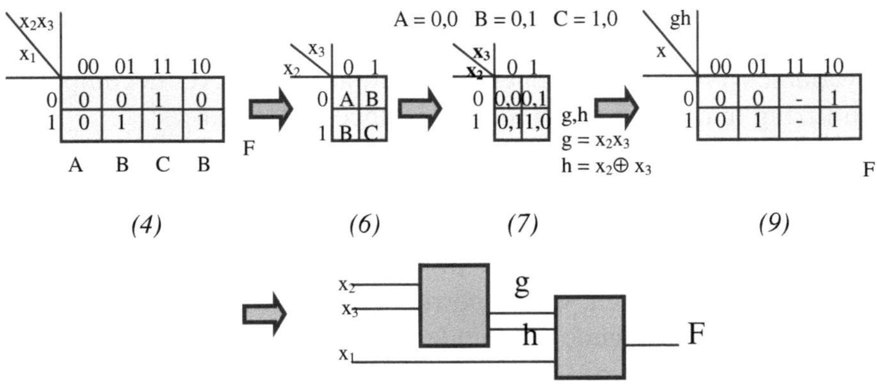

Fig. H.1. AC decomposition. Steps (4), (6), (7), and (9) are discussed in the text.

In Example H.1, the first block of the decomposed structure has two outputs (i.e., intermediate variables g and h). The DFC measure of the decomposed structure is equal to $2{\cdot}2^2 + 1{\cdot}2^3 = 16$, while the DFC of the original LUT is $= 1{\cdot}2^3 = 8$.

The calculation of the DFC complexity measure in Example H.1 shows the inadequacy of DFC as a measure of complexity because the decomposition produces a more complex structure than the non-decomposed LUT (which should be considered as the most complex data model).

By contrast, the log-functionality (LF) complexity measure for the decomposed structure in Fig. H.1 is 8, which does not exceed the complexity of the LUT (which is equal to 8). Therefore, for the AC decomposition of Boolean functions with three variables, if the first block of the decomposed structure has two outputs, then the decomposed structure is at least as complex as the LUT, and consequently, the decomposition is rejected. For other NPN functions, the AC decomposition produces only one output in the first block, and the decomposed structure is less complex than the LUT (which should be considered as the most complex model since it involves the original non-decomposed data), and thus these decompositions are not rejected.

H.2 Multiple-Valued AC decomposition

This Sect. provides the AC decomposition for multiple-valued logic functions. Since every function which is non-decomposable using disjoint decomposition can be decomposed by the process of repetition (sharing) of variables in bound and free sets (i.e., non-disjoint decomposition), AC decomposition, in general, can be found in both its disjoint and non-disjoint variants. In general, the process of repeating a single n-valued variable will produce $\frac{(n-1)}{n}$ don't cares of the total number of cells of the n-valued map of the corresponding n-valued function with repeated variables. As a result of AC decomposition, intermediate constructs are created. The new variables (constructs) correspond to new induced concepts, and therefore learning is achieved.

The following example illustrates the use of multi-valued AC decomposition.

Example H.2. Let us produce the AC decomposition for the following logic function:

x_1 \ x_2x_3	00	01	02	10	11	12	20	21	22
0	0	0	1	1	1	0	0	0	1
1	0	2	2	0	0	2	1	0	2
2	0	2	0	1	1	2	0	0	0

F

For the sub-set of variables $\{x_2, x_3\}$ to be the Bound Set, and the sub-set of variable $\{x_1\}$ to be the Free Set, by applying the previously mentioned AC procedure for the AC decomposition, one obtains the following disjoint AC decomposition of the function F as shown in Fig. H.2.

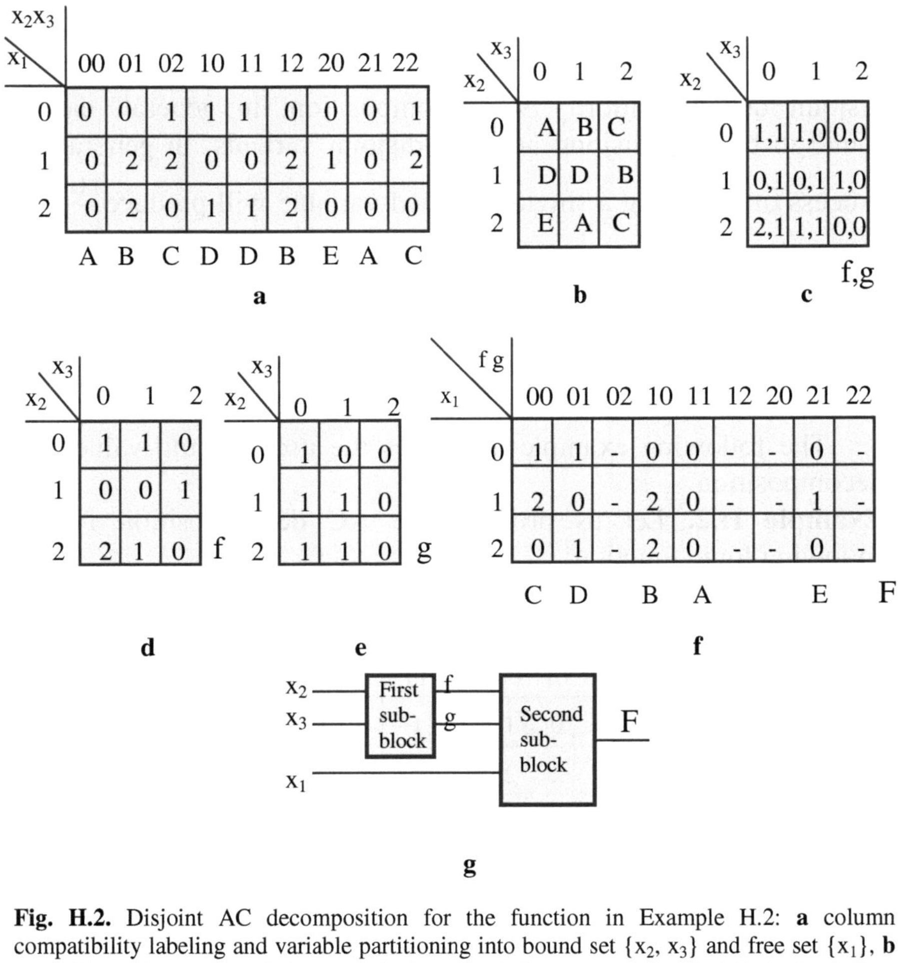

Fig. H.2. Disjoint AC decomposition for the function in Example H.2: **a** column compatibility labeling and variable partitioning into bound set $\{x_2, x_3\}$ and free set $\{x_1\}$, **b** map of bound set, **c** encoded map of the bound set, **d** output function f of first sub-block, **e** output function g of the first sub-block, **f** map of the second decomposed sub-block, and **g** the total resulting structure for AC disjoint decomposition.

The same ternary function F can be equivalently decomposed by using the method of the repetition of variables to produce the corresponding non-disjoint AC decomposition, as follows:

x_1x_2 \ x_2x_3	00	01	02	10	11	12	20	21	22
00	0	0	1	-	-	-	-	-	-
01	-	-	-	1	1	0	-	-	-
02	-	-	-	-	-	-	0	0	1
10	0	2	2	-	-	-	-	-	-
11	-	-	-	0	0	2	-	-	-
12	-	-	-	-	-	-	1	0	2
20	0	2	0	-	-	-	-	-	-
21	-	-	-	1	1	2	-	-	-
22	-	-	-	-	-	-	0	0	0
	A	B	C	A	A	B	B	A	C

a

x_2 \ x_3	0	1	2
0	A	B	C
1	A	A	B
2	B	A	C

b

x_2 \ x_3	0	1	2
0	0	1	2
1	0	0	1
2	1	0	2

g

c

g \ x_1x_2	00	01	02	10	11	12	20	21	22	
0	0	1	0	0	0	0	0	1	0	A
1	0	0	0	2	2	1	2	2	0	B
2	1	-	1	2	-	2	0	-	0	C

d F

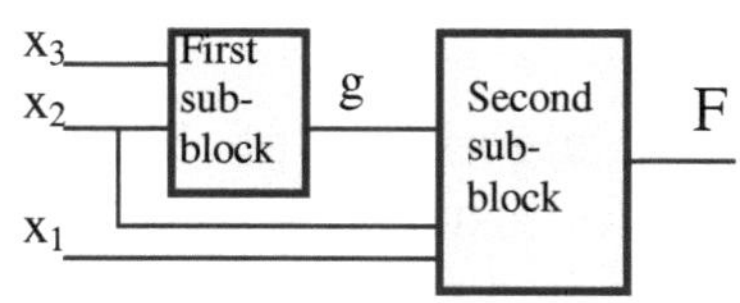

e

Fig. H.3. Non-disjoint AC decomposition for the function in Example H.2: **a** variable repetition, column compatibility labeling, and variable partitioning into bound set $\{x_2, x_3\}$ and free set $\{x_1, x_2\}$, **b** map of bound set, **c** encoded map of the bound set and the output function of the first decomposed sub-block, **d** map of the second decomposed sub-block, and **e** the total resulting structure for AC non-disjoint decomposition.

The decomposition obtained in Fig. H.2 is not the classical Curtis decomposition; it is the generalized Curtis decomposition [96]. This is due to the fact that in classical Curtis decomposition the number of outputs from the predecessor decomposed block must be less than the number of inputs to that block (this includes the case of having only one output, which is the Ashenhurst decomposition as a special case of Curtis decomposition), but in Fig. H.2 the number of outputs of the first sub-block is equal to the number of inputs to the same sub-block (i.e., number of outputs = number of inputs = 2).

H.3 Bi-decomposition

This Sect. presents Bi-decomposition (BD) algorithm that is used later in this Sect. to be compared to MRA decomposition. Bi-decomposition is another type of decomposition, which is widely used in logic synthesis of Boolean functions. Let x $=\{x_1, x_2\}$ be a partition of the variables representing a function f(x) such that $x_1 \cap x_2 = 0$, and x_1 is a set of m variables and x_2 is a set of (n-m) variables. The function, $F(x_1,x_2)=h(g_1(x_1), g_2(x_2))$, is said to be disjoint bi-decomposition if f(x) can be realized as shown in Fig. H.4, where h is a two input gate (e.g., AND, OR, XOR gates) and g_1 and g_2 are either AND, OR, or EXOR gates.

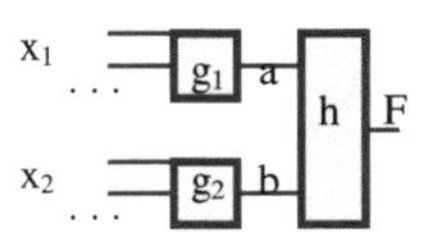

Fig. H.4. Disjoint Bi-decomposition.

A general algorithm of bi-decomposition utilizing K-map representation, for instance, is as follows:

(1) Partition the input set of variables in the K-map into two sets: set1 (bound set) and set2 (free set), and label all the different columns and rows.

(2) Decompose set1 and create a new K-map for the decomposed set (utilizing minimum graph coloring, maximum clique, or some other algorithm to combine similar columns into a single column). Each cell in the new K-map represents a labeled column in the original K-map.

(3) Encode the labels in the cells of the new K-map using minimum number of intermediate binary variables. Express the intermediate variables as functions of the set of variables.

(4) Decompose set2 and create a new K-map for the decomposed set (utilizing minimum graph coloring, maximum clique, or some other algorithm to combine similar columns into a single column). Each

cell in the new K-map represents a labeled row in the original K-map. This step is equivalent to making set2 as bound set and set1 as free set.
(5) Encode the labels in the cells of the new K-map using minimum number of intermediate binary variables. Express the intermediate variables as functions of the set of variables.
(6) Produce the decomposed structure, i.e., a K-map specifying the function (F) in terms of the intermediate variables from steps 3 and 5, respectively.

The following example illustrates the use of BD for decomposition of logic functions. Although the following example is performed for the case of disjoint BD, non-disjoint BD could be also obtained in a manner similar to the method used in Fig. H.3 in Example H.2.

Example H.3. For the logic function F in Fig. H.5 which is represented using the K-map, let the sub-set of variables $\{x_1, x_2\}$ be set2, and the sub-set of variables $\{x_3, x_4\}$ be set1. Figure H.5 is the disjoint Bi-decomposition of F (where {–} means don't care). Note that the decomposed structure in Fig. H.5 has the following logic blocks as a result of BD: g_1 is an AND gate: $g_1 = x_3x_4$, g2 is an AND gate: $g_2 = x_1'x_2$, and h is an AND gate: $h = g_1'g_2$. In general, for a disjoint BD circuit, h can be a two input gate (e.g., AND, OR, XOR gates) and g_1 and g_2 can consist of AND, OR, or EXOR gates.

H.4 Complexity of the Two-Valued Modified Reconstructability Analysis Versus Ashenhurst-Curtis Decomposition and Bi-Decomposition

Utilizing the new methods of 1-MRA and 0-MRA decomposition that were described in Sect. 7.1, one obtains the following results in Table H.1 for the decomposition of 3-variable NPN-classified Boolean functions (from Table G.1) using the different decompositions of the new Modified Reconstructability Analysis (MRA), Ashenhurst-Curtis (AC), and Bi-decomposition (BD), respectively.

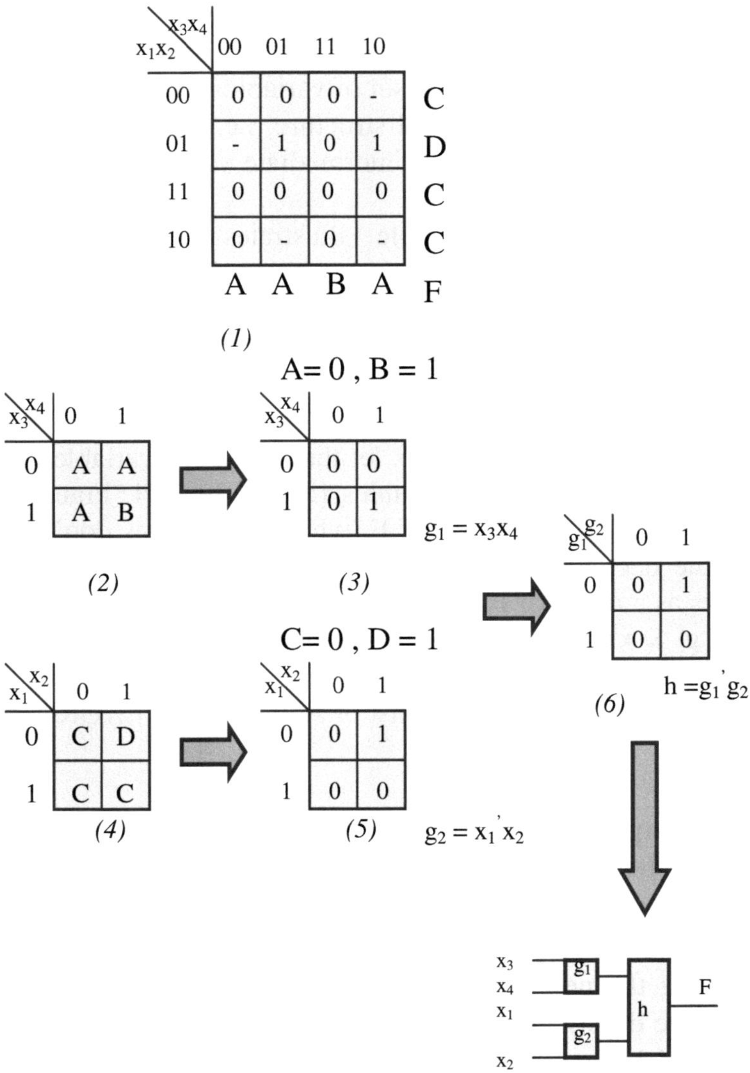

Fig. H.5. Bi-decomposition. Steps (1) - (6) are discussed in the text.

Table H.1. AC and BD decompositions versus MRA for the decomposition of all NPN-classes of 3-variable Boolean functions (See Table G.1 in Appendix G). (Compare the right-most two columns.)

NPN Representative Function	Simplest Modified RA model (0-MRA or 1-MRA)	Simplest AC/BD circuit	Simplest Modified RA circuit	C_{data} (LUT)	C_{LF} (MRA)	C_{LF} (AC)
Class 1 (8) $F = x_1x_2 + x_2x_3 + x_1x_3$	[illegible]	$g = x_2x_3$, $h = x_2 \oplus x_3$, $F = g + x_1h$		8	7.2	8
Class 2 (2) $F = x_1 \oplus x_2 \oplus x_3$	non-decomposable	$g = x_2 \oplus x_3$, $F = x_1 \oplus g$	-	8	8	6.5
Class 3 (16) $F = x_1 + x_2 + x_3$	x_1 f_1 x_2 f_2 x_3 f_3 / 0 0 0 0 0 0 / 1 1 1 1 1 1	$g = x_2 + x_3$, $F = x_1 + g$		8	4.3	6.5
Class 4 (48) $F = x_1(x_2 + x_3)$	[illegible]	$g = x_2 + x_3$, $F = x_1 g$		8	6.5	6.5
Class 5 (8) $F = x_1x_2x_3 + x_1'x_2'x_3'$	[illegible]	$g = x_2x_3$, $h = x_2 \oplus x_3$, $F = x_1'g'h' + x_1g$		8	6.6	8
Class 6 (24) $F = x_1'x_2x_3 + x_1x_2' + x_1x_3'$	non-decomposable	$g = x_2x_3$, $F = x_1 \oplus g$	-	8	8	6.5
Class 7 (24) $F = x_1(x_2x_3 + x_2'x_3')$	[illegible]	$g = x_2 \oplus x_3$, $F = x_1g'$		8	6.5	6.5
Class 8 (24) $F = x_1x_2 + x_2x_3 + x_1'x_3$	[illegible]	$g = x_3$, $h = x_2 \oplus x_3$, $F = x_1g + gh' + x_1'g'h$		8	6.6	8
Class 9 (16) $F = x_1'x_2x_3 + x_1x_2'x_3 + x_1x_2x_3'$	non-decomposable	$g = x_2x_3$, $h = x_2 \oplus x_3$, $F = x_1h + x_1'g$	-	8	8	8
Class 10 (48) $F = x_1x_2'x_3' + x_2x_3$	[illegible]	$g = x_2x_3$, $h = x_2 \oplus x_3$, $F = g + x_1g'h'$		8	6.6	8

Table H.1 shows that, in terms of the log-functionality complexity measure from Appendix G, in three NPN classes (4, 7, 9) MRA and AC and BD decompositions give equivalent complexity decompositions. In two remaining classes (2, 6), which encompass 26 functions, AC and BD decompositions are superior, but in five classes (1, 3, 5, 8, 10), which encompass 104 functions, MRA is superior. Figure H.6 provides a quantitative analysis of the decomposition, of the NPN-classified functions, using MRA, AC, and BD decompositions (from Table H.1), respectively.

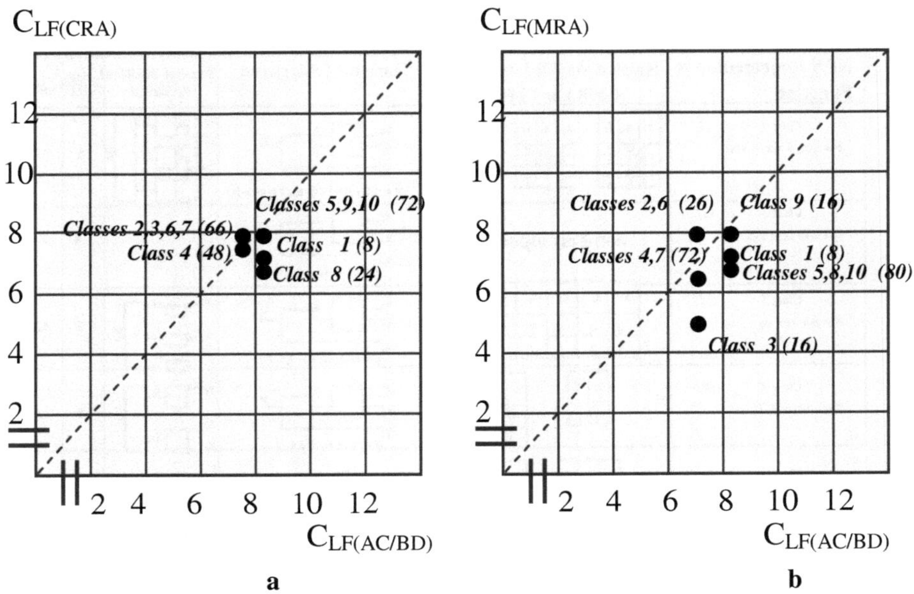

Fig. H.6. a Comparison of the Log-Functionality complexity measure between CRA versus AC/BD decompositions, and **b** MRA versus AC/BD decompositions, of 3-variable NPN-classified Boolean functions.

The analysis, in terms of complexity, of the results in Fig. H.6a is as follows:

Total number of classes that AC and BD is better than CRA: 5 (2,3,4,6,7).
Total number of functions that AC and BD is better than MRA: 114.
Total number of classes that CRA is better than AC and BD: 2 (1,8).
Total number of functions that CRA is better than AC and BD: 32.
Total number of classes for AC and BD is the same as CRA: 3 (5,9,10).
Total number of functions for AC and BD is the same as CRA: 72.

The analysis, in terms of complexity, of the results in Fig. H.6b is as follows:

Total number of classes that AC and BD is better than MRA: 2 (2,6).
Total number of functions that AC and BD is better than MRA: 26.
Total number of classes that MRA is better than AC and BD: 5 (1,3,5,8,10).
Total number of functions that MRA is better than AC and BD: 104.
Total number of classes for AC and BD is the same as MRA: 3 (4,9,7).
Total number of functions for AC and BD is the same as MRA: 88.

We can also summarize the results, from Table H.1 and Fig. H.6, by comparing the decomposability versus non-decomposability for the various approaches. Figure H.7 shows the number of functions decomposable by one method but not by another (upper right and lower left cells).

AC,BD \ MRA	ND	D
ND	1 (9: 16)	4 (1,5,8,10: 88)
D	2 (2,6: 26)	3 (3,4,7: 88)

a

CRA \ AC,BD	ND	D
ND	3 (5,9,10: 72)	4 (2,3,6,7: 66)
D	2 (1,8: 32)	1 (4: 48)

b

CRA \ MRA	ND	D
ND	3 (2,6,9:42)	4 (3,5,7,10: 96)
D	0 (0)	3 (1,4,8: 80)

c

Fig. H.7. a Comparison of the Decomposability (D) versus Non-Decomposability (ND) for AC and BD decompositions versus MRA, **b** CRA versus AC and BD decompositions, and **c** CRA versus MRA, respectively.

Utilizing the results of decomposability from Fig. H.7, one concludes that for NPN-classified 3-variable Boolean functions, MRA decomposition is superior to AC and BD decompositions (88 versus 26), AC and BD decompositions are superior to CRA decomposition(66 versus 32), and MRA decomposition is superior to CRA decomposition (96 versus 0).

While the log-functionality used in Table H.1 and Figs. H.6 and H.7, is a good cost measure for machine learning, it is not a good measure for circuit design. An alternative acceptable cost measure for circuit design will be the count of the total number of 2-input gates in the final circuit ($C_{\#}$). Table H.2 presents an initial comparison between MRA and AC using the $C_{\#}$ complexity measure.

Table H.2 shows that, using the $C_{\#}$ cost measure, in four NPN classes (1,2,6,9) which encompass 50 functions AC/BD is superior to MRA for both including and not including the cost of the inverters. For two NPN classes (4, 8), which encompass 72 functions, AC/BD is equivalent to MRA for both including and not including the cost of the inverters. For four NPN classes (3,5,7,10), which encompass 96 functions, MRA is superior to AC/BD when including the cost of the inverters. For two NPN classes (5, 10), which encompass 56 functions, MRA is superior to AC/BD when not including the cost of the inverters.

While the results in Table H.2 are technology-independent, the results obtained in Table H.2 can be viewed from technology-dependent point of view as well. This is because while the realization of certain two-input logic primitives (gates) from Fig. G.3 need less number of physical primitives (devices) in certain technology, the same gates may need more number of devices in another technology.

Table H.2. Comparison of AC versus MRA using $C_{\#}$ cost measure.

Class	$C_{\#}$ with inverters (AC/BD)	$C_{\#}$ without inverters (AC/BD)	$C_{\#}$ with inverters (MRA)	$C_{\#}$ without inverters (MRA)
1	4	4	5	5
2	2	2	-	-
3	2	2	1	1
4	2	2	2	2
5	9	6	3	3
6	2	2	-	-
7	3	2	2	2
8	4	3	4	3
9	6	5	-	-
10	7	5	3	3

Appendix I

Count for Reversible Nets

This Appendix provides the counts to charaterize the complexity of the two-valued reversible Nets from Chapt. 8.

Theorem I.1. Every positive unate symmetric funtion of n variables can be realized in $1+2+3+...+(n-1)=\frac{n(n-1)}{2}$ MIN/MAX gates.

Proof. Every positive unate (symmetric) function of 2 variables can be realized in 1 gate. Every positive unate function of 3 variables can be realized in 1+ 2 gates. Every positive unate function of four variables can be realized in 1 + 2 + 3 gates, etc. Thus, the total number of gates for n variables will be $\frac{n(n-1)}{2}$. **Q.E.D.**

Theorem I.2. Every single index totally symmetric function of n variables can be realized in $\frac{n(n-1)}{2}$ MIN/MAX gates, (n-2) fan out gates, and (n-1) Feynman gates.

Theorem I.3. Every single-output totally symmetric function of n variables can be realized in $\frac{n(n-1)}{2}$ MAX/MIN gates, (n-2) fan out gates, (n-1) Feynman gates in the second plane, and at most (n-1) Feynman gates in the third plane.

Theorem I.4. Every m-output totally symmetric function of n variables can be realized in $\frac{n(n-1)}{2}$ MAX/MIN gates, (n-2) fan out gates, (n-1) Feynman gates in the second plane, and at most (m · (n-1)) gates in the third plane.

Appendix J

New Optical Realizations for Two-Valued and Multiple-Valued Classical and Reversible Logics

Many optical devices are naturally reversible. When processing light, such devices can operate on the inputs as outputs and the outputs as inputs. One reversibility aspect of light is illustrated in Fig. J.1, which shows that the incident light (I_1) can be totally reconstructed as the transmitted light (T_2), by reversing the input light beams to be outputs and the output light beams to be inputs [261].

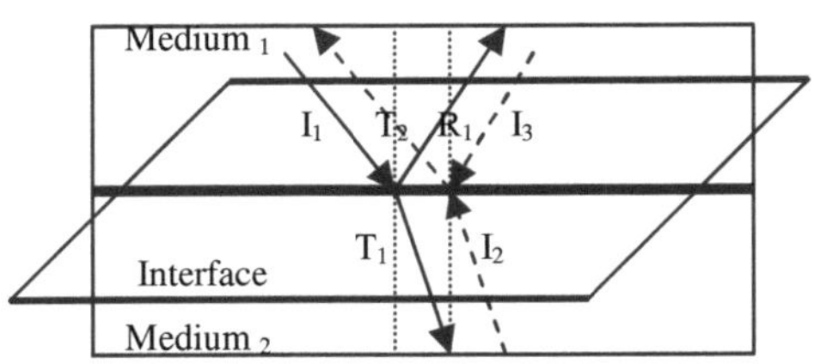

Fig. J.1. Illustration of reversibility of light: I is incident ray, R is reflected ray, and T is transmitted ray. Solid lines are forward rays and dashed lines are reverse rays. Note that $I_1 = T_2$ (Here Absorption (A) is neglected (i.e., A = 0) since in general: T + R + A = 1.0).

Optical realizations of reversible logic circuits are presented in this Appendix. The optical circuits utilize coherent light beams to perform the functionality of basic logic primitives. Three new optical devices are presented that utilize the properties of frequency, polarization, and incident angle that are associated with any light-matter interaction. The hierarchical implementation of such optical reversible primitives results in the synthesis of optical regular reversible structures. The synthesis of reversible lattice structures using such optical devices is described. The concept of optical parallel processing of an array of input laser beams using the new optical devices is also presented. This Appendix reports the synthesis of regular Boolean and multi-valued optical reversible circuits using: (1) Total internal reflection, (2) Optical polarizers, and (3) Optical frequency shifters [17,24].

In general, if the optical properties of the material (i.e., channel or medium) through which light is traveling such as material index of refraction n (that determines the phase of the traveling field) (or material gain or loss (that determines the magnitude of the traveling field), etc) are a function of the electric field then light polarization is a non-linear function of the electric field and this is called non-linear optics (and the medium is called non-linear medium). On the other hand, if the optical properties of the material through which light is traveling such as n (or gain or loss, etc) are constants (i.e., not a function of the electric field) then the light polarization is a linear function (proportional) of electric field and this is called linear optics (and the medium is called linear medium) [261].

On the other hand, if the optical properties of the material through which light is traveling such as n (or gain or loss, etc) are linear function (propotional) of the electric field (or magnetic field) then this is called linear electro-optics (or linear magneto-optics). If the optical properties of the material through which light is traveling such as n (or gain or loss, etc) are non-linear function of the electric field (or magnetic field) then this is called non-linear electro-optics [261] (or non-linear magneto-optics).

This Appendix uses linear optics, but results could also be generalizable and extendable to the non-linear optics domain.

J.1 Optical Realization of Two-Valued and Multiple-Valued Logic

This Sect. presents the implementation of two-valued and multiple-valued logics using optical circuits.

J.1.1 Two-to-one optical multiplexers

A laser beam possesses many properties that may be involved in light-matter interaction. Such properties include [261]: wavelength λ, frequency ν, speed c, polarization, phase front curvature R, spot

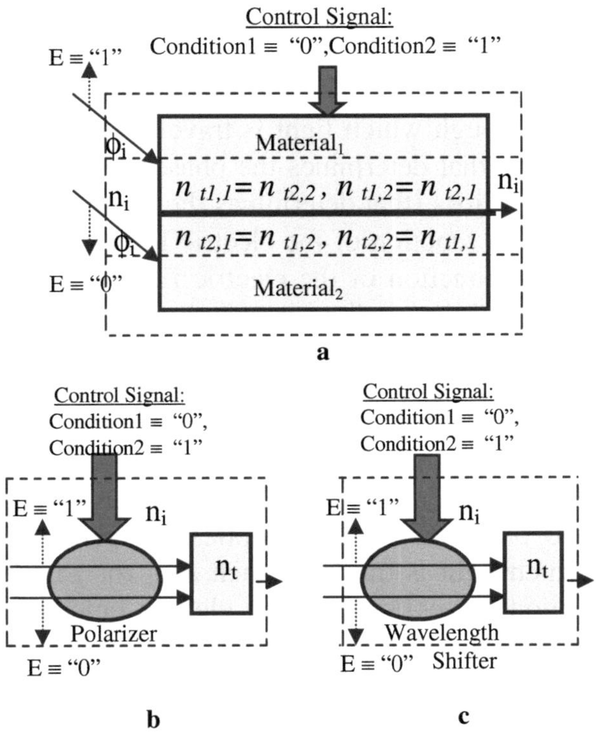

Fig. J.2. Three optical devices that implement the 2-to-1 multiplexer.

size ω_s, and incident angle θ_i. One can construct optical devices using transformations of such properties.

Figure J.2 presents three optical primitives that realize the 2-to-1 multiplexer.

In Fig. J.2, the input control signal that is used to change the electric properties of the devices can be thermal, acoustic, optical, or electrical [261]. While thermal and acoustic control signals can impose slow changes of the electric properties of the material of the device, optical and electrical control signals can impose fast changes [261]. In any of the two-to-one optical devices shown in Fig. J.2, the (orthogonal) polarization of light serves as the label or ID of that light signal. Thus, if the polarization is (↑) then the signal is of value "1", and if the polarization is (↓) then the signal is of value "0". The devices in Fig. J.2 operate as follows:

Device 1 (Fig. J.2a):

In Fig. J.2a, two laser beams with the same frequency (ω) are incident on the device with the same incident angle θ_i. The design specifications of the device are as follows:

$$\theta_i = constant, \tag{J.1}$$

$$\theta_c < \theta_i, \tag{J.2}$$

$$n_{t_{1,1}} = n_{t_{2,2}}, \tag{J.3}$$

$$n_{t_{1,2}} = n_{t_{2,1}}, \tag{J.4}$$

$$n_{t_{1,1}} = n_{t_{2,2}} > n_{t_{1,2}} = n_{t_{2,1}}, \tag{J.5}$$

$$n_{t_{1,1}} = n_{t_{2,2}} > n_i > n_{t_{1,2}} = n_{t_{2,1}}. \tag{J.6}$$

Where n_i is the incidence index of refraction, θ_i is the incidence angle, θ_c is the critical angle, $n_{t1,1}$ is the index of refraction of material 1 when condition 1 is imposed, $n_{t1,2}$ is the index of refraction of material 1 when condition 2 is imposed, $n_{t2,1}$ is the index of refraction of material 2 when condition 1 is imposed, $n_{t2,2}$ is the index of refraction of material 2 when condition 2 is imposed.

The Eqs. that govern the behavior of the laser propagation through an optical device are as follows [17,24], where n_t is the transmitting index of refraction, and θ_t is the transmitting angle.

(1) Snell's law for the angle of transmission (θ_t):

$$n_i \sin\theta_i = n_t \sin\theta_t. \tag{J.7}$$

(2) Critical angle (**θ_c**):

$$\theta_c = \sin^{-1}(\frac{n_t}{n_i}). \tag{J.8}$$

(3) Fresnel's Eqs. for the calculation of the magnitudes and polarizations of the transmitted electric field $\vec{E}_t$ and reflected electric field $\vec{E}_r$:

$$\frac{|\vec{E}_r|_\perp}{|\vec{E}_i|_\perp} = \frac{n_i\cos\theta_i - n_t\cos\theta_t}{n_i\cos\theta_i + n_t\cos\theta_t}, \tag{J.9}$$

$$\frac{|\vec{E}_t|_\perp}{|\vec{E}_i|_\perp} = \frac{2n_i \cos\theta_i}{n_i \cos\theta_i + n_t \cos\theta_t}, \tag{J.10}$$

$$\frac{|\vec{E}_r|_\parallel}{|\vec{E}_i|_\parallel} = \frac{n_t \cos\theta_i - n_i \cos\theta_t}{n_i \cos\theta_t + n_t \cos\theta_t}, \tag{J.11}$$

$$\frac{|\vec{E}_t|_\parallel}{|\vec{E}_i|_\parallel} = \frac{2n_i \cos\theta_i}{n_i \cos\theta_t + n_t \cos\theta_t}. \tag{J.12}$$

(4) The wavelength of a laser beam in a medium is:

$$\lambda_m = \frac{\lambda_0}{n_m}. \tag{J.13}$$

where λ_0 is the light wavelength in vacuum, and n_m is the index of refraction of the medium given by:

$$n_m \approx \sqrt{\frac{\varepsilon_m}{\varepsilon_0}}. \tag{J.14}$$

where ε_m is the material electric permittivity, and ε_o is the vacuum electric permittivity.

A qualitative description of the operation of device 1 in Fig. J.2a is as follows: by imposing a certain value of the control (modulation) signal, the material changes its electric permittivity ε_m. It can be observed from Eq. (J.14), that this change of the electric properties of the material imposes changes on the index of refraction. From Eq. (J.8), changing the index of refraction of the material results in changing the critical angle. This change of the critical angle of the material will make one laser beam totally reflect back and the other laser beam propagate through the material. This device implements the functionality of a 2-to-1 multiplexer ($f = b\bar{c} + ac$).

Example J.1. The following are design specifications for device 1 (Fig. J.2a): $\theta_i = 60^o$, $n_i = 1.5$, $n_{t1,1} = n_{t2,2} = 1.8$, $n_{t1,2} = n_{t2,1} = 1.2$, $\lambda_{green} = 500$ nm. Then, by using Eqs. (J.7) and (J.8), the transmitted angle (θ_t) and the critical angle (θ_c) of the laser beams are: $\theta_t =$

46.24°, and θ_c = 53.13°. Thus when condition1 is imposed on the device, laser beam1 which resembles logic value "1" is transmitted through the device with transmitted angle θ_t = 46.24°, and laser beam2 which resembles logic value "0" is totally reflected back. When condition2 is imposed on the device, then laser beam2 which resembles logic value "0" is transmitted through the device with transmitted angle θ_t = 46.24°, and laser beam1 which resembles logic value "1"is totally reflected back. Using Eqs. (J.9) - (J.12), the magnitudes and directions of the orthogonal and parallel polarizations for the transmitted and reflected laser beams are:

$$|\vec{E}_t|_\perp = 0.752\,|\vec{E}_i|_\perp, |\vec{E}_t|_\parallel = 0.774\,|\vec{E}_i|_\parallel,$$

$$|\vec{E}_r|_\perp = -0.25\,|\vec{E}_i|_\perp, |\vec{E}_r|_\parallel = -0.071\,|\vec{E}_i|_\parallel .$$

Device 2 (Fig. J.2b):

A qualitative description of the operation of device 2 is as follows: by imposing control (modulation) signal, the polarizer changes the polarization of the incident laser beam. This change of polarization of laser beams will make one beam totally absorbed into the incident material where the other beam propagates through the incident material. The polarizations of the transmitted and reflected light beams follow Eqs. (J.9)-(J.12), respectively.

Device 3 (Fig. J.2c):

A qualitative description of the operation of device 3 is as follows: by imposing certain control (modulation) signal, the frequency of the light changes [261]. The up-shift of the frequency by $\Delta\omega$ for both laser beams will make beam 2 to be totally absorbed into the material when its frequency is equal to the resonant frequency of that material (i.e., $\omega_{laser2} = \omega_0$ in Eq. (J.17)), and beam 1 to propagate through the material, and vice versa occurs when the frequency is down-shifted by $\Delta\omega$ for both laser beams according to Eq. (J.16).

$$\omega_{material} = \omega_0, \quad (J.15)$$

$$\omega_{laser1} = \omega_0 + \Delta\omega, \quad (J.16)$$

$$\omega_{laser2} = \omega_0 - \Delta\omega. \quad (J.17)$$

J.1.2 Many-to-one optical multiplexers

The idea of building many-to-one optical multiplexers is possible using the basic two-to-one optical multiplexers from Fig. J.2. For ternary signals one needs two devices to realize the functionality of three-to-one optical multiplexers. This idea is illustrated in Fig. J.3.

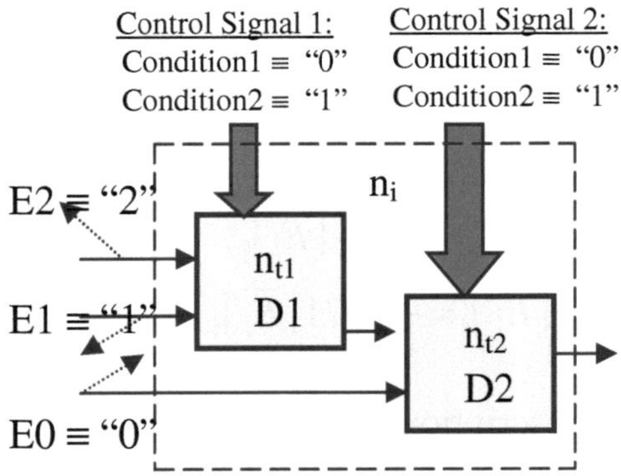

Fig. J. 3. 3-to-1 optical multiplexer.

For N-valued logic signals one needs (N-1) devices to realize the functionality of N-to-1 multiplexer. This idea is illustrated in Fig. J.4. In the 3-to-1 optical device in Fig. J.3, the rotation of light polarization serves as the label of light signal; If polarization is ( ) then this is signal of value "2", if the polarization is ( ) then this is signal of value "1", and if the polarization is ( ) then this is signal of value "0".

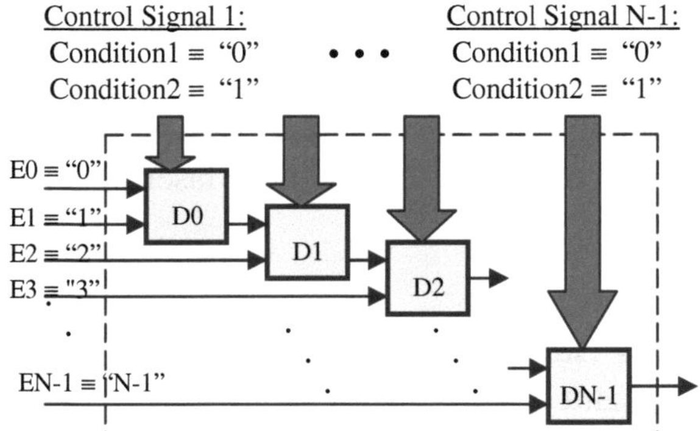

Fig. J.4. Optical realization of an N-to-1 logic multiplexer. Devices D1,..., DN can be any of the three devices in Fig. J.2.

Note that, for instance, in Fig. J.3 that device D1 outputs one signal from two input signals and device D2 outputs one signal from two input signals thus the overall functionality of the device in Fig. J.3 is a three-to-one multiplexer.

J.2 Optical Reversible Lattice Structures

Using the previously introduced three optical devices in Sect. J.1, Figs. J.5a and J.5b illustrate the optical realization of the 2-valued and 3-valued reversible lattice structures (from Chapt. 6), respectively.

For example, in Fig. J.5a, two laser beam sources generate two distinct laser beams with two distinct (orthogonal) polarizations (as was shown in Sect. J.1.1) that correspond to logic values "0" and "1". A beam splitter splits each beam to the desired number of beams. These partitioned beams are then directed by the optical switch to the desired path at the input nodes (or terminals T1-T9) of the two-valued reversible lattice structure. The resulting processed light beams are then measured using optical sensors at output nodes D1-D9.

In Fig. J.5b, the optical switch distributes the logic values of "0", "1", and "2", that result from the beam splitter, at it's output. These outputs are denoted by T1-T11. M is a 3-to-1 optical multiplexer from Fig. J.3. As was shown in Fig. J.5a, the propagation of the laser beam can be guided using a fiber-optic cable or propagate freely in the medium (i.e., channel or material), and the resulting processed light beams are then measured using optical sensors at output nodes D1-D11.

Various beam splitters can be used for the purpose of splitting the light beam into many light beams. Figure J.6 illustrates three methodologies of splitting the light beam [See [24] and references therein].

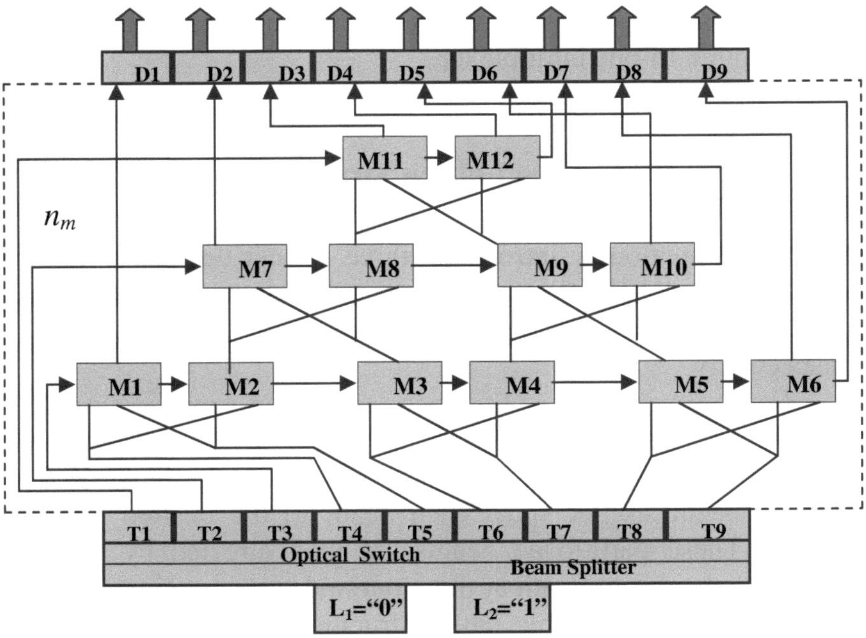

Fig. J.5a. Optical realization of two-valued reversible lattice structure.

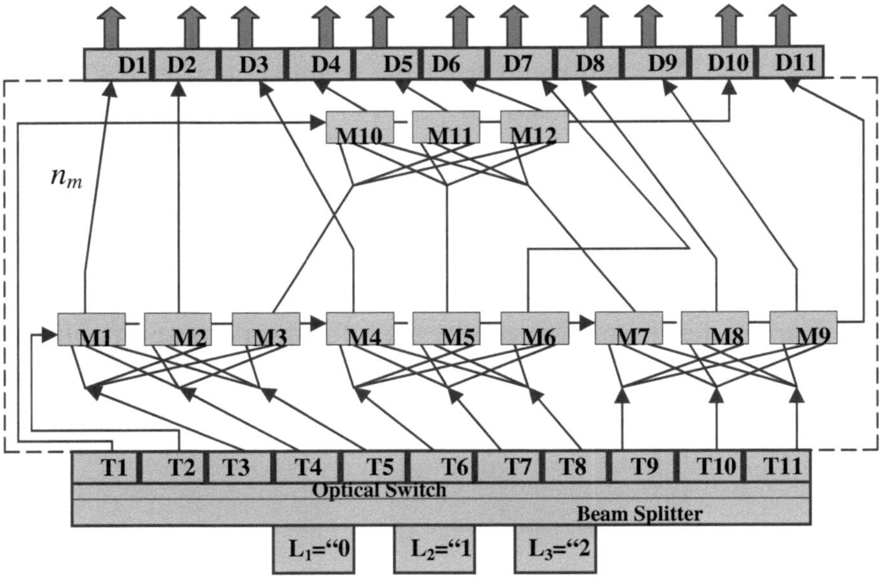

Fig. J.5b. Optical realization of three-valued reversible lattice structure.

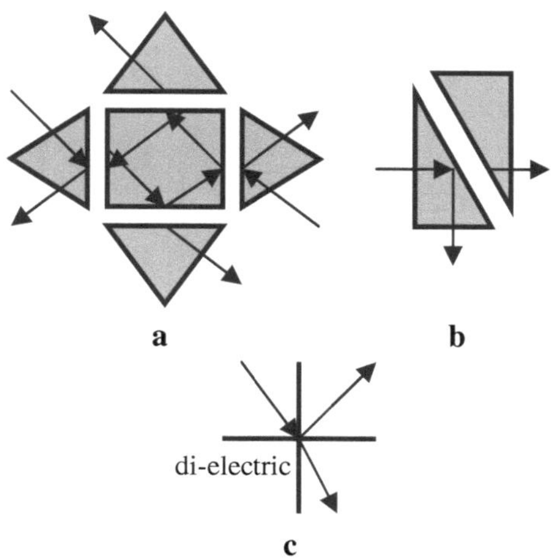

Fig. J.6. Three possible beam splitters.

The beam splitter in Figs. J.6a and J.6b utilize the phenomenon of evanescent surface waves [261] that result when the incident angle of the light beam is greater than the critical angle. On the other hand, the beam splitter in Fig. J.6c utilizes the transmission and reflection of light at dielectric boundaries governed by Eq. (J.7). The compound layer of the beam splitter and the optical switch sub-layers is used repetitively whenever it is needed for splitting and directing the propagating laser beams. The optical switch in Fig. J.5 can be constructed using an array of reflecting micro-mirrors [See [24] and references therein] that direct the laser beam to the desired destination.

J.3 Optical parallel processing

One major advantage of optical computing is that optical signals don't experience cross talk to the extent that electrical signals do, and thus the optical "signal integrity" is high. This advantage is proposed for the optical parallel processing of an array of input laser beams as in Fig. J.7.

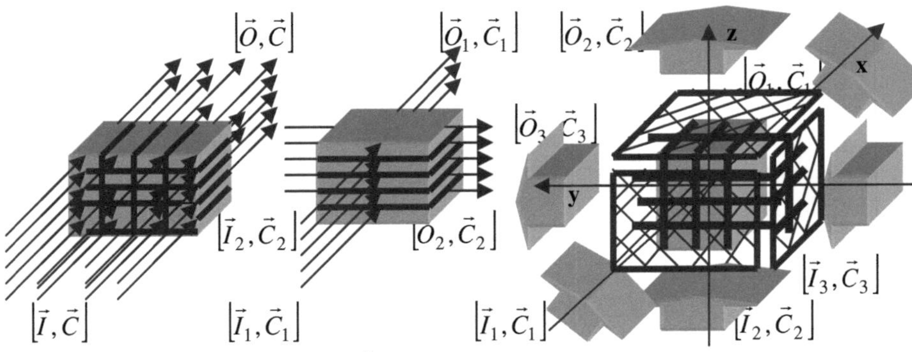

Fig. J.7. Three 3-D optical devices for parallel processing of logic signals.

The parallel processing of many input laser beams is done using a single optical cell to produce an array of outputs as in Fig. J.7a, where $\vec{C}$ is the vector of control signals, $\vec{I}$ is the vector of inputs, and $\vec{O}$ is the vector of outputs. While the optical cell in Fig. J.7a processes a 1-D input array, the cell in Fig. J.7b processes a 2-D input array. Each element of the input and output arrays in Fig. J.7b contains two signals: one is the input signal and the other is the control signal. Each layer in Fig. J.7a, and each cube in Fig. J.7b, represents a single device from Fig. J.2 for two-valued logic, or Fig. J.4 for any N-valued logic. Note that each layer and each cube of the 3-D cells, in Figs. J.7a and J.7b respectively, is an internal node that lays in the plane of a single lattice structure. Thus, if one has N streams of distinct inputs to each cell then one can construct N lattice structures to process N logic functions at the same time. Distinct laser beams in Fig. J.7 propagate freely and their propagation paths can intersect since no cross talk exists between different laser beams.

The only constraint is that insulation sheets must be presented between the neighbor layers in Fig. J.7a, and neighbor cubes in Fig. J.7b to isolate the effects of adjacent non-optical control signals.

Further extensions can include the processing of laser beams that propagate in three dimensions in an anistropic material [261] controlled by a control signal as shown in Fig. J.7c where six sheets surround each optical cell to perform reversible beam splitting, optical switching, and providing control signals. Further extension can include the application of Fizeau's effect to double the number of processed laser beams that propagate in opposite directions within each dimension in a 3-D optical cell from Fig. J.7c.

Appendix K

Artificial Neural Network Implementation Using Multiple-Valued Quantum Computing

This Appendix provides another implementation of multiple-valued quantum computing (QC) that was presented in Chapt. 11. Quantum neural network (NN) implementation using the general scheme of multiple-valued QC is presented in this Appendix. The proposed method uses the multiple-valued orthonormal computational basis states, that were presented in Chapt. 11, to implement such computations. Physical implementation of NNQC is performed by controlling the potential V to yield specific wavefunction as a result of solving Shrodinger Eq. that governs the dynamics of QC. The main contributions of this Appendix are: (1) quantum implementation of NNs using multiple-valued QC. This is achieved via the use of discrete-grid weight space and making an assignment (encoding; mapping) of points on the grid to individual components of a many-valued orthonormal set of quantum basis states, (2) show the underlying mathematical methodology and formalisms for such many-valued QC of NNs, and (3) propose a "reverse engineering" method to develop look-up tables (LUTs) for potential functions associated with specified many-valued logic functions to be performed.

K.1 Neural Networks

The importance of neural networks in application is their ability to learn to perform functions in a problem domain, based on interacting with data from that domain [See [249] and references therein]. A key role in the process is performed by the training set (i.e., a collection of input-output pairs from the problem domain). The training set can be said to provide problem domain "constraints."

The role of learning in the classical domain can be implemented in the quantum domain by the dynamics of a physical system governed by the Schrodinger Eq. (Eqs. (11.1) and (11.2)). The role of training set in classical NN learning can be implemented in the quantum domain by the potential function V (also considered as "constraints"). Thus, there is motivation to establish a mechanism for converting the training set (i.e., collection of input-output pairs) from the problem domain into an appropriate V function in the quantum domain. (An approach for this is presented in Sect. K.3.)

As was shown in Fig. 11.3d in Chapt. 11, a physical system comprising trapped ions under multiple-laser excitations can be used to reliably implement MVQC. A physical system in which an atom (particle) is exposed to a specific potential field (function) can also be used to implement MVQC with two-valued being a special case (cf. Fig. 11.3a in Chapt. 11). In such an implementation, the (resulting) distinct energy states are used as the orthonormal basis states which can be further used for neural network realization. The latter is illustrated in Example K.1 below.

Example K.1. We assume the following *constraints*: (1) spring potential $V(x) = (1/2)\, kx^2$, where m is a particle, $k = m\omega^2$ is spring constant, and ω is angular frequency ($= 2\pi\cdot$frequency), and (2) boundary conditions. Also, assuming the solution of SE for these constraints is of the following form (i.e., the Gaussian function):

$$\psi(x) = Ce^{-\alpha\frac{x^2}{2}},$$

where $\alpha = m\omega/(h/2\pi)$. The general solution for the wavefunction $|\psi\rangle$ (for a spring potential) is:

$$C = \left[\frac{\alpha}{\pi}\right]^{1/4} \frac{1}{\sqrt{2^n n!}} H_n(\sqrt{\alpha}x),$$

where $H_n(x)$ are the Hermite polynomials. This solution leads to the sequence of evenly spaced energy levels (eigenvalues) E_n characterized by a quantum number (n) as follows:

$$E_n = (n + \frac{1}{2})(h/2\pi)\omega.$$

The distribution of the energy states (eigenvalues) and their associated probabilities are shown in Fig. K.1.

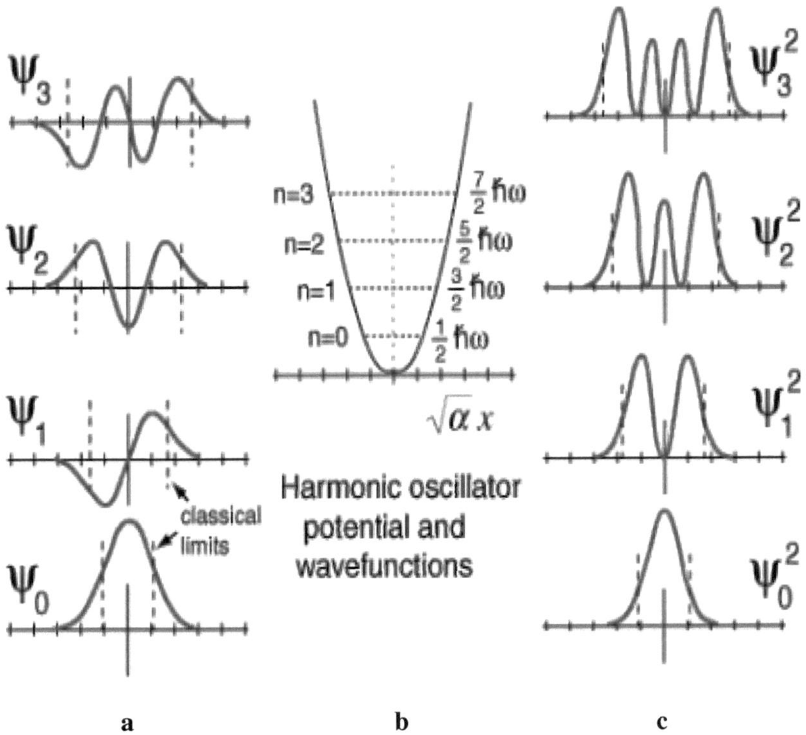

Figure K.1. Harmonic oscillator potential and wavefunctions: **a** wavefunctions for various energy levels (subscripts), **b** spring potential V(x) and the associated energy levels E_n, and **c** probabilities for measuring particle (m) in each energy state (E_n).

K.2 Multiple-Valued Quantum Implementation of Neural Networks: Methodology and Notation

A classic shortcoming of single-neuron neural networks (e.g., perceptron, Adeline) is their inability to implement the XOR function [See [249] and references therein]. In [249] on the other hand, a single quantum neuron is shown to be capable of solving this not-linearly separable (NLS) function. In this Sect., we develop a methodology and formalisms for dealing with multiple-valued logic functions using multiple-valued quantum neurons.

The notion of linearly separable (LS) and not-linearly separable (NLS) mappings in the two-valued context generalizes to the many-valued case.

The following steps describe the notion of associating a quantum state to a point in the weight space of a neural network (NN) using a two-weight NN as an example (extension to any number of weights is straightforward). (See Fig. K.2.)

(1) Assume each weight can take a finite (discrete) set of values (i.e., 0, 1, 2, etc).

(2) Form a two-dimensional grid of all possible combinations of weight value pairs (2-tuples). (In the case of three weight NN this generalizes to three-dimensional grid of all possible combinations of weight value tuples, and in general for N weight NN this results in N-dimensional grid.)

(3) Assign each of the grid points (i.e., each 2-tuple) to be a quantum basis state (in quantum state space). These quantum basis states are orthogonal (cf. Chapt. 11), and thus unique operations can be implemented on such representation. (For higher dimension quantum space, the Kronecker (tensor) product is used to produce all possible combinations of grid points as was shown in Sects. 11.1 and 11.4 in Chapt. 11.)

(4) If each of the two weights (w_1 and w_2) can take m values, then there will be m^2 quantum basis states, each with dimension m^2 (to yield an orthonormal basis set). (Thus, for the general case of N-weight NN, if each weight can take m values then there will be m^N quantum basis states each with dimension m^N to yield a unique (orthonormal) quantum basis set.)

(5) Let:

$$\vec{w}_{i\,j} = \begin{bmatrix} w_{1i} \\ w_{2j} \end{bmatrix}, \qquad i, j = 0,1,2,\ldots, m-1. \tag{K.1}$$

represent the m^2 points in the 2-dimensional weight space, where (i,j) are position indices for the vector w_{ij} and the components of w_{ij} are weight values at the corresponding positions (i,j).

(6) Then define:

$$\vec{w}^{2,m} = \begin{bmatrix} \vec{w}_{00}{}^{T} \\ \vec{w}_{01}{}^{T} \\ \cdots \\ \vec{w}_{0,m-1}{}^{T} \\ \vec{w}_{10}{}^{T} \\ \cdots \\ \vec{w}_{m-1,m-1}{}^{T} \end{bmatrix}, \tag{K.2}$$

where the superscript refers to 2 dimensions, with m (discrete) values in each dimension. (To reference a subset of all these possibilities, an appropriate subscript may be provided.)

For example, by letting each weight take three values from the set {a,b,c} where a, b, c are any discrete real values (i.e., m = 3) then one would have nine grid points {00,01,02,10,11,12,20,21,22}, and Eq. (K.2) becomes:

$$\vec{w}^{2,3} = \begin{bmatrix} w_{00}{}^{T} \\ w_{01}{}^{T} \\ w_{02}{}^{T} \\ w_{10}{}^{T} \\ w_{11}{}^{T} \\ w_{12}{}^{T} \\ w_{20}{}^{T} \\ w_{21}{}^{T} \\ w_{22}{}^{T} \end{bmatrix} = \begin{bmatrix} aa \\ ab \\ ac \\ ba \\ bb \\ bc \\ ca \\ cb \\ cc \end{bmatrix}.$$

Thus, one can perform MVQC by making the following assignments (maps; encodings) between the weight space and the quantum space:

$$\vec{w}^{2,3} = \begin{bmatrix} aa \\ ab \\ ac \\ ba \\ bb \\ bc \\ ca \\ cb \\ cc \end{bmatrix} \quad \begin{matrix} \longrightarrow \\ \longleftarrow \end{matrix} \quad \left|\vec{w}^{2,3}\right\rangle = \begin{bmatrix} |00\rangle \\ |01\rangle \\ |02\rangle \\ |10\rangle \\ |11\rangle \\ |12\rangle \\ |20\rangle \\ |21\rangle \\ |22\rangle \end{bmatrix}. \tag{K.3}$$

(Left: **MV Weight Space**; right: **MV Quantum Space**.)

Using the notation from Chapt. 11, the above may be written as follows:

$$\begin{aligned} |\psi_1\psi_2\rangle &= |\psi_1\rangle \otimes |\psi_2\rangle, \\ &= (\alpha_1|0\rangle + \beta_1|1\rangle + \gamma_1|2\rangle) \otimes (\alpha_2|0\rangle + \beta_2|1\rangle + \gamma_2|2\rangle) \\ &= \alpha_1\alpha_2|00\rangle + \alpha_1\beta_2|01\rangle + \alpha_1\gamma_2|02\rangle + \beta_1\alpha_2|10\rangle + \\ &\quad \beta_1\beta_2|11\rangle + \beta_1\gamma_2|12\rangle + \gamma_1\alpha_2|20\rangle + \gamma_1\beta_2|21\rangle + \gamma_1\gamma_2|22\rangle. \end{aligned} \tag{K.4}$$

Note that each component of the tensor product (in Eq. (K.4) is associated with a product of two probabilities. The coefficients of the quantum basis functions (i.e., probabilities) are the system parameters, obtained by solving the waveequation with the specified potential function V applied. We note that different Vs will (normally) result in different solutions (i.e., different probabilities) for each of the quantum basis states. Upon measurement of an observable variable in a physical quantum implementation, by definition, the highest probability state is the most likely one to occur. In the context of neural networks (NNs) with an assignment such as the one given in Eq. (K.3), each basis state corresponds to a particular combination of weight values. These weight values determine the mapping performed (e.g., logic function) by the NN. (See Fig. K.2.)

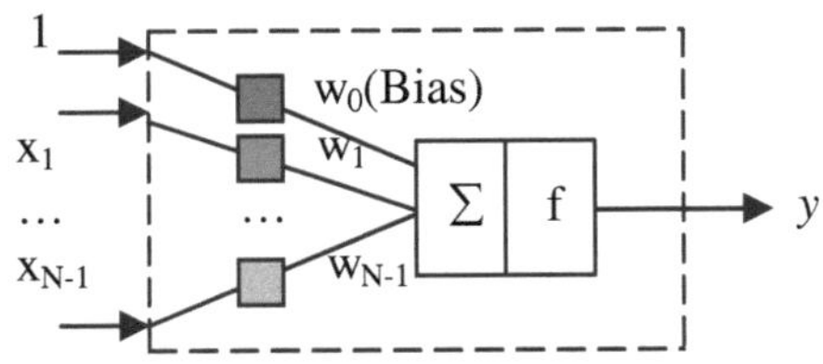

Fig. K.2. A simple neuron.

Where f is the activation (transfer) function, and:

$$y = f\,(w_0 \cdot 1 + w_1 x_1 + w_2 x_2 + \ldots + w_{N-1} x_{N-1}). \tag{K.5}$$

and f can be an appropriate mapping (such as threshold function, sigmoid, etc). The manner in which the MVQC is implemented for neural computing (NC) is illustrated in Fig. K.3.

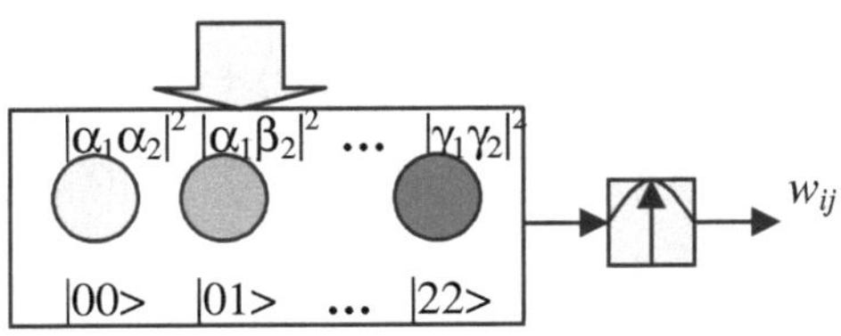

Fig. K.3. MVQC scheme to implement a NN using a ternary 2-qubit QC system. (All possible quantum states are shown in different colors.)

Example K.2. For a quantum neuron, let the following unitary ternary quantum operator **A** [15,23] perform a function analogous to the activation function (AF) and summing junction (SJ) in classical artificial neurons.

$$A = \begin{bmatrix} 1 & 0 & 0 \\ 0 & 0 & 1 \\ 0 & 1 & 0 \end{bmatrix}.$$

Let us denote a ternary 2-weight quantum neuron as in Fig. K.4.

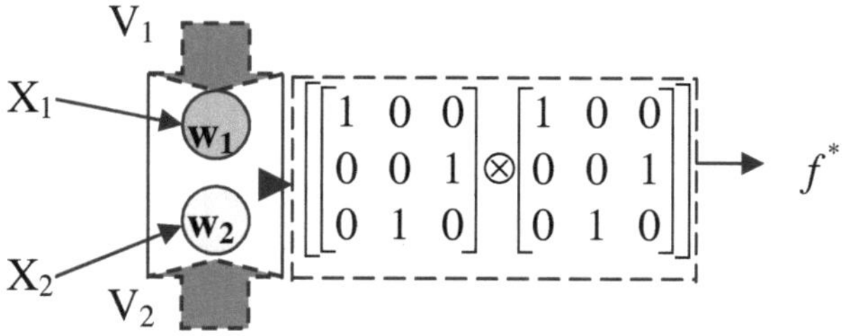

Fig. K.4. Ternary quantum neuron.

Then, for instance, for binary inputs $\{x_1,x_2\}$, the MVQC would proceed as follows to produce the following ternary function f^*:

x_1	x_2	f^*
0	0	0
0	1	2
1	0	0
1	1	2

The quantum weights will be determined by a suitable learning algorithm utilizing the operator **A** (e.g., the algorithm using bipolar quantum Fourier operater in [249]), which is equivalent to solving the TISE with an appropriate potential V. In the notation of Fig. K.4, an example result would be $\{w_1 = |2>, w_2 = |1>\}$, where in the MV quantum space (as shown in Chapt. 11):

$$|w_1 w_2\rangle \to |21\rangle = \begin{bmatrix} 0 \\ 0 \\ 1 \end{bmatrix} \otimes \begin{bmatrix} 0 \\ 1 \\ 0 \end{bmatrix} = [0 \quad 0 \quad 0 \quad 0 \quad 0 \quad 0 \quad 0 \quad 1 \quad 0]^T .$$

Then the MVQC (in Fig. K.4) is performed in the following manner: the matrix of inputs $\{x_1,x_2\}$ is transformed, before being processed by the activation function (AF), to a new matrix of inputs by multiplying the set of inputs by the values of the corresponding weights $\{w_1 = 2, w_2 = 1\}$ as follows:

$$[\mathrm{x}_1 \ \ \mathrm{x}_2] \to [\mathrm{w}_1\mathrm{x}_1 \ \ \mathrm{w}_2\mathrm{x}_2]$$
$$\begin{bmatrix} 0 & 0 \\ 0 & 1 \\ 1 & 0 \\ 1 & 1 \end{bmatrix} \to \begin{bmatrix} 0 & 0 \\ 0 & 1 \\ 2 & 0 \\ 2 & 1 \end{bmatrix}.$$

Encoding the new matrix of inputs in the MV quantum space **H** will lead to:

$$\begin{bmatrix} 0 & 0 \\ 0 & 1 \\ 2 & 0 \\ 2 & 1 \end{bmatrix} \to \begin{bmatrix} |00\rangle \\ |01\rangle \\ |20\rangle \\ |21\rangle \end{bmatrix} = \begin{bmatrix} [1 & 0 & 0 & 0 & 0 & 0 & 0 & 0 & 0]^T \\ [0 & 1 & 0 & 0 & 0 & 0 & 0 & 0 & 0]^T \\ [0 & 0 & 0 & 0 & 0 & 0 & 1 & 0 & 0]^T \\ [0 & 0 & 0 & 0 & 0 & 0 & 0 & 1 & 0]^T \end{bmatrix}.$$

By using the 2-qubit ternary operator:

$$[A] \otimes [A] = \left[\begin{bmatrix} 1 & 0 & 0 \\ 0 & 0 & 1 \\ 0 & 1 & 0 \end{bmatrix} \otimes \begin{bmatrix} 1 & 0 & 0 \\ 0 & 0 & 1 \\ 0 & 1 & 0 \end{bmatrix} \right],$$

$$= \begin{bmatrix} \begin{bmatrix} 1 & 0 & 0 \\ 0 & 0 & 1 \\ 0 & 1 & 0 \end{bmatrix} & & \\ & & \begin{bmatrix} 1 & 0 & 0 \\ 0 & 0 & 1 \\ 0 & 1 & 0 \end{bmatrix} \\ & \begin{bmatrix} 1 & 0 & 0 \\ 0 & 0 & 1 \\ 0 & 1 & 0 \end{bmatrix} & \end{bmatrix}.$$

the matrix of the output functions will be obtained from the matrix of the weighted inputs as follows:

$$\begin{bmatrix} \begin{bmatrix} 1 & 0 & 0 \\ 0 & 0 & 1 \\ 0 & 1 & 0 \end{bmatrix} & & \\ & & \begin{bmatrix} 1 & 0 & 0 \\ 0 & 0 & 1 \\ 0 & 1 & 0 \end{bmatrix} \\ & \begin{bmatrix} 1 & 0 & 0 \\ 0 & 0 & 1 \\ 0 & 1 & 0 \end{bmatrix} & \end{bmatrix} \begin{bmatrix} [1 & 0 & 0 & 0 & 0 & 0 & 0 & 0 & 0]^T \\ [0 & 1 & 0 & 0 & 0 & 0 & 0 & 0 & 0]^T \\ [0 & 0 & 0 & 0 & 0 & 0 & 1 & 0 & 0]^T \\ [0 & 0 & 0 & 0 & 0 & 0 & 0 & 1 & 0]^T \end{bmatrix} =$$

$$\begin{bmatrix} [1 & 0 & 0 & 0 & 0 & 0 & 0 & 0 & 0]^T \\ [0 & 0 & 1 & 0 & 0 & 0 & 0 & 0 & 0]^T \\ [0 & 0 & 0 & 1 & 0 & 0 & 0 & 0 & 0]^T \\ [0 & 0 & 0 & 0 & 0 & 1 & 0 & 0 & 0]^T \end{bmatrix},$$

but in MV quantum space, the matrix of outputs correspond to the following values:

$$\begin{bmatrix} \begin{bmatrix} 1 \\ 0 \\ 0 \\ 0 \\ 0 \\ 0 \\ 0 \\ 0 \\ 0 \end{bmatrix} & \begin{bmatrix} 0 \\ 0 \\ 1 \\ 0 \\ 0 \\ 0 \\ 0 \\ 0 \\ 0 \end{bmatrix} & \begin{bmatrix} 0 \\ 0 \\ 0 \\ 1 \\ 0 \\ 0 \\ 0 \\ 0 \\ 0 \end{bmatrix} & \begin{bmatrix} 0 \\ 0 \\ 0 \\ 0 \\ 0 \\ 1 \\ 0 \\ 0 \\ 0 \end{bmatrix} \end{bmatrix} = \left[|00\rangle \quad |02\rangle \quad |10\rangle \quad |12\rangle \right] \rightarrow \left| f_1 f^* \right\rangle.$$

Then by measuring the second output one obtains the function f^*. (Note that in this example another function f_1 is naturally obtained, and thus this adds a possibility of utilizing such additional output in a separate computation.) The generation of other multiple-valued logic functions, at the output of the quantum neuron in Fig. K.4, is performed using the same topology and same AF (and SJ) by changing the values of the weights.

Other varieties of quantum operators from [15,23] could be used as well to perform the functionality of the AF (and SJ) in Fig. K.4.

For instance, one could use the quantum Chrestenson operator (that was introduced in Theorem 11.2 in Chapt. 11 [15,23]):

$$C_{(1)}^{(3)} = \frac{1}{\sqrt{3}}\begin{bmatrix} 1 & 1 & 1 \\ 1 & d_1 & d_2 \\ 1 & d_2 & d_1 \end{bmatrix},$$

where the superscript indicates the radix, the subscript indicates number of variables, and d_1 and d_2 are complex numbers. The quantum Chrestenson operator used here is the quantum multiple-valued Fourier operator (which is equivalently called the quantum multiple-valued Walsh-Hadamard operator), which is the generalization (extension) of the quantum bipolar Fourier operator (which is equivalently called the quantum (two-valued) Walsh-Hadamard operator). Consequently, the (quantum) learning algorithm proposed in [249] for the quantum Walsh operator could be extended to be used for the quantum Chrestenson operator as well.

K.3 Further NN Implementations Using MVQC

As noted earlier, the quantum analog of a training set in the classical NN context is the potential function V, and the quantum analog for the training process are the dynamics described by the SE. An approach to implement a quantum NN suggested here is as follows (cf. Fig. K.5a): (1) specify a set of functions, F_i, and train a separate neural network (in the first stages of this work, think in terms of a single-neuron NN, e.g., perceptron) for each function; (2) construct a table that associates the trained NN weight vector for each function F_i; (3) construct a separate wavefunction ψ_i in the MV quantum space for each F_i such that its highest probability is at the weight vector in the table, and relatively low at all other weight values as illustrated in Fig. K.5b; (4) substitute this ψ_i into the TISE (Eq. (11.2) in Chapt. 11) and solve for V_i. After the above information has been generated and tabulated (as a look-up table) as indicated in Fig. K.5c, one could implement a full quantum NN as shown in Fig. K.6.

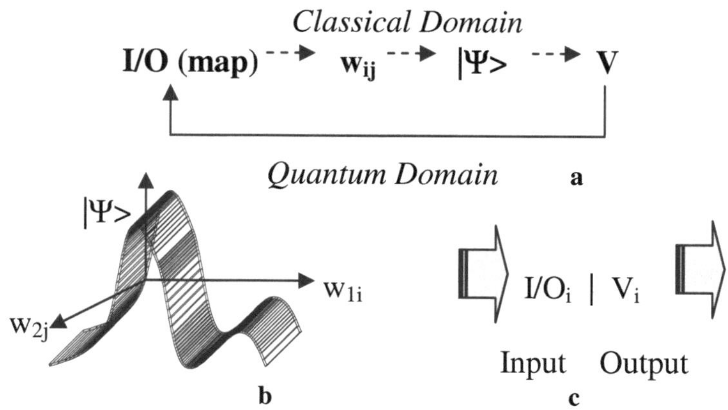

Fig. K.5. **a** Possible MVQC strategy to implement a NN, **b** MV Quantum-Weight space to obtain |Ψ>, and **c** Look-Up-Table to implement a NN for all logic functions.

In this more general case (in Fig. K.6), we specify a single V for an entire weight vector going into a single quantum neuron (corresponding to a single neural element of a NN). Such a quantum neuron (QN) is here represented as shown in Fig. K.6a. A full network would be a collection of such QNs, connected in a specified topology, as in Fig. K.6b.

So far in this Sect., we proposed a methodology of implementing a NN using a multiple-valued quantum computation (MVQC). This method uses the encoding of multiple-valued orthonormal computational basis states in the quantum space to be the weights in a NN. The potential plays the role of a training (I/O) set and the dynamics of the solution of SE to be the training process. Future work will involve (1) simulations for various designs of potential distributions (Vs) that correspond to specific logic functions, (2) determine MVQC equivalents of (a) supervised, (b) reinforcement, and (c) unsupervised learning strategies, and (3) for storing given number of patterns $\mathbf{S_i}$ in Auto-Associative memory (as in a Hopfield NN), where $i = 1, 2, \ldots, N$, and the pattern vector $\mathbf{S_i}$ is of dimension D.

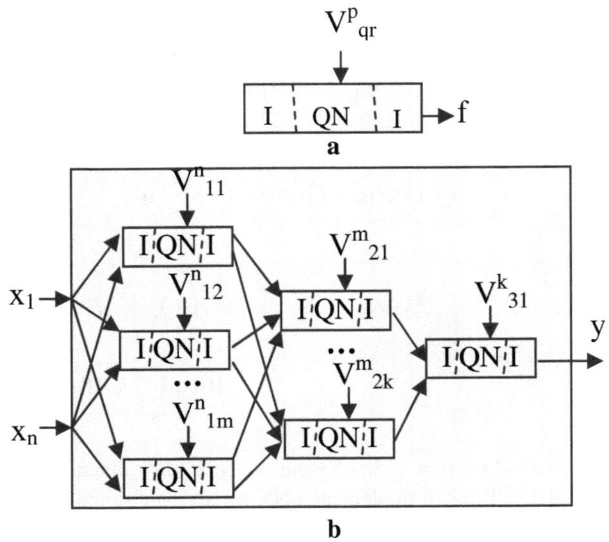

Fig. K.6. An MVQC implementation of a NN: **a** a quantum neuron (QN), which is a dynamical system governed by TISE constrained by V^p_{qr}, where I is the interface mechanism, superscript p is the number of incoming weights, subscript q is the layer number, and subscript r is the element number in layer q, and **b** a 3-layer NN.

This is done conceptually as follows:

(1) Construct a *state-space grid* (equivalent to weight-space grid discussed in Sect. K.2). Each point on the grid corresponds to a specific pattern $\mathbf{S}_i$.

(2) Design a wavefunction ψ_i for each given pattern to be stored. Then solve the TISE for the corresponding potential function V_i.

(3) For a query that is a "dirty" version of a stored pattern $\mathbf{S}_i^*$, construct a corresponding ψ_i^* and V_i^*, where the designed ψ_i^* corresponds to the query pattern, and V_i^* is obtained by solving TISE.

(4) If the original ψ_i was crafted such that probability is maximum at $\mathbf{S}_i$ and gradually decreases for nearby patterns, then the application of V_i^* should yield the quantum state $\mathbf{S}_i$ (i.e., the "clean"/complete pattern).

Bibliography

[1] S. B. Akers, "A Rectangular Logic Array," *IEEE Trans. on Comp.*, Vol. C-21, pp. 848-857, August 1972.

[2] S. B. Akers, "Binary Decision Diagrams," *IEEE Trans. on Comp.*, Vol. C-27, No. 6, pp. 509-516, June 1978.

[3] A. Al-Rabadi, "A Novel Counting Triangle and Expression for Characterizing the Complexity of S/D Trees for Arbitrary Radix Logic and Arbitrary Number of Variables," *Technical Report #2002/02*, ECE Department, PSU, 25th May 2000.

[4] A. Al-Rabadi, and M. Perkowski, "An Extended Green/Sasao Hierarchy of Canonical Ternary Galois Field Decision Diagrams and Forms," *submitted to special issue of Multiple-Valued Logic (MVL) Journal*, 1st August 2000.

[5] A. Al-Rabadi, "New Three-Dimensional Invariant MVL Shannon and Davio Families of Spectral Transforms and their Lattice Structures," *Technical Report #2001/002*, ECE Department, PSU, 9th January 2001.

[6] A. Al-Rabadi, "New Reversible Invariant Multi-Valued Families of Spectral Transforms For Three-Dimensional Layout," *Technical Report #2001/003*, ECE Department, PSU, 3rd March 2001.

[7] A. Al-Rabadi, and M. Perkowski, "New Families of Reversible Expansions and their Lattice Structures," *Submitted to MVL Journal*, 26th March 2001.

[8] A. Al-Rabadi, "Towards a Fully Reversible 2-valued Nano-Processor: A Library of Binary Reversible Hardware Components," *Technical Report #2001/004*, ECE Department, PSU, 3rd April 2001.

[9] A. Al-Rabadi, and M. Perkowski, "Multiple-Valued Galois Field S/D Trees for GFSOP Minimization and their Complexity," *Proc. Int. Symp. on Multiple-Valued Logic (ISMVL) '01*, pp. 159-166, Warsaw, Poland, 22-24th May 2001.

[10] A. Al-Rabadi, "A Novel Reconstructability Analysis For the Decomposition of Logic Functions," *Technical Report #2001/005*, ECE Department, PSU, 1st July 2001.

[11] A. Al-Rabadi, "On the Characterization of Multi-Valued Equally Input-Output Reversible Galois Logic Primitives," *Technical Report #2001/006*, ECE Department, PSU, 1st August 2001.

[12] A. Al-Rabadi, and M. Perkowski, "Families of New Multi-Valued Reed-Muller-Based Spectral Transforms," *Proc. Int. Workshop on Applications of the Reed-Muller Expansion in Circuit Design (Reed-Muller)'2001 workshop*, pp. 226-241, Starkville, Mississippi, 10-11th August 2001.

[13] A. Al-Rabadi, and M. Perkowski, "Shannon and Davio Sets of New Lattice Structures for Logic Synthesis in Three-Dimensional Space," *Proc. Reed-Muller'01*, pp. 165-184, Starkville, Mississippi, 10-11th August 2001.

[14] A. Al-Rabadi, and M. Perkowski, "New Classes of Multi-Valued Reversible Decompositions for Three-Dimensional Layout," *Proc. Reed-Muller'01*, pp. 185-204, Starkville, Mississippi, 10-11th August 2001.

[15] A. Al-Rabadi, "Synthesis and Canonical Representations of Equally Input-Output Binary and Multiple-Valued Galois Quantum Logic: Decision Trees, Decision Diagrams, Quantum Butterflies, Quantum Chrestenson Gate, and Multiple-Valued Bell-Einstein-

Podolsky-Rosen Basis States," *Technical Report #2001/007*, ECE Department, PSU, 22nd August 2001.
[16] A. Al-Rabadi, "Novel Semi-Regular Conventional and Reversible Decompositions of Logic Circuits," *Technical Report #2001/008*, ECE Department, PSU, 3rd September 2001.
[17] A. Al-Rabadi, "Novel Optical Gates and Structures for Reversible Logic," *Technical Report #2001/009*, ECE Department, PSU, 13th October 2001.
[18] A. Al-Rabadi, and M. Perkowski, "Three-Dimensional Regular Realization of Galois Field Logic Circuits," *Submitted to IEE Series-E*, 3rd December 2001.
[19] A. Al-Rabadi, L. W. Casperson, and M. Perkowski, "Multiple-Valued Quantum Computing," *Submitted to Quantum Computers and Computing Journal*, 8th March 2002.
[20] A. Al-Rabadi, M. Zwick, and M. Perkowski, "A Comparison of Enhanced Reconstructability Analysis and Ashenhurst-Curtis Decomposition of Boolean Functions," *Book of Abstracts of the 12th international World Organization for Systems and Cybernetics (WOSC) Congress and the 4th International Institute for General Systems Studies (IIGSS) workshop*, Pittsburgh, Pennsylvania, p. 12, 24-26th March 2002.
[21] A. Al-Rabadi, and M. Zwick, "Modified Reconstructability Analysis for Many-Valued Logic Functions," *Book of Abstracts of the WOSC/IIGSS'02*, Pittsburgh, Pennsylvania, p. 90, 24-26th March 2002.
[22] A. Al-Rabadi, and M. Zwick, "Reversible Modified Reconstructability Analysis of Boolean Circuits and its Quantum Computation," *Book of Abstracts of the WOSC/IIGSS'02*, Pittsburgh, Pennsylvania, p. 90, 24-26th March 2002.
[23] A. Al-Rabadi, L. W. Casperson, M. Perkowski, and X. Song, "Multiple-Valued Quantum Logic," *Booklet of the 11th Post-Binary Ultra Large Scale Integration (ULSI)'02 workshop*, pp. 35-45, Boston, Massachusetts, 15th May 2002.
[24] A. Al-Rabadi, and L. W. Casperson, "Optical Realizations of Reversible Logic," *Proc. International Workshop on Logic and Synthesis (IWLS)'02*, pp. 21-26, New Orleans, Louisiana, 4-7th June 2002.
[25] A. Al-Rabadi, "Symmetry as a Base for a New Decomposition of Boolean Logic," *Proc. IWLS'02*, pp. 273-278, New Orleans, Louisiana, 4-7th June 2002.
[26] A. Al-Rabadi, L. W. Casperson, M. Perkowski, and X. Song, "Canonical Representations for Two-Valued Quantum Computing," *accepted to the 5th International Workshop on Boolean Problems (WBP)'02*, Freiberg, Germany, 19-20th September 2002.
[27] A. Al-Rabadi, M. Zwick, and M. Perkowski, "A Comparison of Modified Reconstructability Analysis and Ashenhurst-Curtis Decomposition of Boolean Functions," *submitted to Kybernetes.*
[28] A. Al-Rabadi, and M. Zwick, "Modified Reconstructability Analysis for Many-Valued Logic Functions," *submitted to Kybernetes.*
[29] A. Al-Rabadi, and M. Zwick, "Reversible Modified Reconstructability Analysis of Boolean Circuits and its Quantum Computation," *submitted to Kybernetes.*
[30] A. Al-Rabadi, and M. Zwick, "Enhancements to Crisp Possibilistic Reconstructability Analysis," *submitted to the International Journal of General Systems (IJGS).*
[31] E. Andersson, and S. Stenholm, "Quantum Logic Gate with Microtraps," *Optics Communications*, 188, pp. 141-148, 1 February 2001.
[32] R. L. Ashenhurst, "The Decomposition of Switching Functions," *Bell Laboratories Report*, Vol. 1, pp. II-1-II-37, 1953.
[33] R. L. Ashenhurst, "The Decomposition of Switching Functions," *Bell Laboratories Report*, Vol. 16, pp. III-1-III-72, 1956.
[34] R. L. Ashenhurst, "The Decomposition of Switching Functions," *International Symposium on the Theory of Switching Functions*, pp. 74-116, 1959.
[35] W. C. Athas, and L. J. Svensson, "Reversible Logic Issues in Adiabatic CMOS," Exploratory Design Group, University of Southern California, Information Sciences Institute, Marina del Rey, CA 90292-6695.

[36] A. Barenco et al., "Elementary Gates for Quantum Computation," *Physical Review A*, Vol. 52, pp. 3457-3467, 1995.

[37] C. H. Bennett, "Logical Reversibility of computation," *IBM Journal of Research and Development,* Vol. 17, pp.525-532, 1973.

[38] C. H. Bennett, and R. Landauer, "The Fundamental Physical Limits of Computation," *Scientific American*, 253, pp. 38-46, July 1985.

[39] C. H. Bennett, "Notes on the History of Reversible Computation," *IBM Journal of Research and Development*, Vol. 32, pp. 16-23, 1988.

[40] C. H. Bennett, and D. P. DiVincenzo, "Progress Toward Quantum Computation," *Nature*, September 1995.

[41] S. Bettelli, L. Serafini, and T. Calarco, "Toward an Architecture for Quantum Programming," *IRST Technical Report #0103-010*, March 8, 2001.

[42] A. Blotti, S. Di Pascoli, and R. Saletti, "A Comparison of Some Circuit Schemes for Semi-Reversible Adiabatic Logic," *Int. J. Electronics*, Vol. 89, No. 2, pp. 147-158, 2002.

[43] G. Boole, *An Investigation into the Laws of Thought on which are Founded the Mathematical Theories of Logic and Probabilities*, 1854.

[44] G. Brassard, S. L. Braunstein, and R. Cleve, "Teleportation as Quantum Computation", *Physica D*, Vol. 120, pp. 43-47, 1998.

[45] R. E. Bryant, "Graph-based Algorithms for Boolean Functions Manipulation," *IEEE Trans. on Comp.,* Vol. C-35, No.8, pp. 667-691, 1986.

[46] J. T. Butler, K. A. Schueller, "Worst Case Number of Terms In Symmetric Multiple-Valued Functions," *Proc. ISMVL'91,* pp. 94-101, Victoria, B. C., Canada, May 26-29,1991.

[47] J. T. Butler, and T. Sasao, "On the Properties of Multiple-Valued Functions that are Symmetric in Both Variable Values and Labels," *Proc. ISMVL'98,* pp. 83-88, Fukuoka, Japan, May 27-29, 1998.

[48] A. Calderbank, E. Rains, P. Shor, and N. Sloane, "Quantum Error Correction Via Codes Over GF(4)," *AT&T Labs Research*, 5 March 1998.

[49] L. W. Casperson, "Synthesis of Gaussian Beam Optical Systems," *Applied Optics*, Vol. 20, No. 13, pp. 2243-2249, 1 July 1981.

[50] M. Chrzanowska-Jeske, Z. Wang, and Y. Xu, "Regular Representation for Mapping to Fine-Grain, Locally-Connected FPGAs," *Proc. Int. Symp. on Circuits and Systems (ISCAS)'97*, Vol. 4, pp. 2749-2752, 1997.

[51] M. Chrzanowska-Jeske, Y. Xu, M. A. Perkowski, "Logic Synthesis for a Regular Layout," *VLSI Design: An International Journal of Custom-Chip Design, Simulation, and Testing*, Vol. 10, No. 1, pp. 35-55, 1999.

[52] M. Chrzanowska-Jeske, A. Mishchenko, and M. Perkowski, "A Family of Canonical AND/EXOR Forms That Includes the Exact Minimum ESOPs," *accepted to VLSI Design Journal, 2000.*

[53] I. Chuang, and Y. Yamamoto, "A Simple Quantum Computer", ERATO Quantum Fluctuation Project, Edward L. Ginzton Laboratory, Stanford University, 28 March 1995.

[54] J. I. Cirac, and P. Zoller, "Quantum Computations with Cold Trapped Ions," *Phys. Rev. Lett.*, Vol. 74, No. 20, pp. 4091-4094, 15 May 1995.

[55] R. Cleve, and J. Watrous, "Fast Parallel Circuits for the Quantum Fourier Transform," *Proc. Symp. on the Theory of Computing*, pp. 526-535, 2000.

[56] M. Cohn, *Switching Function Canonical Form over Integer Fields*, Ph.D. Dissertation, Harvard University, Cambridge, Massachusetts, 1960.

[57] W. Cooley, and J. Tukey, "An Algorithm for the Machine Calculation of Complex Fourier Series," *Mathematics of Computation*, Vol. 19, pp. 297-301, 1965.

[58] Concurrent Logic Inc., "CLI 6000 Series Field Programmable Gate Arrays," *Prelimin. Inform.*, Dec. 1, 1991, Rev. 1.3.

[59] H. A. Curtis, "Generalized Tree Circuit," *ACM,* pp. 484-496, 1963.

[60] H. A. Curtis, "Generalized Tree Circuit-The Basic Building Block of an Extended Decomposition Theory," *ACM,* Vol. 10, pp. 562-581, 1963.
[61] H. A. Curtis, *A New Approach to the Design of Switching Circuits*, Princeton, Van Nostrand, NJ, 1962.
[62] R. Cuykendall, and D. R. Andersen, "Reversible Computing: All-Optical Implementation of Interaction and Priese Gates," *Optics Communications*, Vol. 62, No. 4, pp. 232-236, May 1987.
[63] R. Cuykendall, and D. McMillin, "Control-Specific Optical Fredkin Circuits," *Applied Optics*, Vol. 26, No. 10, pp. 1959-1963, May 1987.
[64] R. Cuykendall, and D. R. Andersen, "Reversible Optical Computing Circuits," *Optics Letters*, Vol. 12, No. 7, pp. 542-544, July 1987.
[65] R. Cuykendall, "Three-Port Reversible Logic," *Applied Optics*, Vol. 27, No. 9, pp. 1772-1779, May 1988.
[66] C. Darwin, *On the Origin of Species by Means of Natural Selection, or the Preservation of Favoured Races in the Struggle for Life*, 1859.
[67] J. Denes and A. D. Keedwell, *Latin Squares and their Applications*, Academic Press, New York, 1974.
[68] B. Desoete, A. De Vos, M. Sibinski, and T. Widerski, "Feynman's Reversible Logic Gates Implemented in Silicon," *Proc. 6th Int. Conference on MIXDES*, pp. 497-502, 1999.
[69] D. Deutsch, "Quantum Theory, the Church-Turing Principle and the Universal Quantum Computer," *Proc. Royal Society of London A*, Vol. 400, pp. 97-117, 1985.
[70] A. De Vos, "Proposal for an Implementation of Reversible Gates in c-MOS," *Int. J. of Electronics,* Vol.76, pp. 293-302, 1994.
[71] A. De Vos, "Reversible Computing in c-MOS," *Proc. Advanced Training Course on Mixed Design of VLSI Circuits,* pp.36-41, 1994.
[72] A. De Vos, "A 12-Transistor c-MOS Building-Block for Reversible Computers," *Int. J. of Electronics*, Vol.79, pp. 171-182, 1995.
[73] A. De Vos, "Reversible and Endoreversible Computing," *Int. J. of Theoretical Physics*, Vol.34, pp. 2251-2266, 1995.
[74] A. De Vos, "Towards Reversible Digital Computers," *Proc. European Conf. on Circuit Theory and Design*, pp. 923-931, Budapest 1997.
[75] A. De Vos, "Reversible Computing," *Progress in Quantum Electronics,* 23, pp. 1-49, 1999.
[76] K. Dill, and M. Perkowski, "Minimization of Generalized Reed-Muller Forms with a Genetic Algorithm," *Proc. Genetic Programming'97 Conf.*, p. 362, Stanford University, CA, July 1997.
[77] K. Dill, J. Herzog, and M. Perkowski, "Genetic Programming and its applications to the synthesis of Digital Logic," *Proc. PACRIM'97*, vol. 2, pp. 823-826, Canada, 20-22 August, 1997.
[78] K. Dill, K. Ganguly, R. Safranek, and M. Perkowski, "A new Linearly-Independent, Zhegalkin Galois Field Reed-Muller Logic," *Proc. Reed-Muller'97,* pp. 247-257, Oxford Univ. U.K., September 1997.
[79] K. Dill, and M. Perkowski, "Evolutionary Minimization of Generalized Reed-Muller Forms," *Proc. ICCIMA'98 Conf.*, pp. 727-733, Australia, February 1998, published by *World Scientific.*
[80] K. M. Dill, and M. A. Perkowski, "Baldwinian Learning Utilizing Genetic and Heuristic Algorithms for Logic Synthesis and Minimization of Incompletely Specified Data With Generalized Reed-Muller (AND-EXOR) Forms," *J. Systems Architecture*, Vol. 47, pp. 477-489, 2001.
[81] P. Dirac, *The Principles of Quantum Mechanics,* first edition, Oxford University Press, 1930.

[82] R. Drechsler, A. Sarabi, M. Theobald, B. Becker, and M. A. Perkowski, "Efficient Representation and Manipulation of Switching Functions Based on Ordered Kronecker Functional Decision Diagrams," *Proc. DAC'94,* pp. 415-419, 1994.
[83] R. Drechsler, "Pseudo-Kronecker Expressions for Symmetric Functions," *IEEE Trans. on Comp.*, Vol. 48. No. 8, pp. 987-990, September 1999.
[84] B. T. Drucker, C. M. Files, M. A. Perkowski, and M. Chrzanowska-Jeske, "Polarized Pseudo-Kronecker Symmetry with an Application to the Synthesis of Lattice Decision Diagarms," *Proc. ICCIMA'98*, pp. 745-755, 1998.
[85] M. Escobar, and F. Somenzi, "Synthesis of AND/EXOR Expressions via Satisfiability," *Proc. Reed-Muller'95,* pp. 80-87, 1995.
[86] D. Etiemble, and M. Israel, "Comparison of Binary and Multivalued ICs According to VLSI Criteria," *IEEE Trans. on Comp.*, pp. 28-42, April 1988.
[87] B. Falkowski, *Spectral Methods for Boolean and Multiple-Valued Input Logic Functions*, Ph.D. Dissertation, Electrical and Computer Engineering Department, Portland State University, 1991.
[88] B. Falkowski, and S. Rahardja, "Efficient Algorithm for the Generation of Fixed Polarity Quaternary Reed-Muller Expansions," *Proc ISMVL'95,* pp.158-163, Bloomington, Indiana, 1995.
[89] B. J. Falkowski, and S. Rahardja, "Classification and Properties of Fast Linearly Independent Logic Transformations," *IEEE Trans. on Circuits and Systems-II: Analog and Digital Signal Processing*, Vol. 44, No. 8, pp. 646-655, August 1997.
[90] B. Falkowski, and L.-S. Lim, "Gray Scale Image Compression Based on Multiple-Valued Input Binary Functions, Walsh and Reed-Muller Spectra," *Proc. ISMVL'00*, pp. 279-284, 23-25 May 2000.
[91] P. Farm, and E. Dubrova, "Technology Mapping for Chemically Assembeled Electronic Nanotechnology," *Proc. IWLS'02*, pp. 121-124, New Orleans, Louisiana, 4-7th June 2002.
[92] R. Feynman, "Simulating Physics with Computers," *Int. J. of Theoretical Physics*, Vol. 21, Nos. 6/7, pp. 467-488, 1982.
[93] R. Feynman, "Quantum Mechanical Computers," *Optics News*, 11, pp. 11-20, 1985.
[94] R. Feynman, "There is Plenty of Room at the Bottom: an Invitation to Enter a New Field of Physics," *Nanotechnology*, edited by B. C. Crandal and J. Lewis, MIT Press, pp. 347-363, 1992.
[95] R. Feynman, *Feynman Lectures on Computation*, Addison Wesley, 1996.
[96] C. M. Files, *A New Functional Decomposition Method as Applied to Machine Learning and VLSI Layout*, Ph.D. Dissertation, Electrical and Computer Engineering Department, Portland State University, Portland, Oregon, 2000.
[97] M. P. Frank, *Reversibility for Efficient Computing*, Ph.D. Dissertation, Massachusetts Institute of Technology, 1999.
[98] E. Fredkin, and T. Toffoli, "Conservative Logic," *Int. J. of Theoretical Physics,* 21, pp.219-253, 1982.
[99] H. Fujiwara, *Logic Testing and Design for Testability*, The MIT Press, 1985.
[100] Y. Z. Ge, L. T. Watson, and E. G. Collins, "Genetic Algorithms for Optimization on a Quantum Computer," *Int. Conf. Unconventional Models of Computation (UMC)'98*, pp. 218-227, Auckland, New Zealand, 5-9 January 1998.
[101] N. A. Gershenfeld, and I. L. Chuang, "Bulk Spin Resonance Quantum Computation," *Science*, 275, pp. 350-356, 1997.
[102] D. E. Goldberg, *Genetic Algorithms in Search, Optimization, and Machine Learning*, Addison Wesley, 1989.
[103] S. Goldstein, and M. Budiu, "Nanofabrics: Spatial Computing Using Molecular Electronics," *Proc. Int. Symp. on Computer Architecture*, Gothenborg, Sweden, June 2001.

[104] D. H. Green, "Families of Reed-Muller Canonical Forms," *Int. J. of Electronics,* No. 2, pp. 259-280, February 1991.

[105] L. K. Grover, "A Fast Quantum-Mechanical Algorithm for Database Search," *Proc. Symp. on Theory of Computing (STOC)'96*, pp. 212-219, 1996.

[106] L. K. Grover, "A Framework for Fast Quantum Mechanical Algorithms," *Proc. STOC'98*, pp. 53-62, May 1998.

[107] J. Gruska, *Quantum Computing*, McGraw-Hill, 1999.

[108] S. Grygiel, *Decomposition of Relations as a new Approach to Constructive Induction in Machine Learning and Data Mining*, Ph.D. Dissertation, Electrical and Computer Engineering Department, Portland State University, Portland, Oregon, 2000.

[109] D. Hammerstrom, "Computational Neurobiology Meets Semiconductor Engineering," *Proc. ISMVL'00,* pp. 3-12, Portland, Oregon, 23-25 May 2000.

[110] K. Han, K. Park, C. Lee, and J. Kim, "Parallel quantum-inspired genetic algorithm for combinatorial optimization problems," *Proc. of the Congress on Evolutionary Computation (CEC),01,* volume 2, pp. 1422-1429, 2001.

[111] H. Hasegawa, "Quantum Devices and Integrated Circuits Based on Quantum Confinement in III-V Nanowire Networks Controlled by Nano-Schottky Gates," *ECS Joint Int. Meeting and Sixth Int. Symp. on Quantum Confinement*, San Fransisco, 2-7th September 2001.

[112] H. Hasegawa, A. Ito, C. Jiang, and T. Muranaka, "Atomic Assisted Selective MBE Growth of InGaAs Linear and Hexagonal Nanowire Networks for Novel Quantum Circuits," *Proc. of the 4th Int. Workshop on Novel Index Surfaces (NIS)'01*, Apset, France, 16-20 September 2001.

[113] S. Hassoun, T. Sasao, and R. Brayton (editors), *Logic Synthesis and Verification*, Kluwer Academic Publishers, November 2001.

[114] M. Helliwell, M. A. Perkowski, "A Fast Algorithm to Minimize Multi-Output Mixed-Polarity Generalized Reed-Muller Forms," *Proc. Design Automation Conference (DAC)'88*, pp. 427-432, 1988.

[115] M. Hirvensalo, *Quantum Computing*, Springer, 2001.

[116] T. Hogg, C. Mochon, W. Polak, and E. Rieffel, "Tools for Quantum Algorithms," *Int. J. of Modern Physics C*, Vol. 10, No. 7, pp. 1347-1361, 1999.

[117] R. J. Hughes, D. F. V. James, E. H. Knill, R. Laflamme and A. G. Petschek, "Decoherence Bounds on Quantum Computation with Trapped Ions," *Los Alamos National Laboratory,* Los Alamos, New Mexico, 8 April 2002.

[118] S. L. Hurst, *Logical Processing of Digital Signals*, Crane Russak and Edward Arnold, London and Basel, 1978.

[119] S. L. Hurst, "Multiple-Valued Logic – Its Status and Its Future," *IEEE Trans. on Comp.*, Vol. C-33, pp. 1160-1179, December 1984.

[120] S. L. Hurst, D. M. Miller, and J. C. Muzio, *Spectral Techniques in Digital Logic,* Academic Press Inc., 1985.

[121] K. Iwama, Y. Kambayashi, and S. Yamashita, "Transformation Rules for Designing CNOT-based Quantum Circuits," *Proc. DAC'2002*, pp. 419-424, New Orleans, Louisiana, 10-14th June 2002.

[122] T. P. Johnson, "Reversible Fuzzy Topological Spaces," *J. Fuzzy Mathematics*, Vol. 1, No. 3, p. 491, September 1993.

[123] R. Josza, "Quantum Algorithms and the Fourier Transform," *Proc. Royal Society of London*, Vol. 454, pp. 323-337, 1997.

[124] U. Kalay, *Highly Testable Quasigroup-Based Combinational Logic Circuits*, Ph.D. Dissertation, Electrical and Computer Engineering Department, Portland State University, Portland, Oregon, 2001.

[125] M. G. Karpovski, *Finite Orthogonal Series in the Design of Digital Devices,* Wiley, New York, 1976.

[126] P. Kerntopf, "A Comparison of Logical Efficiency of Reversible and Conventional Gates," *Proc. Symp. on Logic, Design and Learning (LDL)'00,* Portland, Oregon, 2000.
[127] P. Kerntopf, "Logic Synthesis Using Reversible Gates," *Proc. LDL'00, Portland,* Oregon, 31 May 2000.
[128] P. Kerntopf, "On Efficiency of Reversible Logic (3,3)-Gates," *Proc. Int. Conf. on Mixed Design of Integrated Circuits and Systems (MIXDES)'00*, pp. 185-190, June 2000.
[129] P. Kerntopf, "An Approach to Designing Complex Reversible Logic Gates," *Proc. IWLS*'02, pp. 31-36, New Orleans, Louisiana, 2002.
[130] R. Keyes, and R. Landauer, "Minimal Energy Dissipation in Logic," *IBM J. Research and Development*, 14, pp. 153-157, 1970.
[131] S. Kim, C. H. Ziesler, and M. C. Papaefthymiou, "A True Single-Phase 8-bit Adiabatic Multiplier," *DAC'01*, pp. 758-763, 2001.
[132] K. Kinoshita, T. Sasao, and J. Matsuda, "On Magnetic Bubble Logic Circuits," *IEEE Trans. on Comp.*, Vol. C-25, No. 3, pp. 247-253, March 1976.
[133] G. Klir, *Architecture of Systems Problem Solving*, Plenum Press, New York, 1985.
[134] G. Klir (editor), "Reconstructability Analysis Bibliography," *Int. J. of General Systems*, Vol. 24, pp. 225- 229, 1996.
[135] G. Klir, and M. J. Wierman, *Uncertainty-Based Information: Variables of Generalized Information Theory*, Physica-Verlag, New York, 1998.
[136] Z. Kohavi, *Switching and Finite Automata Theory,* McGraw-Hill Inc., 1978.
[137] J. R. Koza, *Genetic Programming: On the Programming of Computers by Means of Natural Selection*, MIT Press, Cambridge, MA, 1992.
[138] K. Krippendorff, *Information Theory: Structural Models for Qualitative Data*, Sage Publications, Inc., 1986.
[139] R. Landauer, "Irreversibility and heat generation in the computational process," *IBM J. of Research and Development,* 5, pp. 183-191, 1961.
[140] R. Landauer, "Fundamental Physical Limitations of the Computational Process," *Ann. N.Y. Acad. Sci.*, 426, 161, 1985.
[141] R. Landauer, "Computation and Physics: Wheeler's Meaning Circuit," *Found. Phys.*, 16, 551, 1986.
[142] C. Y. Lee, "Representation of Switching Circuits by Binary Decision Diagrams," *Bell Syst. Tech. J.*, Vol. 38, pp. 985-999, 1959.
[143] J. Lim, D. Kim, and S. Chae, "Reversible Energy Recovery Logic Circuits and Its 8-Phase Clocked Power Generator for Ultra-Low-Power Applications," *IEICE Trans. Electron*, OL. E82-C, No. 4, April 1999.
[144] P. Lindgren, R. Drechsler, and B. Becker, "Synthesis of Pseudo-Kronecker Lattice Diagrams," *Proc. Reed-Muller'99,* pp. 197-204, Victoria, B. C., Canada, 1999.
[145] S. Loloyd, "A Potentially Realizable Quantum Computer," *Science*, 261, pp. 1569-1571, 1993.
[146] S. MacLane and G. Birkoff, *Algebra*, Macmillan Company, New York, 1967.
[147] F. J. MacWilliams and N. J. A. Sloane, *The Theory of Error-Correcting Codes*, North-Holland, Amsterdam, 1977.
[148] K. K. Maitra, "Cascaded Switching Networks of Two-Input Flexible Cells," *IRE Trans. Electron Comput.*, pp. 136-143, 1962.
[149] C. Marand, and P. Townsend, "Quantum Key Distribution Over Distances as Long as 30 km," *Optics Letters*, Vol. 20, No. 16, pp. 1695-1697, 15 August 1995.
[150] N. Margolus, *Physics and Computation*, Ph.D. Dissertation, Massachusetts Institute of Technology, 1988.
[151] R. J. McEliece, *Finite Fields for Computer Scientists and Engineers*, Kluwer Academic Publishers, Boston, 1987.
[152] R. C. Merkle, and K. Drexler, "Helical Logic," *Nanotechnology*, 7, pp. 325-339, 1996.

[153] R. C. Merkle, "Reversible Electronic Logic Using Switches," *Nanotechnology*, 4, pp. 21-40, 1993.
[154] R. C. Merkle, "Two Types of Mechanical Reversible Logic," *Nanotechnology*, 4, pp. 114-131, 1993.
[155] L. J. Micheel, A. H. Taddiken, and A. C. Seabaugh, "Multiple-Valued Logic Computation Using Micro- and Nanoelectronic Devices," *Proc. ISMVL'93*, pp. 164-169, 1993.
[156] G. J. Milburn, "Quantum Optical Fredkin Gate," *Phys. Rev. Lett.*, Vol. 62, No. 18, pp. 2124-2127, 1 May 1989.
[157] A. Mishchenko, M. Perkowski, "Fast Heuristic Minimization of Exclusive-Sum-Of-Products," *Proc. Reed-Muller'01*, pp. 242-250, Starkville, Mississippi, August 2001.
[158] A. Mishchenko, and M. Perkowski, "Logic Synthesis of Reversible Wave Cascades," *Proc. IWLS*'02, pp. 197-202, New Orleans, Louisiana, 2002.
[159] C. Moraga, "Ternary Spectral Logic," *Proc. ISMVL'77,* pp. 7-12, 1977.
[160] J. N. Mordeson, "Fuzzy Galois Theory," *J. Fuzzy Mathematics,* Vol. 1, No. 3, 659, September 1993.
[161] A. Mukherjee, R. Sudhakar, M. Marek-Sadowska , and S. I. Long, "Wave Steering in YADDs: A Novel Non-Iterative Synthesis and Layout Technique," *Proc. DAC'99,* pp. 466-471, New Orleans, June 1999.
[162] J. Mullins, "Quantum Physics Spins Off Marketable Products: Uncrackable Encryption Key is One of First New Devices," *IEEE Spectrum Magazine*, pp. 21-22, May 2002.
[163] J. Mullins, "Making Unbreakable Code: The Quantum Properties of Photons Could Make Encrypted Messages Absolutely Secure," *IEEE Spectrum Magazine*, pp. 40-45, May 2002.
[164] S. Muroga, *Logic Design and Switching Theory,* Wiley, New York, 1979.
[165] A. Muthukrishnan, and C. R. Stroud, "Multivalued Logic Gates for Quantum Computation," *Physical Review A*, Vol. 62, 052309, 2000.
[166] J. C. Muzio, and T. Wesselkamper, *Multiple-Valued Switching Theory*, Adam-Hilger, 1985.
[167] M. A. Nielsen, and I. L. Chuang, *Quantum Computation and Quantum Information,* Cambridge University Press, 2000.
[168] M. A. Nielsen, and Isaac L. Chuang, "Programmable Quantum Gate Arrays," *Submitted to Phys. Rev. Lett.*, 12th May 1998.
[169] M. Oskin, F. T. Chong, and I. L. Chuang, "A Practical Architecture for Reliable Quantum Computers," *IEEE Computer Magazine*, pp. 79-87, January 2002.
[170] A. Peres, "Reversible Logic and Quantum Computers," *Physical Review A*, 32, pp. 3266-3276, 1985.
[171] M. Perkowski, and M. Chrzanowska-Jeske, "An Exact Algorithm to Minimize Mixed-Radix Exclusive Sums of Products for Incompletely Specified Boolean Functions," *Proc. ISCAS'90,* Intern. Symposium on Circuits and Systems, pp. 1652-1655, New Orleans, 1-3 May 1990.
[172] M. Perkowski, "The Generalised Orthonormal Expansion of Functions with Multiple-Valued Inputs and Some of its Applications," *Proc. ISMVL'92,* pp. 442-450, 1992.
[173] M. Perkowski, "A Fundamental Theorem for Exor Circuits," *Proc. Reed-Muller'93,* pp. 52-60, 1993.
[174] M. Perkowski, A. Sarabi, and F. Beyl, "Fundamental Theorems and Families of Forms for Binary and Multiple-Valued Linearly Independent Logic," *Proc. Reed-Muller'95,* pp. 288-299, Chiba, Japan, August 1995.
[175] M. Perkowski, T. Ross, D. Gadd, J. A. Goldman, and N. Song, "Application of ESOP Minimization in Machine Learning and Knowledge Discovery," *Proc. Reed-Muller'95,* pp. 102-109, Chiba, Japan, August 1995.

[176] M. Perkowski, M. Marek-Sadowska, L. Jozwiak, T. Luba, S. Grygiel, M. Nowicka, R. Malvi, Z. Wang, and J. Zhang, "Decomposition of Multiple-Valued Relations," *Proc. ISMVL'97*, pp. 13-18, Halifax, Nova Scotia, Canada, 28-30th May 1997.
[177] M. Perkowski, E. Pierzchala, and R. Drechsler, "Layout-Driven Synthesis for Submicron Technology: Mapping Expansions to Regular Lattices," *Proc. First International Conference on Information, Communications, and Signal Processing, (ICICS)'97,* Session 1C1: Spectral Techniques and Decision Diagrams, Singapore, 9-12th September 1997.
[178] M. A. Perkowski, E. Pierzchala, and R. Drechsler, "Ternary and Quaternary Lattice Diagrams for Linearly-Independent Logic, Multiple-Valued Logic and Analog Synthesis," *Proc. ISIC'97,* Singapore, Vol. 1, pp. 269-273, 10-12th September 1997.
[179] M. Perkowski, M. Chrzanowska-Jeske, and Y. Xu, "Lattice Diagrams using Reed-Muller Logic," *Proc. Reed-Muller'97*, Oxford Univ., U.K., pp. 85-102, 19-20th September 1997.
[180] M. Perkowski, L. Jozwiak, and R. Drechler, "A Canonical AND/EXOR Form that includes both the Generalized Reed-Muller Forms and Kronecker Reed-Muller Forms," *Proc. Reed-Muller'97,* Oxford Univ., U.K., pp. 219-233, 19-20th September 1997.
[181] M. Perkowski, L. Jozwiak, and R. Drechler, "Two Hierarchies of Generalized Kronecker Trees, Forms, Decision Diagrams, and Regular Layouts," *Proc. Reed-Muller'97,* Oxford Univ., U.K., pp. 115-132, 19-20th September 1997.
[182] M. Perkowski, A. Al-Rabadi, P. Kerntopf, A. Mishchenko, and M. Chrzanowska-Jeske, "Three-Dimensional Realization of Multiple-Valued Functions using Reversible Logic," *Booklet ULSI'01*, pp. 47-53, Warsaw, Poland, 21st May 2001.
[183] M. Perkowski, P. Kerntopf, A. Buller, M. Chrzanowska-Jeske, A. Mishchenko, X. Song, A. Al-Rabadi, L. Jozwiak, and A. Coppola, "Regularity and Symmetry as a Base for Efficient Realization of Reversible Logic Circuits," *Proc. IWLS'01*, pp. 90-95, Lake Tahoe, California, 12-15th June 2001.
[184] M. Perkowski, L. Jozwiak, P. Kerntopf, A. Mishchenko, A. Al-Rabadi, A. Coppola, A. Buller, X. Song, M. Khan, S. Yanushkevich, V. Shmerko, and M. Chrzanowska-Jeske, "A General Decomposition for Reversible Logic," *Proc. Reed-Muller'01*, pp. 119-138, Starkville, Mississippi, 10-11th August 2001.
[185] M. Perkowski, P. Kerntopf, A. Buller, M. Chrzanowska-Jeske, A. Mishchenko, X. Song, A. Al-Rabadi, L. Jozwiak, A. Coppola, and B. Massey, "Regular Realization of Symmetric Functions using Reversible Logic," *Proc. Euro-Micro'01*, pp. 245-252, Warsaw, Poland, September 2001.
[186] M. Perkowski, D. Foote, Q. Chen, A. Al-Rabadi, and L. Jozwiak, "Learning Hardware Using Multiple-Valued Logic, Part 1: Introduction and Approach," *IEEE Micro*, pp. 41-51, May/June 2002.
[187] M. Perkowski, D. Foote, Q. Chen, A. Al-Rabadi, and L. Jozwiak, "Learning Hardware Using Multiple-Valued Logic, Part 2: Cube Calculus and Architecture," *IEEE Micro*, pp. 52-61, May/June 2002.
[188] M. Perkowski, and A. Mishchenko, "Logic Synthesis for Regular Layout using Satisfiability," *accepted to International Workshop on Boolean Problems (WBP)'02*, Freiberg, Germany, 2002.
[189] M. Perkowski, B. Falkowski, M. Chrzanowska-Jeske, and R. Drechsler, "Efficient Algorithms for Creation of Linearly-Independent Decision Diagrams and their Mapping to Regular Layout," *accepted to VLSI Design.*
[190] P. Picton, "Optoelectronic Multi-Valued Conservative Logic," *Int. J. of Optical Computing*, Vol.2, pp. 19-29, 1991.
[191] P. Picton, "Modified Fredkin Gate in Logic Design," *Microelectronics J.*, 25, pp. 437-441, 1994.

[192] P. Picton, "Multi-Valued Sequential Logic Design Using Fredkin Gates," *MVL J.*, Vol. 1, pp. 241-251, 1996.
[193] P. Picton, "A Universal Architecture for Multiple-Valued Reversible Logic," *MVL J.*, 5, pp. 27-37, 2000.
[194] E. Pierzchala, M. A. Perkowski, and S. Grygiel, "A Field Programmable Analog Array for Continuous, Fuzzy, and Multiple-Valued Logic Applications," *Proc. ISMVL'94*, pp. 148-155, Boston, 25-27th May 1994.
[195] E. Pierzchala, M. A. Perkowski, "Programmable analog array circuit," *U. S. Patent US5959871*. Issued/Filed Dates: Sept. 28, 1999/ Dec. 22, 1994.
[196] H. W. Postma, T. Teepen, Z. Yao, M. Grifoni, and C. Dekker, "Carbon Nanotube Single-Electron Transistors at Room Temperature," *Science*, Vol. 293, No. 5527, pp. 76-79, 6th July 2001.
[197] A. J. Poustie, and K. J. Blow, "Demonstration of an All-Optical Fredkin Gate," *Optics Communications*, 174, pp. 317-320, 2000.
[198] D. K. Pradhan, "Universal Test Sets for Multiple Fault Detection in AND-EXOR Arrays," *IEEE Trans. on Comp.*, Vol. 27, pp. 181-187, February 1978.
[199] D. K. Pradhan, *Fault-Tolerant Computing: Theory and Techniques*, Vol. I, Prentice-Hall, New Jersey, 1987.
[200] L. Priese, "On a Simple Combinatorial Structure Sufficient for Sublying Nontrivial Self-Reproduction," *J. Cybernet.*, 6, 101, 1976.
[201] S. Rahardja, and B. J. Falkowski, "Family of Unified Complex Hadamard Transforms," *IEEE Trans. on Circuits and Systems-II: Analog and Digital Signal Processing*, Vol. 46, No. 8, pp. 1094-1100, August 1999.
[202] K. R. Rao, and D. F. Elliott, *Fast Transforms Algorithms, Analyses, Applications,* Academic Press Inc., 1982.
[203] M. R. Rayner, and D. J. Newton, "On the Symmetry of Logic," *J. of Phys. A: Mathematical and General*, 28, pp. 5623-5631, 1995.
[204] S. M. Reddy, "Easily Testable Realizations of Logic Functions," *IEEE Trans. on Comp.,* C-21, pp. 1183-1188, Nov. 1972.
[205] R. Rovatti, and G. Baccarani, "Fuzzy Reversible Logic," *Proc. Int. Conference on Fuzzy Systems (FUZZ-IEEE)'98*, 1998.
[206] K. Roy, and S. Prasad, *Low-Power CMOS VLSI Circuit Design,* John Wiley & Sons Inc., 2000.
[207] B. I. P. Rubinstein, "Evolving quantum circuits using genetic programming," *Proc. Congress on Evolutionary Computation (CEC)'01,* pp. 144-151, 2001.
[208] J. J. Sakurai, *Modern Quantum Mechanics*, Addison-Wesley, Reading, Massachusetts, 1995.
[209] A. Sarabi, N. Song, M. Chrzanowska-Jeske, and Marek A. Perkowski, "A Comprehensive Approach to Logic Synthesis and Physical Design for Two-Dimensional Logic Arrays," *Proc. DAC'94*, pp. 321-326, 1994.
[210] T. Sasao, and K. Kinoshita, "Cascade Realization of 3-Input 3-Output Conservative Logic Circuits," *IEEE Trans. on Comp.*, Vol. C-27, No. 3, pp. 214-221, March 1978.
[211] T. Sasao, and K. Kinoshita, "Realization of Minimum Circuits with Two-Input Conservative Logic Elements," *IEEE Trans. on Comp.*, Vol. C-27, No. 8, pp. 749-752, August 1978.
[212] T. Sasao, and K. Kinoshita, "Conservative Logic Elements and Their Universality," *IEEE Trans. on Comp.*, Vol. C-28, No. 9, pp. 682-685, September 1979.
[213] T. Sasao (editor), *Logic Synthesis and Optimization*, Kluwer Academic Publishers, January 1993.
[214] T. Sasao, "EXMIN2: A Simplified Algorithm for Exclusive-OR-Sum-Of-Products Expressions for Muliptle-Valued Input Two-Valued Output Functions," *IEEE Trans. on Computer Aided Design*, Vol. 12, No. 5, pp. 621-632, May 1993.

[215] T. Sasao, "Representation of Logic Functions using EXOR Operators," *Proc. Reed-Muller'95,* pp. 11-20, 1995.
[216] T. Sasao, and J. T. Butler, "Planar Multiple-Valued Decision Diagrams," *Proc. ISMVL'95,* pp. 28-35, 1995.
[217] T. Sasao, and M. Fujita (editors), *Representations of Discrete Functions*, Kluwer Academic Publishers, April 1996.
[218] T. Sasao, "Easily Testable Realizations for Generalized Reed-Muller Expressions," *IEEE Trans. on Comp.*, Vol. 46, pp. 709-716, June 1997.
[219] T. Sasao, *Switching Theory for Logic Synthesis*, Kluwer Academic Publishers, February 1999.
[220] A. C. Seabaugh, "Multi-Valued Logic and the Esaki Tunnel Diode," *Booklet ULSI'02*, p. 46, Boston, Massachusetts, 15th May 2002.
[221] D. Shah, *Self-Repairable Field Programmable Gate Array*, M.S. Thesis, Electrical and Computer Engineering Department, Portland State University, November 2000.
[222] J. Shamir, H. J. Caulfield, W. Micelli, and R. Seymour, "Optical Computing and the Fredkin Gates," *Applied Optics*, Vol. 25, p.1604, 1986.
[223] C. E. Shannon and W. Weaver, *A Mathematical Theory of Communication,* University of Illinois Press, 1949.
[224] V. V. Shende, A. K. Prasad, I. L. Markov, and J. P.Hayes, "Reversible Logic Circuit Synthesis," *Proc. IWLS'02*, pp. 125-130, New Orleans, Louisiana, 2002.
[225] Z. Shi, and R. Lee, "Bit Permutation Instructions for Accelerating Software Cryptography," *Int. Conference on Application-Specific Systems*, Architectures, and Processors, pp. 138-148, July 2000.
[226] P. W. Shor, "Algorithms for Quantum Computation: Discrete Logarithms and Factoring," *Proc. Symp. Foundations of Computer Science*, pp. 124-134, 1994.
[227] P. W. Shor, "Fault-tolerant quantum computation," *Proc. Symp. Foundations of Computer Science*, 1996.
[228] P. W. Shor, "Polynomial Time Algorithms for Prime Factorization and Discrete Logarithm," *SIAM J. of Computing*, 26(5), pp. 1484-1509, 1997.
[229] K. C. Smith, "Prospects for VLSI Technologies in MVL," *Booklet ULSI'02*, p. 4, Boston, Massachusetts, 15th May 2002.
[230] J. A. Smolin, and D. P. DiVincenzo, "Five Two-Bit Quantum Gates are Sufficient to Implement the Quantum Fredkin Gate," *Phys. Rev. A*, 53, pp. 2855-2856, 1996.
[231] F. Somenzi, *CUDD Package*, Release 2.3.1.
[232] N. Song, and M. Perkowski, "EXORCISM-MV-2: Minimization of Exclusive Sum of Products Expressions for Multiple-Valued Input Incompletely Specified Functions," *Proc. ISMVL'93*, pp. 132-137, 24 May, 1993.
[233] N. Song, and M. Perkowski, "Minimization of Exclusive Sum of Products Expressions for Multi-Output Multiple-Valued Input Incompletely Specified Functions," *IEEE Trans. on Computer Aided Design*, Vol. 15, pp. 285-395, April 1996.
[234] N. Song, and M. Perkowski, "Fast Look-Ahead Algorithm for Approximate ESOP Minimization of Incompletely Specified Multi-Output Boolean Functions," *Proc. Reed-Muller'97*, pp. 61-72, 1997.
[235] N. Song, *A New Design Methodology for Two-Dimensional Logic Cell Arrays*, Ph.D. Dissertation, Electrical and Computer Engineering Department, Portland State University, 1997.
[236] L. Spector, H. Barnum, H. J. Bernstein, and N. Swamy, "Finding a better-than-classical quantum AND/OR algorithm using genetic programming," *Proc. CEC'99,* volume 3, pp. 2239-2246, Washington D.C, 6-9. July 1999.
[237] L. Spector, H. Barnum, H. J. Bernstein, and N. Swamy, "Quantum Computing Applications of Genetic Programming," In *Advances in Genetic Programming*, Vol. 3, pp. 135-160, 1999.

[238] R. S. Stankovic, "Functional Decision Diagrams for Multiple-Valued Functions," *Proc. ISMVL'95*, pp. 284-289, 1995.
[239] R. S. Stankovic, *Spectral Transform Decision Diagrams in Simple Questions and Simple Answers*, Nauka, Belgrade, 1998.
[240] R. S. Stankovic, C. Moraga, and J. T. Astola, "Reed-Muller Expressions in the Previous Decade," *Proc. Reed-Muller'01*, pp. 7-26, Starkville, Mississippi, 10-11th August 2001.
[241] A. Steane, "Quantum Error Correction," *in Introduction to Quantum Computation and Information,* H-K. Lo, T. Spiller, and S. Popescu (editors), World Scientific, 1998.
[242] B. Steinbach, and A. Mishchenko, "A New Approach to Exact ESOP Minimization," *Proc. Reed-Muller'01*, pp. 66-81, Starkville, Mississippi, August 2001.
[243] L. Storme, A. De Vos, and G. Jacobs, "Group Theoretical Aspects of Reversible Logic Gates," *J. Universal Computer Science,* Vol. 5, No. 5, pp. 307-321, May 1999.
[244] V. P. Suprun, "Fixed Polarity Reed-Muller Expressions of Symmetric Boolean Functions," *Proc. Reed-Muller'95,* pp.246-249.
[245] T. Toffoli, "Reversible Computing," *Tech. Memo. 151*, Laboratory for Computer Science, MIT, Cambridge, MA, 1980.
[246] T. Toffoli, "Reversible Computing," *Automata, Languages and Programming*, pp. 632-644, Springer Verlag, 1980.
[247] R. Tosic, I. Stojmenovic, and M. Miyakawa, "On the Maximum Size of the Terms in the Realization of Symmetric Functions," *Proc. ISMVL'91,* pp. 110-117, Victoria, B.C., Canada, May 26-29,1991.
[248] V. I. Varshavsky, "Logic Design and Quantum Challenge," *Preprint from the author.*
[249] D. Ventura, *Quantum and Evolutionary Approaches to Computational Learning*, Ph.D. Dissertation, Computer Science Department, Brigham Young University, 1998.
[250] P. Wayner, "Silicon in Reverse," *Byte,* p. 67, August 1994.
[251] A. Wiles, "*Modular Elliptic Curves and Fermat's Last Theorem", Annals of Mathematics*, 141 (3), May 1995.
[252] C. P. Williams, and A. Gray, "Automated Design of Quantum Circuits," *Proc. of first NASA International Conference on Quantum Computing and Quantum Communications,* Palm Springs, CA, vol. 1509, Springer Verlag lecture notes in computer science, 1998.
[253] C. P. Williams, and S. H. Clearwater, *Explorations in Quantum Computing*, Springer-Verlag, New York, 1998.
[254] C. P. Williams, and S. H. Clearwater, *Ultimate Zero and One- Computing at the Quantum Frontier*, Springer, 2000.
[255] C. P. Williams, and A. G. Gray, "Automated Design of Quantum Circuits," *Quantum Computing and Quantum Communications (QCQC)'98,* pp. 113-125, Palm Springs, California, 17-20th February 1998.
[256] M. R. Williams, *A History of Computing Technology*, second edition, 1997.
[257] H. Wu, M. A. Perkowski, X. Zheng, and N. Zhuang, "Generalized Partially-Mixed-Polarity Reed-Muller Expansion and Its Fast Computation," *IEEE Trans. on Comp.*, Vol. 45, No. 9, pp. 1084-1088, 1996.
[258] T. Yabuki and H. Iba. "Genetic algorithms and quantum circuit design, evolving a simpler teleportation circuit," *Late Breaking Papers at the 2000 Genetic and Evolutionary Computation Conference*, pp. 421-425, 2000.
[259] T. Yamada, Y. Kinoshita, S. Kasai, H. Hasegawa, Y. Amemiya, "Quantum Dot Logic Circuits Based on Shared Binary-Decision Diagram," *Japanese J. Applied Physics*, Vol. 40, Part 1, No. 7, pp. 4485-4488, July 2001.
[260] S. N. Yanushkevich, *Logic Differential Calculus in Multi-Valued Logic Design*, Technical University of Szczecin Academic Publisher, Poland, 1998.
[261] A. Yariv, *Quantum Electronics*, John Wiley, 1989.

[262] S. G. Younis, *Asymptotically Zero Energy Computing Using Split-Level Charge Recovery Logic*, Ph.D. Dissertation, MIT, 1994.
[263] S. G. Younis, and T. F. Knight, "Asymptotically Zero Energy Split-Level Charge Recovery Logic," *Workshop on Low Power Design'94*, pp. 177-182, 1994.
[264] X. Zeng, M. Perkowski, K. Dill, and A. Sarabi, "Approximate Minimization of Generalized Reed-Muller Forms," *Proc. Reed-Muller'95,* pp. 221-230.
[265] I. I. Zhegalkin, "On the Techniques of Calculating Sentences in Symbolic Logic," *Math. Sb.*, Vol. 34, pp. 9-28, 1927 (in Russian).
[266] I. I. Zhegalkin, "Arithmetic Representations for Symbolic Logic," *Math. Sb.*, Vol. 35, pp. 311-377, 1928 (in Russian).
[267] Z. Zilic, and Z. G. Vranesic, "Current-Mode CMOS Galois Field Circuits," *Proc. ISMVL'93*, pp. 245-250, 1993.
[268] Z. Zilic, and K. Radecka, "The Role of Super-Fast Transforms in Speeding Up Quantum Computations," *Proc. ISMVL'02*, pp. 129-135, Boston, Massachusetts, 15-18th May 2002.
[269] G. W. Zobrist (editor), *Progress in Computer-Aided VLSI Design: Tools*, Volume 1, 1985.
[270] G. Zorpette, "The Quest for the Spin Transistor," *IEEE Spectrum Magazine*, 2002.
[271] M. Zwick, "Control Uniqueness in Reconstructibility Analysis," *Int. J. General Systems*, 23(2), 1995.
[272] M. Zwick, and H. Shu, "Set-Theoretic Reconstructability of Elementary Cellular Automata," *Advances in System Science and Application, Special Issue I*, pp. 31-36, 1995.
[273] M. Zwick, *Wholes and Parts in General Systems Methodology*, *In: The Character Concept in Evolutionary Biology*, edited by G. Wagner, Academic Press, 2001.
[274] M. Zwick, "Reconstructability Analysis with Fourier Transforms," *Book of Abstracts of the WOSC/IIGSS'02*, Pittsburgh, Pennsylvania, p. 82, 24-26th March 2002.
[275] M. Zwick, *Elements and Relations*, Book in preparation.

Index

C

F

G

M

N

O

P

Q

MIX
Papier aus verantwortungsvollen Quellen
Paper from responsible sources
FSC® C105338

If you have any concerns about our products,
you can contact us on
ProductSafety@springernature.com

In case Publisher is established outside the EU,
the EU authorized representative is:
Springer Nature Customer Service Center GmbH
Europaplatz 3, 69115 Heidelberg, Germany

Printed by Libri Plureos GmbH
in Hamburg, Germany